T0321942

ION EXCHANGE ADVANCES

Proceedings of IEX '92

Papers presented at the SCI Conference IEX '92—Ion Exchange Advances, held at Churchill College, Cambridge, UK, 12–17 July 1992.

ION EXCHANGE ADVANCES

Proceedings of IEX '92

Edited by

M. J. SLATER

Department of Chemical Engineering,
University of Bradford, UK

Published for SCI
by
ELSEVIER APPLIED SCIENCE
LONDON and NEW YORK

ELSEVIER SCIENCE PUBLISHERS LTD
Crown House, Linton Road, Barking, Essex IG11 8JU, England

WITH 90 TABLES AND 188 ILLUSTRATIONS

© 1992 SCI
© 1992 UNITED KINGDOM ATOMIC ENERGY AUTHORITY—pp. 214–221, 272–278

British Library Cataloguing in Publication Data

Ion Exchange Advances: Proceedings of IEX
'92
I. Slater, M. J.
541.3723

ISBN 1-85166-882-9

Library of Congress CIP data applied for

Preface

This volume contains the papers presented at the Sixth International Ion Exchange Conference organised by the SCI and held at Churchill College, Cambridge, UK, in July 1992. As on previous occasions, most recently in 1988, the organising committee did not engage plenary speakers but decided to solicit state-of-the-art contributions from the ion exchange community. This book contains the refereed papers presented at the meeting, whether in poster or oral form. Extra papers were presented at the meeting as posters because they were not available in time for refereeing purposes.

The subject matter of the meeting and therefore the contents of the book is subdivided into seven separate topic areas as follows: resin developments; water treatment; fundamentals; biotechnology, food and pharmaceuticals; environmental and pollution control; membranes, inorganic materials and nuclear; and hydrometallurgy. The coverage of the meeting is similar to 1988 although there are fewer subdivisions on this occasion. The more restricted coverage this time reflects the smaller number of papers offered by authors. This is probably due to the world-wide industrial recession which has affected commercial development and exploitation of the technology and restricts the ability of practitioners and academics to contribute to and attend international meetings. Nevertheless, the advances in biotechnology, growing concern about the environment and the need for novel separation processes have provided sufficient impetus to stimulate a sufficient number of workers in the field. Original contributions in the fields of hydrometallurgy, nuclear technology and inorganic materials are less evident at this conference and this is possibly because these topics have reached a level of maturity. There is a striking absence of papers concerning chemical applications and catalysis. This is surprising since it is a dynamic field but possibly rather too close to the market and therefore subject to confidentiality. That is unfortunate if it is the case.

Three outstanding contributors to the subject of ion exchange were honoured by the SCI Solvent Extraction and Ion Exchange Group at the meeting: Dr T. V. Arden was recognised for his contribution to the science and industrial application of ion exchange, Dr F. Martinola for his work on resin developments and Dr D. E. Weiss for his innovative contribution to ion exchange technology.

A meeting of the size of IEX '92 can only be organised with the willing help of a committee of volunteers supported by their companies and universities. The Chairmen wish to express their gratitude to the following: Kevin Blaxall, Howard Chase, Harry Eccles, Jim Greig, Michael Sadler, Bob Scott, George Solt and

Michael Verrall. We are also indebted to many other people who helped to make the work-load bearable, in particular our session chairmen and referees who facilitated the production of this book. The staff at SCI, in particular Anne Potter and Monique Heald, have played a large part in creating IEX '92.

D. C. SHERRINGTON
University of Strathclyde

M. J. SLATER
University of Bradford

M. STREAT
Loughborough University of Technology

Co-Chairmen, IEX '92

Acknowledgements

The SCI and the Conference Committee are most grateful for the support of the following companies in the organisation of IEX '92 and the sponsoring of events held during the meeting:

Corning Process Systems, Corning Ltd
Davy McKee (Stockton) Ltd
Dewplan (WT) Ltd
Dow Separation Systems, Dow Chemical Company Ltd
Ecolochem International Ltd
Ion Exchange India Ltd
Mitsubishi Kasei Corporation
NEI Thompson Ltd
Purolite International Ltd
Rohm and Haas (UK) Ltd

Contents

Fundamentals

Biotechnology, Food and Pharmaceuticals

Environmental and Pollution Control

Membranes, Inorganic Materials and Nuclear

Hydrometallurgy

DESIGN OF POLYMERIC ADSORBENTS FOR FOOD AND CHEMICAL PROCESSING

R.T.STRINGFIELD, P.A.ANDREOZZI, and R.R.STEVENS

The Dow Chemical Company, Midland, Michigan, USA

ABSTRACT

The usage of polymeric adsorbents for purification and bulk
separations is increasing in the food and chemical industries
due to the reliability, improved efficiency and simplicity of
adsorption processes. The large number of permutations
relating to the adsorbent and the solution matrix often makes
it difficult to select the appropriate adsorbent or optimize
the adsorbent properties without first evaluating numerous
adsorbents under the actual process conditions. Styrenic
adsorbents with various pore structures were designed to
achieve good selectivity, high adsorption capacity and optimal
adsorption rates. Mathematical models based on
phenomenological correlations of adsorption behavior are
presented that aid in selecting an adsorbent or identifying
the adsorbent structural modifications required to meet the
process specifications.

1. INTRODUCTION

The interplay of molecular forces between the adsorbate,
adsorbent and solution coupled with nearly unlimited three-
dimensional geometric configurations makes it difficult to
predict accurately the adsorption selectivities of complex
matrices. Typically, the equilibrium adsorption capacity and
the adsorption rate are measured for the components of
interest in a given process using various adsorbents. This

procedure can be tedious when designing adsorbents for specific separations. The phenomenological correlations approach to adsorbent design which we call Quantitative Structure Property Relationships (QSPR) can be very useful in mapping an adsorbent's characteristics. The QSPR approach is similar to the QSAR[1] approach used to correlate chemical structure relationships with biological activity in drug design. QSPR correlates the adsorbent's structure and physico-chemical properties of the adsorbates with the adsorption characteristics of probe compounds. Empirical adsorption measurements and physico-chemical parameters are incorporated into a mathematical model via statistical analysis and then compared to the measurable properties of the adsorbent. Once a database is generated for a class of adsorbents it is possible to determine quickly the adsorbents with the best properties for a particular separation.

This paper presents mathematical models and phenomenological correlations describing the adsorption behavior of several polymeric adsorbents. The models allow estimation of the adsorption rates and adsorption capacities of low molecular weight organic compounds for the adsorbents presented.

2. EXPERIMENTAL

The SAS system[2] and RS/1[3] were the general data manipulation programs used for statistical analysis, analytical modeling and graphical display. Physico-chemical properties for the adsorbate compounds were determined using molecular mechanics via Chemlab II[4] and Medchem[5] software packages. Several common low molecular weight compounds that cover a wide range of physico-chemical parameters were selected as probe compounds. Fifteen adsorbate physico-chemical properties that describe the adsorbate's size, shape, polarity and solubility in aqueous and organic solution were initially evaluated in this study. Six of these properties, including the octanol-water partition coefficient (LOG P), the solvation energies[6] in octanol (SEO) and acetic acid (SEA), molecular volume (MVOL), equilibrium diameter (EQD), and molecular weight were found to be useful in modeling

Table 1.
Properties of Polymeric Adsorbents

Adsorbent Identity	Water Retention Capacity (%)	Pore Volume (wet) (mL/g)	Exclusion Value (nm)
A	34.4	0.52	0.80
B	42.8	0.75	0.90
C	47.1	0.89	1.05
D	53.3	1.14	5.10
E	58.0	1.38	27
XUS-40285	55.0	1.22	150

adsorption behavior. A series of styrenic adsorbents were prepared using conventional synthesis methods. The degree of crosslinking ranged from 50 to 70 percent. Both swelling and non-swelling porogens were used to control the pore size and pore volume. Properties of these adsorbents are shown in Table 1. The size exclusion values for the adsorbents were determined using the method by Crispin[7]. The size exclusion value of the smallest carbohydrate not penetrating the adsorbents is reported in Table 1.

The overall adsorption rate (OAR) was determined by measuring the change in concentration of the solute in contact with the adsorbent as a function of time. Each wet adsorbent (2.3 g on a dry basis) was agitated in 200 grams of solution containing 4.0 g/L of each probe compound. Linearization of the data was obtained by plotting the amount adsorbed per unit weight of adsorbent versus time ($h^{1/2}$) for the initial fraction of the reaction. The overall adsorption rates determined from the slope of the line with units mg/g-adsorbent/$h^{1/2}$, are not true reaction rates but relative rates useful for comparative purposes and are given in Table 2.

Dynamic mini-column adsorption experiments were performed to determine the dynamic adsorption capacity (DAC) of each adsorbent for the probe compounds. Methods for determining capacity from breakthrough curves are described in the Chemical Engineers' Handbook[8]. The mini-column feed concentration was 4.0 g/L for each probe compounds. The dynamic adsorption capacity data are given in Table 3.

3. RESULTS AND DISCUSSION

The correlation coefficient (γ^2) is a value between 0 and 1 that measures the portion of the total variation in the response variable explained by the multiple linear regression model equation. A γ^2 value of 1.0 generally indicates a good fit. As shown in Table 4 the molecular volume is highly correlated with the Log(OAR) of the probe compounds for the adsorbents A and B which have the lower size exclusion values. The molecular size of the adsorbate is close to the adsorbent size exclusion value so diffusion of the adsorbate into the pores of the adsorbent is hindered. The size exclusion value of the adsorbents C and D are sufficiently large to allow less hindered access into the adsorbents' interior, hence, the LOG(OAR) is less well correlated with the molecular volume. The two parameter models of LOG P/SEO and LOG P/SEA best describe the LOG(OAR) for the adsorbents C and D. The LOG P describes the adsorbate's relative aqueous and organic solubility while the solvation energies in octanol and acetic acid give a qualitative estimate of the adsorbate-adsorbent potential. The overall adsorption rate into the larger pores is dependent on the rate of diffusion within the pore-filled liquid and migration along the solid surface of the pore.

Other investigators[9] have shown that the size of the adsorbate should affect the overall adsorption rate when the

Table 2.
Overall adsorption rates (OAR) of probe compounds
(mg/g-adsorbent/h$^{1/2}$)

| | Adsorbent | | | |
Compound	A	B	C	D
Methylene chloride	852	1100	1074	1164
Dioxane	202	290	294	352
Phenol	522	646	742	965
Methylal	196	478	502	676
Catechol	300	435	655	905
Cyclohexanol	118	328	500	629
Caffeine	58.6	112	193	635
Chlorogenic acid	0	18	73.8	307

Table 3.
Dynamic adsorption capacities (DAC)

PROBE COMPOUND	ADS. A (mL/g)	ADS. B (mL/g)	ADS. C (mL/g)	ADS. D (mL/g)	ADS. E (mL/g)
Methylene chloride	0.36	0.491	0.304	-	0.483
Dioxane	0.092	0.132	0.071	0.069	0.122
Phenol	0.389	0.483	0.344	0.324	0.429
Methylal	0.141	0.224	0.144	0.134	0.189
Catechol	0.262	-	0.264	0.223	0.33
Cyclohexanol	0.309	0.567	0.288	0.283	0.493
Caffeine	0.311	0.576	0.301	0.331	0.504
Triethylamine	0.477	0.6	0.464	-	-
t-butylcatechol	-	-	0.435	0.542	0.758

molecular size of the adsorbate is close to the size of the pores of the adsorbent. Kuga[10] and Crispin[7] tabulated the molecular weights, radii of equivalent spheres and size exclusion values for several polymers, dextrans and sugar compounds. The radii of equivalent spheres are the compounds' molecular sizes when swollen in a good solvent. The size exclusion values plotted in Figure 1 are the adsorbent pore sizes required for unhindered diffusion of non-adsorbing compounds. Molecules of molecular weight less than 500 diffuse unhindered in pore sizes 1.2 to 1.5 times the equilibrium diameter. The overall adsorption rate modeling observations are in agreement with the results reported by Kuga and Crispin. The adsorbent XUS-40285 contains macropores for the rapid diffusion of low molecular weight adsorbates to the smaller micropores. Adsorbents A to E contain micropores predominantly and are useful in adsorption processes where exclusion of higher molecular components is preferred.

The dynamic adsorption capacity was correlated to the molecular weight and the LOG P of the adsorbate as shown in Table 5. The DAC of the adsorbent increases with increasing LOG P and increasing molecular weight as shown in Figure 2. This is in good agreement with results obtained by other investigators[11,12]. The experimental adsorbate solution concentrations were within the linear portion of the Langmuir equilibrium adsorption isotherms which allows the estimation of the DAC at lower solution concentration.

Table 4.
Correlation Coefficients (γ^2) For Log(OAR)

Adsorbate physico-chemical properties
used in linear regression models

	MVOL & LOG P	MVOL	LOG P	SEO & LOG P	SEO	SEA & LOG P	SEA
Ads. A	0.99	0.97	0.73	0.91	0.83	0.73	0.23
Ads. B	0.98	0.95	0.75	0.96	0.90	0.76	0.19
Ads. C	0.94	0.84	0.84	0.99	0.87	0.84	0.24
Ads. D	0.77	0.48	0.77	0.92	0.53	0.91	0.65

Table 5.
Correlation Coefficients (γ^2)
For Dynamic Adsorption Capacity

Adsorbate physico-chemical properties
used in linear regression models

	MW & LOG P	MW	LOG P
Adsorbent A	0.86	0.07	0.48
Adsorbent B	0.96	0.33	0.23
Adsorbent D	0.98	0.48	0.75
Adsorbent E	0.92	0.41	0.70

Table 6.
Multiple linear regression models
for predicting LOG(OAR) and DAC

Ads. A	LOG(OAR) = 0.163 X LOG P - 0.011 X MVOL + 3.32
Ads. B	LOG(OAR) = 0.124 X LOG P - 0.0063 X MVOL + 3.14
Ads. C	LOG(OAR) = 0.161 X LOG P + 0.105 X SEO + 3.13
Ads. D	LOG(OAR) = 0.110 X LOG P - 0.193 X SEA + 2.80
Ads. A	(DAC) = 0.00176 X MW + 0.153 X LOG P + 0.003
Ads. B	(DAC) = 0.00347 X MW + 0.198 X LOG P - 0.058
Ads. D	(DAC) = 0.00174 X MW + 0.100 X LOG P - 0.016
Ads. E	(DAC) = 0.00227 X MW + 0.140 X LOG P + 0.031

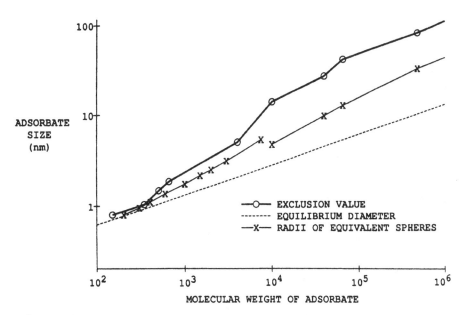

Figure 1. Exclusion values of adsorbate compounds required for unhindered diffusion. Kuga[12], Crispin[5]

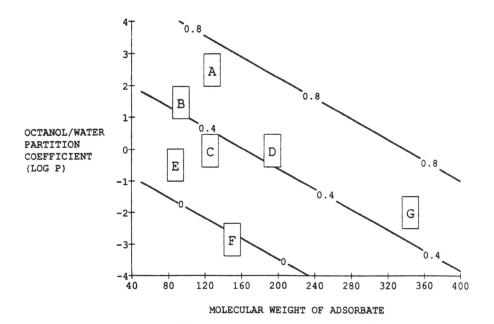

Figure 2. Contourplot of Adsorbent E DAC predictions.
(A) p-chlorophenol, (B) phenol, (C) hydroxymethylfurfural,
(D) **caffeine**, (E) dioxane, (F) **ribose**, (G) chlorogenic acid

4.0 SUMMARY

The ability of the QSPR modeling approach to distinguish the various adsorption behavior of a diverse series of probe compounds has been demonstrated. The physico-chemical parameters describing the probe compounds' solubility and molecular size were used to model adsorption behavior. The multiple linear regression equations given in Table 6 can be used to predict the adsorption characteristics of other compounds for the series of adsorbents presented. The QSPR approach has been used to aid development of adsorbents for several applications including sugar decolorization[13], orange juice debittering[14] and coffee decaffeination[15].

5.0 REFERENCES

1. Hansch, C. J., <u>Med. Chem.</u>, 19, 1 (1976)
2. SAS Institute Inc., Cory, NC, USA
3. BBN Software Products, 10 Fawcett ST, Cambridge, MA 02238,USA
4. Molecular Design Limited, 2132 Farallon Drive, San Leandro, CA 94577, USA
5. DAYLIGHT Chemical Information Systems, Inc. Irvine, CA USA
6. Hopfinger, A.J., <u>Conformational Properties of Macromolecules,</u> Academic Press, New York (1973)
7. Crispin T., Halasz, I., <u>J. of Chromatography,</u> 239 (1982) p. 351-362
8. Perry, R.H., Chilton, C.H., <u>Chemical Engineers' Handbook,</u> 5th ed., McGraw-Hill Book Co., (1973), section 16-24
9. Weber, W.J. Jr., and J.C. Morris, <u>J. Sanit. Eng. Div. ASCE SA2,</u> 31 (1963)
10. Kuga, S., <u>J. of Chromatography,</u> 206 (1981) p. 449-461
11. Hildebrand, J.H., and R.L. Scott, <u>The Solubility of Nonelectrolytes,</u> 3rd ed. Dover, New York (1964)
12. McGuire, M.J., and I.H. Suffet. "The Calculated Net Adsorption Energy Concept," in <u>Activated Carbon Adsorption of Organics From the Aqueous Phase,</u> Vol. 1. I.H. Suffet and M.J. eds. Ann Arbor Science Publishers, Ann Arbor, MI (1980), p. 91.
13. Stringfield, R.T., et. al., U.S. Pat. 4,950,332, (1990)
14. Norman, S.I., et. al., U.S. Pat. 4,965,083, (1990)
15. Dawson-Ekeland, K.R., et. al., U.S. Pat. 5,021,253, (1991)

COMB-LIKE POLYETHYLENE OXIDE POLYMERS
AS CHELATING AGENTS FOR MERCURIC (II) CHLORIDE:
STUDY OF THE COMPLEX FORMATION BY SOLID-STATE ^{1}H-NMR

Y.FRERE and Ph. GRAMAIN

Institut Charles Sadron, CNRS,
6, rue Boussingault, 67083 Strasbourg Cedex, France

ABSTRACT

Polyethylene oxide (PEO) forms with mercuric chloride a great variety of stable complexes of different stoichiometry, depending on the molecular weight, the type of end groups and the solvent. To take advantage of this unique property, the synthesis of hydrogels by polymerization of acrylates or methacrylates of short PEO chains (macromers) is carried out and the formation of mercuric complexes is discussed. The stoichiometry of the different complexes formed is compared and discussed in relation to the results of solid state ^{1}H-NMR experiments using the free induction decay technique. It is shown that, (i) the binding properties of the gels are similar to those of PEO in solution, (ii) the first two ethylene oxyde units next to the grafting points are rather rigid and there-fore are unable to participate in the complexes formation.

INTRODUCTION

Polyethyleneoxide, polyethylene imine and polyethylene sulfide are among the simplest polymers and are structurally similar. Due to the presence in the backbone of donor atoms, the three polymers are potential chelating agents, very attractive from a fundamental point of view considering their structural simplicity. The development of such polymers as chelating materials supposes the immobilisation of the chains and the study of the induced effects. Among the three structures, we have chosen to study polyethylene oxide (PEO). In the first part of this paper, we describe the spectacular binding properties of PEO towards mercuric chloride in water. In the second part, we analyse the properties of hydrogels obtained by polymerization of PEO acrylate or methacrylate macromers.

COMPLEXATION OF HgCl$_2$ BY PEO IN WATER

S=O/Hg	R-[O-CH$_2$-CH$_2$]$_n$-O-R' in water		Mp in °C
n=4	4<n<14		n>14
R=H R'=H R=CH$_3$ R'=H or CH$_3$	R=CH$_3$ R'=CH$_3$	R=H R'=H or CH$_3$	R=H or CH$_3$ R'=H or CH$_3$

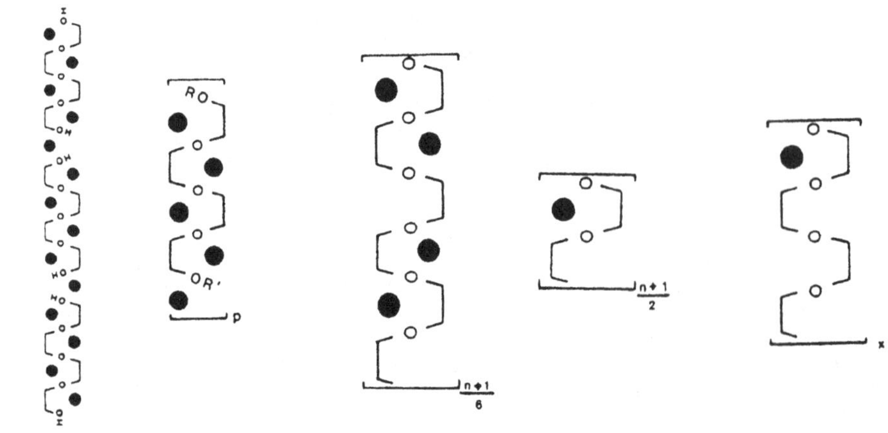

(1)	(2)	(3)	(4)	(5)
S=15/14 tricomplex	S=5/5 polycomplex	S=1.5/1 monocomplex if (n+1)/3 whole number, if not tricomplex	S=2/1 monocomplex with odd n. dicomplex with even n	S=4/1
Mp 125	Mp 150 R'=H Mp 160 R'=CH$_3$	160<Mp<175	175<Mp<195	110<Mp<150

Figure 1 - Complexes obtained in water with poly(ethylene oxide)s and HgCl$_2$.

The formation of complexes between PEO and mercuric chloride in methanol or ether has been known since the work of Blumberg (1) and Iwamoto (2), but curiously no study has been published on the formation and characterization of the complexes obtained from water solution. This is especially surprising, since, according to our observations, the mixture of PEO and $HgCl_2$ in water leads immediately to the formation of a precipitate. The fascinating aspect is that the characteristics of the complexes formed (melting point, stability, stoechiometry and structure) depend strongly on the degree of polymerization (DP) n of the PEO chains and to a lesser extent, on the nature of the terminal groups (OH or OCH_3 for example) (3). The lower the DP, the richer in $HgCl_2$ are the complexes. With n=4, which is the minimum in order to observe precipitation in water, the stoichiometry is one oxygen atom for one mercury (S=1/1), while it becomes four oxygens for one mercury (S=4/1) with n>14. Moreover, when the number of oxygen atoms in the PEO chains is even, di-,tri-or polycomplexes are formed by association of PEO chains. The structure of the complexes and the range of melting points of each type are schematically represented in Fig. 1. They are purely covalent and can be easily decomposed either by heating (sublimation of $HgCl_2$) or by addition of nitric acid (formation of uncomplexed ionic mercuric nitrate).

The strong influence of the PEO DP on the stoichiometry of the complexes is easily understood by considering the three types of energetic contributions involved; the polymer-solvent and polymer-mercury interactions and the change of conformational energy of the polymer chain due to the complex formation. In particular, the results show that the loss of conformational energy of the chain becomes more important as the degree of polymerization increases. This is of importance when the use of polymers as chelating agents is considered.

COMPLEXATION OF $HgCl_2$ BY COMB-LIKE POLY(PEO-ACRYLATES)

In order to take advantage of the unique binding property of PEO, polymers containing PEO as side-chains are interesting structures. From a fundamental point of view, such a material may give important indications on the effect not only of the attachment of ligands but also of its local density.

Water soluble polymers with PEO as side-chains (comb-like structure) can be easily obtained by radical polymerization of acrylate or methacrylate of PEO (macromers) of different lengths (4,5):

$$CH_2=CR-\underset{\underset{O}{\|}}{C}-O[CH_2-CH_2-O]_n-CH_3$$

Due to transfer reactions, the polymers prepared in concentrated solutions are obtained as slightly crosslinked particles which swell strongly in water (the water content varies from 70 to 95 weight %). We have prepared and studied the $HgCl_2$ binding properties of PEO-acrylate polymers with n=3, 6.4, and 14, together with polydispersed macromers with n= 43.5, and 111.7 (calculated according to the average molecular

poly(PEO-acrylates) n	Maximum capacity mol $HgCl_2$/mol PEO chain	S=O/Hg	type of complex (see Fig.1) (S)	
3	0.54 $\pm$ 0.04	5.6 $\pm$ 0.4	-	-
6.4	2.65 $\pm$ 0.05	2.41 $\pm$ 0.05 2.0 with n=5.3	4	(2)
14	6.25 $\pm$ 0.3	2.25 $\pm$ 0.1 2.0 with n=12.5	4	(2)
43.5 av.	12.0 $\pm$ 2	3.7 $\pm$ 0.5	5	(4)
111.7 av.	30.9 $\pm$ 0.5	3.6 $\pm$ 0.5	5	(4)

Table 1 - **Complexation of $HgCl_2$ by hydrogels of poly(PEO-acrylates).**

Conditions : 3 to 100 mg of gel in 10 mL $HgCl_2$ 0.1M.

Capacities after 24h determined according to ref. 3.

weight given by the manufacturer). The maximum uptake of mercuric chloride by PEO chain, supposing all the chains accessible, is given in table 1 for each polymer.

Analysis of these results shows that for the two longer polydispersed samples, a stoichiometry of the complex of one mercury for four EO units is obtained as for the free PEO of the same length in solution. For the other samples and in particular for $n=6.4$ and 14, a stoichiometry $S=2$ as obtained in solution is quite probable, although the experimental S values are higher. The gel with $n=3$ shows very poor binding properties difficult to analyse. A probable reason for the observation of higher S values for the three last gels is that we have considered that all the chains or the entire chains were able to form a complex. The numbers of the theoretical EO units n leading to complexes with the same stoichiometry as in solution are shown in the same table (column 3). According to this, 1.3 ± 0.2 EO units are unable to be engaged in a complex. In order to obtain information on the validity of this hypothesis, solid state ^{1}H-NMR experiments were carried out.

STUDY BY SOLID STATE ^{1}H-NMR SPECTROSCOPY

One of the problems often encountered with insolubilized complexing phases is the difficulty of determining the exact amount of ligand participating in the complexation process. The preceding study well illustrates this difficulty, since in this case various structures can be formed. It is possible to determine the type of complex formed only by knowing the amount of PEO side-chains or of EO units accessible. To analyse this, wide-line ^{1}H NMR using free induction decay (FID) analysis is a powerful tool. The FID signals are indicative of the mobility of the hydrogen in the samples. A signal showing a slow return to equilibrium (long T_2) characterizes an important proton mobility as in the case of liquids. When several phases of different mobility are present, as in the case of many polymeric materials, a separate analysis of each component of the FID signal allows one to distinguish the mobile and the rigid parts of the samples and to evaluate their proportion. Applied to dry gels of poly(PEO-methacrylate) prepared under the same conditions as the preceding poly(PEO-acrylate), the method demonstrates the presence in the samples of an important amount of rigid protons at 25°C, detected as a Gaussian component of the FID signal with a relaxation time of 13 ms $\pm$ 1. Fig.2 presents the evolution of the content of rigid protons with the length of macromers. The shorter the PEO chain, the higher the content in rigid protons. In particular, it has to be noted that the gel with n=3 contains 51% of rigid protons.

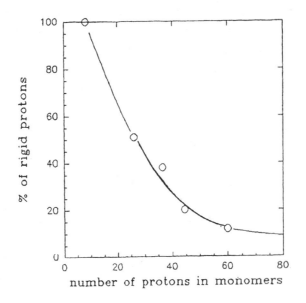

Figure 2 - Proton rigid fractions in PEO gels at 25°C.

With the very likely hypothesis that the samples are homogeneous and that the rigid part is composed of the backbone and of the EO units located near the ester function, it can be calculated that the rigid part for all the samples is constituted of the protons of the backbone plus the protons of 1.7 $\pm$ 0.3 EO units. This value is in good agreement with the preceding calculated value of 1.3 $\pm$ 0.2 units and confirms the hypothesis.

In particular, it is clear that the n=3 gel containing only one EO mobile unit, has very poor binding properties and forms an undefined complex. The remarkable agreement between the FID results and the binding results demonstrates the usefulness of the FID method. It shows that the rigid nature of a segment of the chains inhibits the participation of this part in complexation due to its inability to adopt an adequate conformation. This conclusion is of great importance in understanding the binding properties of insolubilized phases. Once the contribution of the rigid part is analysed, the results demonstrate that the binding properties of the PEO ligands grafted on a backbone are quite similar to that of free PEO in solution.

The stoichiometries observed contrast with the behaviour of the same chains grafted onto macroporous polystyrene beads (PSDVB) for which the type 5 complex was always observed (6). It is quite probable that the crowded micro-environment inside the PSDVB beads and the limitation of the swelling are responsible for such a change of stoichiometry by restricting the global mobility of the chains and the room available for the formation of mercury rich complexes. Such a crowding effect has recently been well demonstrated with polythioethers grafted onto macroporous PS matrices (7).

CONCLUSIONS

The variety of complexes formed with PEO and mercury chloride together with the solid state ^{1}H-NMR technique allow a better understanding of the binding properties of grafted ligands. In particular, the observation of the rigidity of the first protons of the ligands near the grafting points justifies the empirical use of an arm or spacer between the phase and the ligand.

ACKNOWLEDGMENT

The authors thank Dr. N. Parizel for the ^{1}H-NMR experiments on gels.

REFERENCES

1- Blumberg A.A. and Pollack S.S., J. Polym. Sci., Part A, 2, 2499 (1964)

2- Yokuyama Y., Ishihara H., Iwamoto R. and Tadokoro H., Macromolecules, 2, 184 (1969)

3- Frère Y., Guilbert Y. and Gramain Ph., New J. Chem. 12, 773 (1988)

4- Gramain Ph. and Frère Y., Polym. Commun., 27, 16 (1986)

5- Yan F., Déjardin Ph., Frère Y. and Gramain Ph., Makromol. Chem., 191, 1197 (1990)

6- Lauth M., Frère Y., Meurer B. and Gramain Ph., Reactive Polymers, 12, 155 (1990)

7- Lauth M., Frère Y., Prévost M. and Gramain Ph., Reactive Polymers, 13, 73 (1990)

PREPARATION OF CHELATING HOLLOW FIBERS BASED ON CHLOROSULFONATED POLYETHYLENE AND ITS USES FOR METAL EXTRACTION

S. BELFER, S. BINMAN and E. KORNGOLD

The Institutes for Applied Research, Ben-Gurion University of the Negev,

Beer-Sheva, Israel

ABSTRACT

Different anion-exchange and chelating fibers were prepared from chlorosulfonated polyethylene, to which amine groups were attached. The chemical reactions of amination, quaternization and carboxylation were carried out under different conditions, and the relationship between the properties of fibers and their structure was established.

INTRODUCTION

Ion-exchange materials in a fibrous form offer much higher rates of diffusion than particulate materials and therefore represent a real alternative to the latter. Chelating ion-exchange resins have been widely described in the literature (1a,b,c). A considerable number of publications, mainly by Russian investigators, have been devoted to the preparation and application of fibrous (but not hollow) ion exchangers (2a,b,c). Recently, a small number of papers, mainly by Japanese researchers, have been published on hollow-fiber chelating polymers (3). Such chelating hollow fibers have been produced by radiation grafting of vinyl monomers to polyethylene.

Our approach is based on the use of chlorosulfonated polyethylene hollow fibers (4). Interaction of this starting material with different diamines opens a wide range of possibilities for the production of chelated hollow fibers.

Moreover, the introduction of amino groups facilitates further chemical modifications and the production of aminoacetic and aminophosphoric chelating groups. The present paper describes the results of our research in this direction.

EXPERIMENTAL

Materials

A range of chlorosulfonated polyethylene hollow fibers was prepared in our Institute in the laboratory headed by Dr. Korngold (5). The characteristics of these fibers are given in Table 1.

Table 1
Characteristics of the Chlorosulfonated Hollow Fibers

	PECl 1	PECl 2	PECl 3	PECl 4
S content (%)	7.17	6.4	6.08	7.03
Cl content (%)	10.0	16.0	10.0	12.0
Swelling in DClE (%)	37.8	20.0	16.0	17.4
Outer diameter (mm)	1.559	0.891	0.824	1.351
Inner diameter (mm)	1.147	0.517	0.509	0.892
Thickness of the wall	0.206	0.187	0.158	0.229

The following diamines were used for amination: ethylenediamine (EDA), propylenediamine (PDA), dimethylaminopropylamine (DMAPA), and tetra-ethylenepentamine (TEPA). All amines were supplied by Merck and were employed without purification except for the few cases when DMAPA was used. In this case the amine was azeotropically distilled with toluene in order to remove the water as completely as possible. Chloracetic ($ClCH_2COOH$) and bromacetic ($BrCH_2COOH$) acids were supplied by Merck and were used as received. All metal salts were analytical grade. Dioxan was dried over KOH. Dimethylformamide was used as received.

The amination and carboxylation reactions were carried out according to refs. (6) and (7), respectively, and capacity was determined according to standard procedure. A batch equilibrium technique was used to obtain extraction characteristics of chelating fibers.

RESULTS

The characteristics of the initial chlorosulfonated polyethylene hollow fibers are presented in Table I. From the table we can see that the fibers with almost the same wall thickness differ each from other by their swelling and S and Cl content. Those data can be explained by the fact that chlorosulfonation is accompanied by chlorination (higher Cl content), crosslinking (higher S content and lower swelling) and other side reactions.

Amination

In the first series of experiments aliphatic amines with two or more amino groups were used. The amines differ in the length of the chain (EDA and PDA), the number of amine groups (TEPA) and their basicity (DMAPA). The results of amination are given in Table 2. H^+ capacity could be a result of anion exchange between the anion exchanger in the OH^- form and the Cl^- of HCl (a decrease of

Table 2
Amination of Chlorosulfonated Polyethylene

Exp no.	Fiber no.	Amine	T (°C)	Solvent	Time	Capacity (meq/g)		Water content (%)
						H+	Cl-	
1	PECl III	EDA	22	Dioxane	3 days	0.92	0.97	18.5
2	PECl III	"	22	Dioxane	3 days	0.9	0.98	20.2
3	PECl III	"	60	DMF	5 h	0.64	0.58	12.7
4	PECl IV	"	22	–	3 days	0.95	1.04	22
5	PECl I	PDA	22	Decalin	24 h	0.89	0.86	28.7
6	PECl I	"	22	Dioxane	5 h	0.9	1.1	35.8
7	PECl III	"	22	Dioxane	24 h	0.83	0.86	28.5
8	PECl IV	"	22	–	24 h	1.2	1.2	22
9	PECl I	DMAPA	22	Decalin	5 h	1.3	0.8	49.3
10	PECl I		22	Decalin	24 h	0.9	0.75	45.6
11	PECl I		40	Dioxane	3 days	0.64	0.46	26.8
12	PECl I		50	Dioxane	7 h	0.83	0.86	48.3
13	PECl II		60	Dioxane	5 h	1.4	0.95	36
14	PECl III		22		3 days	1.05	1.1	31
15	PECl II	TEPA	22	DClE	24 h	0.63	0.54	18.9
16	PECl II		22	"	24 h	0.67	0.56	21.5
17	PECl III		22	"	24 h	1.41	1.2	48
18	PECl III		22	"	24 h	1.10	0.8	42

the acidity of the HCl due to the presence of OH in solution). The decrease of acidity could also be a result of cation exchange between Na$^+$ attached to the sulfamide group on the fiber and the H$^+$ present in the external solution (8). The H$^+$ capacity may thus be an expression of the presence both cation- and anion-exchange sites in the fiber. On the other hand, the decrease of Cl$^-$ in solution results solely from OH$^-$/Cl$^-$ exchange and is therefore a measure of anion-exchange sites alone.

From Table 2 we see that the two amines EDA and PDA gave aminated fibers with almost the same anion-exchange capacity, 0.9-1.2 meq/g, but the water content of fibers differed significantly depending on the amine used. EDA produced fibers with a low water content, 18-20%, while in fibers aminated with PDA the values reached 28 and 36%. The solvent had no effect on the capacity; moreover, when pure amine was used (exp. no. 4) the value of the capacity was the same as in most cases in which solvent was used. In contrast, the temperature had a significant effect on the reaction: for example, the capacity (0.6 meq/g) and the water content (12%) decreased markedly when amination with EDA was carried out at 60°C. The amination of chlorosulfonated polyethylene is possibly accompanied by additional crosslinking, a situation analogous with amination of chloromethylated polystyrene (9,10). Usually, high temperature and high amine concentrations enhanced crosslinking.

In the first set of experiments with TEPA, anion-exchangers with low capacities and low water contents were obtained. It should be noted here that the solution of TEPA in batches N15-16 has high viscosity, which increased significantly during the amination. Therefore, in the next set of experiments (no. 17-18) TEPA was diluted significantly and the tertiary amine (triethylamine) was added into the reaction mixture to the [6]. As a result, the capacity was doubled and the water content increased from 20 to 48%.

The amination with DMAPA has a special goal, because this amine possesses a pendant tertiary amino group, which could create a strong base anion exchanger after quaternization reaction. We can see that amination at room temperature resulted in fibers with a high capacity (0.86-1.3 meq/g) and a high water content (45.6-49.3%). However, the capacity and water content dropped drastically when the reaction was carried out at about 40°C. Increasing the temperature to 60°C leads to the formation of fibers with a higher capacity and a higher water content (exp. no. 11,13).

Quaternization

Quaternization was performed as described in (5) (Table 3). We can see that the degree of quaternization is strongly dependent on the type of amino group. Tertiary amino groups undergo reaction easily. Generally, more than 50% of

Table 3
Quaternization of Aminated Fibers

Fiber	Amine	Conditions of quaternization		Strong base capacity (meq/g)	Degree of quaternization (%)
		(°C)	Time, h		
PEQ-1	DMAPA	0	48	10.8	70
PEQ-2	EDA	0	48	0.10	10
PEQ-3	PDA	0	48	0.05	4
PEQ-4	DMAPA	0	48	0.55	55
PEQ-5	"	0	48	0.41	41
PEQ-6	PDA	25	24	0.06	4.1
PEQ-7	"	25	48	0.05	5.2
PEQ-8	ADA	25	48	0.16	12.0
PEQ-9	DMAPA	25	7	1.13	66
PEQ-10	"	0	48	1.02	60
PEQ-11	"	25	48	0.76	45

Table 4
Alkyl Carboxylation of Aminated Polyethylenes with Chloracetic Acid
and Bromacetic Acid

Fiber no.	Amine	Capacity (meq/g)				Co uptake (meq/g)
		Amino		Amino carboxy		
		H+	Cl-	H+	Cl-	
		Chloracetic acid				
28	EDA	0.92	0.97	1.2	0.4	0.93
27	"	0.64	0.57	0.77	0.07	0.33
6	PDA	0.91	1.1	1.03	-	0.9
22	"	0.47	0.67	0.48	-	0.22
CB-1	TEPA	1.4	0.82	0.82	-	0.96
34	"	0.63	0.56	0.84	0.08	0.45
		Bromacetic acid				
33	EDA	0.90	0.98	1.4	-	0.68
23	EDA	0.54	0.64	0.42	0.08	0.4
CB-6	PDA	1.2	0.75	0.82	-	0.9
30	"	0.8	0.69	0.58	0.07	0.35
63	TEPA	1.28	0.28	1.14	-	1.4
29	"	0.56	0.58	0.73	0.04	0.61

weak base groups were converted to the strong amino group over a period of
7 h at 25°C or 48 h at 0°C. The degree of quaternization fell drastically for fibers

containing a primary amine group. In this case only 5-12% of quarternization was achieved.

Introduction of amino-acetic groups

Aminated hollow fibers were allowed to react with chloracetic or bromacetic acids in order to obtain chelating groups. After reaction, the fibers were converted into the Na^+ form, and H^+ and Cl^- capacities were determined. The results are summarized in Table 4. Two different types of aminated fiber were used, which differed from one another in the value of their H^+ capacity. From the Table we can see that the higher the H^+ capacity of the aminated fiber, the higher the H^+ capacity of the resulting aminocarboxylated fiber, and in most cases the Cl^- capacity disappeared completely.

Chromate extraction

The ability of EDA- aminated fibers to extract Cr^{+6} from a bichromate solution was investigated (11). In the adsorption isotherms given in Fig. 1 Cr^{6+} concentration in the fibers is plotted as a function of its concentration in solution at different pH. We can see that extraction by weak base anion-exchangers was strongly dependent on the pH: the lower pH, the better the extraction. At pH 2 the fibers showed a very high capacity, being able to take up about 9 meq/g of Cr^{6+} from a 0.23 M solution of dichromate. The elution with 1.25 M NaOH, 5% $NaHSO_3$, and 0.1 M NaOH was studied as a function of time (Fig. 2). We can see that very fast elution was obtained with 1.25 M NaOH: more than 50% of the loaded Cr^{6+} was eluted within 2 min.

Gold extraction

The extraction of gold from a cyanide solution was studied for the weak and strong base anion- exchangers and hollow fibers containing the both weak and strong base groups. The results are given in Fig. 3. We can see that for weak base fibers the gold extraction was reduced sharply as pH was raised. The strong base PEQ-1 exchanger shows no dependence on pH, and gold extraction remained very high over the whole pH range. Very interesting results were obtained for the samples prepared with PDA and EDA partly quaternized. Even though the samples have a low content of strong base groups (6 and 12.5%), they were able to extract gold, at much higher values of pH.

The results of our experiments demonstrate the unique properties of the fibrous material. However a more comprehensive investigation should be done in order to reveal all the advantages and to assign this separation system to the most suitable applications.

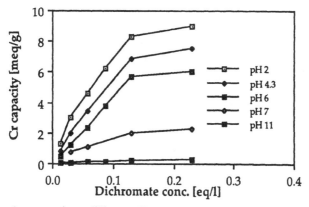

Fig. 1. Chromium capacity at different pH as a function of dichromate concentration in solution.

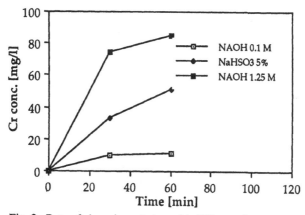

Fig. 2. Rate of chromium elution with different eluents.

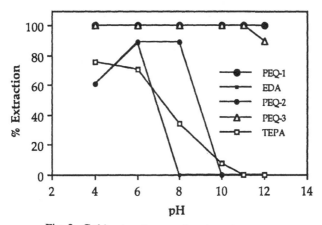

Fig. 3. Gold extraction as a function of pH.

24

REFERENCES

1. a) A. Warshawsky. Chelating ion exchangers, in "Ion Exchange and Sorption Processes, in Hydrometallurgy, Ed. M. Streat, D. Naden, Wiley, 1987.
   b) Y. Zhu, E. Millon, A.K. Sengupta. Towards separation of toxic metal (11) cations by chelating polymers: some noteworthy observations. React. Polymers, 13,3,241,1990.
   c) D. Lindsay, D.S. Sherrington. Copper selective chelating resins. Batch Extractions. React. Polymers, 12, 10, 59 (1990).

2. a) V.S. Soldatov, A.A. Shunkevich, G.I. Sergeev. Synthesis, structure and properties of new fibrous ion exchangers. React. Polym, Ion Exch, Sorbents, 7, 159-72, 1988.
   b) N.E. Nemilov, O.V. Prisekina, S.I. Besmykh, L.V. Emets, L.A. Volfi Fibrous sulfocationite. USSR SU 1 131 884.
   c) A.A. Lysenko, O.V. Astashikina, L.V. Emets, L.A. Volfi. Study of protolytic properties of ion-exchanger fibers. Zh. Prikl. Khim 62, 10, 2287, 1989.
   d) V.S. Soldatov and G.I. Sergeev. Fibrous ion exchangers - prospective sorbents for separating heavy metal ions from aqueous solutions. Zh. Vses Khim. O va in Dm Mendeleeva, 35, 1, 101, 1990.

3. a) S. Tsumeda, K. Saito, Sh. Furusaki, T. Sugo, J. Okamoto. Metal collection using chelating hollow fiber membrane. J. Memb. Sci. 58, 221, 1991.
   b) C.H. Yamagishi, K. Saito, Sh. Furusaki, T. Sugo, I. Ishigaki. Introduction of a High-Density Chelating Group into a Porous Membrane without Lowering the Flux. Ind. Eng. Chem. 30, 2234, 1991.

4. E. Korngold and D. Vofsi. Water Desalination by Ion Exchange Hollow Fibers, Proceeding of the 12th Int. Symp on Desalination and Water Re-use, Malta, April 1991.

5. E. Korngold. Manufacture of improved anion exchange polyethylene and bipolar membranes for electrodialysis of electroless copper solution. Report No. BGUN ARI-49-87 1987.

6. T.M. Suzuki, T. Yokoyama, Preparation and complication properties of polystyrene resins containing diethylentetramine derivatives. Polyhedron, 3, 8, 933, 1984.

7. T. Sata, R. Izuo. Modification of Transport Properties of Ion Exchange Membrane. XI. Electrodialytic Properties of Cation Exchange Membrane Having Polyethylenimine Layer Fixed by Acid-Amide Bonding. J. Appl. Pol. Sci. 41, 2349 (1990).

8. F. de Körösy. An amphoteric ion-permselective membrane, Desalination 16, 85, 1975.

9. S. Belfer and R. Glozman. Anion-exchange resin prepared from polystyrene Crosslinked via a Friedel-Crafts reaction. J. Appl. Polym. Sci. 24, 2147, 1979.

10. A.B. Pashkov, V.A. Vakulenko, I.V. Samborsky, M.K. Makarov. Production of High-Molecular Amphoteric compounds by means of polymer analogous conversions. J. Pol. Sci., Polymer Symposia, 47, 147 (1974)

11. S. Belfer and S. Binman, Removal of heavy metals from aqueous solutions by means of hollow fibers. Report no. BGUN-ARI-12-91, Beer-Sheva, The Institutes of Applied Research, 1991.

12. B.R. Green and A.H. Potgieter. Nonconventional weak-base anion exchange resins useful for the extraction of metals, especially gold, in: "Ion Exchange Technology," Ed. D. Naden, H. Streat, London, 1984, p. 626.

13. A.M. Schwellnus and R.R. Green. Structural factors influencing the extraction of gold cyanide by weak-base resins, in "Ion-Exchange for Industry" Ed. M. Streat,. London, 1988, p. 207.

PREPARATION AND CHROMATE SELECTIVITY OF WEAKLY BASIC ION EXCHANGERS BASED ON MACROPOROUS POLYACRYLONITRILE

R J ELDRIDGE and S VICKERS
CSIRO, Division of Chemicals and Polymers,
Private Bag 10, Clayton, Victoria 3168, Australia

ABSTRACT

Macroporous crosslinked polyacrylonitrile beads were prepared by suspension copolymerization. Heating with oligo-amines converted the beads to weak base ion exchange resins. Capacities up to 7.6 meq/g were obtained with diethylenetriamine. The resins are resistant to attack by acidic or alkaline chromate solutions. They are selective for Cr(VI) over sulfate and chloride, although less so than commercial strong base resins. Because of their stability and ready regeneration with alkali they are well suited to recovery of chromate from industrial wastewaters.

INTRODUCTION

Chromic acid and chromates are still widely used in anodising, electroplating and corrosion control, despite the toxicity of Cr(VI) in the environment. Stringent limits on Cr(VI) concentrations in industrial wastewaters are usually met by reduction to Cr(III) followed by precipitation and landfilling, although recovery of Cr(VI) is clearly preferable in principle and has often been demonstrated [1].

In 1976 Kunin [2] described the use of a weakly basic ion exchanger to recover chromate corrosion inhibitors from cooling tower blowdown. The strong base resins which had been used previously suffered two major drawbacks—they were readily fouled and their very high chromate selectivities meant that caustic brine was needed for regeneration, so that the recovered chromate was contaminated with large amounts of salt. Weak base resins are easily regenerated with alkali, but those available before the mid 1970s were severely degraded by Cr(VI). Even modern, oxidation-resistant weak base resins have not completely displaced strong base resins in chromate recovery applications: both strong and weak base anion exchangers are used to treat chrome plating rinsewaters [1,3], with the choice depending on the relative importance of low leakage, ease of regeneration and resin lifetime. It should be noted that even highly selective strong base resins show gradual breakthrough of Cr(VI), necessitating the use of a polishing column. The reason for this behaviour has been elucidated by Sengupta and Clifford [4,5].

Soldatov et al [6,7] have described Cr(VI) recovery on fibrous anion exchangers. Modification of polypropylene-graft-acrylonitrile fibres with a series of oligo-amines [8,9] yields weakly basic materials having high Cr capacities, unusually high basicities and excellent oxidation resistance. Green et al [10-12] used similar chemistry—the reaction of crosslinked polyacrylonitrile (PAN) beads with 1,2-diaminoethane—to prepare a weak base resin in bead form. We have now prepared a range of resins from macroporous PAN beads by reaction with oligo-amines $H(NHCH_2CH_2)_nNH_2$, where n = 1–4, and evaluated their potential for chromate recovery. For resins with n = 2 and different porosities we measured ion exchange (IX) rates and equilibrium chromate uptakes in chromate-sulfate and chromate-chloride systems. To facilitate comparison with commercial resins, column runs were done with the same feed compositions used by Sengupta and Clifford [4,5].

EXPERIMENTAL

Macroporous PAN beads crosslinked with a combination of divinylbenzene (DVB) and ethylene dimethacrylate were made by suspension polymerization [13]. Loss of acrylonitrile monomer to the aqueous phase could be reduced but not entirely eliminated by salting-out with calcium chloride. IX groups were introduced by heating the PAN beads, together with a catalyst, in an excess of neat oligo-amine for 6-32 h. Products (designated "CRO" resins) were eluted with 2 M HCl, dried and their IX capacities measured by shaking a weighed portion in excess 0.1 M NaOH and back-titrating. Internal surface areas were determined (by adsorption of nitrogen in a Micromeritics 2205 instrument) and ir spectra recorded on the free-base form. Selectivities were measured by equilibrating 0.15 g (1.0 meq) of resin in the sulfate form with varying amounts of potassium dichromate in 200 mL of sodium sulfate solution. Cr remaining in solution was determined after 48 h by plasma emission spectroscopy. Loading curves were established by pumping solutions containing dichromate and chloride and/or sulfate at 100 mL/h through 1 cm diameter columns containing 6.5-8 mL of resin. Fractions were collected and analyzed for Cr. IX rates were determined by converting 0.3 g of 400-800 μm beads to the sulfate form and adding them to 400 mL of a stirred solution containing 2 or 4 g/L of sulfate and 65, 130 or 260 mg/L of Cr(VI). Samples were withdrawn at intervals, filtered to remove resin and analyzed for Cr. Stabilities were assessed by allowing 0.5 g lots of resin in the salt form to stand, with occasional shaking, in 1 L of 0.01 M sulfuric acid containing 200 mg/L Cr(VI), and similarly exposing 0.5 g lots of base form resin to 20 mL of 3 M NaOH containing 90 g/L Cr(VI). After seven days the resins were eluted with HCl, then shaken in excess NaOH and the supernatant solution analyzed for both OH⁻ consumed and Cl⁻ released.

RESULTS AND DISCUSSION

The best IX capacities were obtained with diethylenetriamine (DETA) at a temperature of 105 °C, an amine:bead ratio $\geq$ 5:1 w/w and a catalyst:bead ratio $\geq$ 0.08. Under these conditions PAN beads containing 14% nitrogen gave a capacity of 7.6 meq/g after a reaction time of 24 h. Rubeanic acid (RA) was the usual catalyst [14], but sulfur and mercaptoacetic acid gave similar results. Figure 1 shows the capacity obtained (at 100 °C) with four different oligo-amines. Infrared bands due to nitrile at 2234-2240 cm⁻¹ and ester at 1723-1732 cm⁻¹ were greatly reduced. A new band appeared at about 1650 cm⁻¹, attributable to >C=N— and/or —C(O)N< groups and therefore consistent with the 2-aminoethylimidazolinyl and substituted amide groups expected to be among the products [7-12, 14]. Amination reduced the porosity

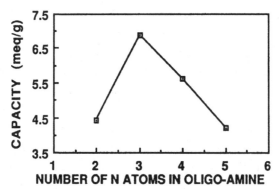

FIGURE 1. IX capacity of CRO resins made by heating 1g PAN beads with 50 mmol oligo-amine and 80 mg catalyst for 24 h at 100 °C.

of the beads, but increasing the DVB content increased both the surface area of the PAN precursor and the fraction retained after amination, although at some cost in capacity. Thus as the final surface area increased from 19 to 67 m^2/g the capacity obtained with DETA fell from 6.5 to 2.7 meq/g.

The amount of chromium(VI) adsorbed by DETA beads from acidic solution was equal to the chloride capacity of the resin, indicating that Cr(VI) is adsorbed as $HCrO_4^-$ and/or $Cr_2O_7^{2-}$. Figure 2 is a titration curve for a resin having a capacity of 6.5 meq/g, in the presence of 2 g/L sulfate. The sloping nature of the curve is typical of polyelectrolytes containing closely-spaced ionogenic groups.

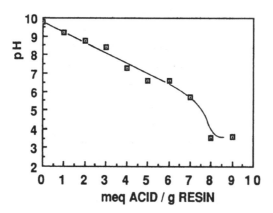

FIGURE 2. Titration of a DETA resin (capacity 6.5 meq/g) with sulfuric acid. 0.1g resin / 200mL of 2 g/L sulfate solution.

The DETA resins are selective for Cr(VI) over both chloride and sulfate. Selectivity over sulfate is illustrated in Figure 3, where the equivalent fraction Cr(VI) in the resin phase is plotted against the residual Cr concentration in sodium sulfate solution at pH 4.5. The highest Cr concentration corresponds to an equivalent fraction Cr(VI) in the solution phase of 0.07 at 2 g/L sulfate and 0.04 in 4 g/L sulfate. Some data of Sengupta and Clifford (5) are included, and show that the selectivity of the commercial strong base resin Amberlite IRA 900 (at pH 4.0) is

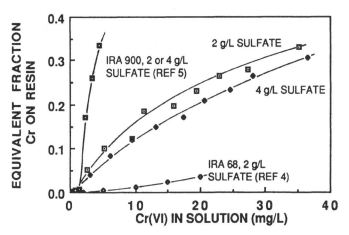

FIGURE 3. Chromate-sulfate selectivity of a CRO resin compared with IRA 68 and IRA 900 (Refs. 4, 5) in acidic solution.

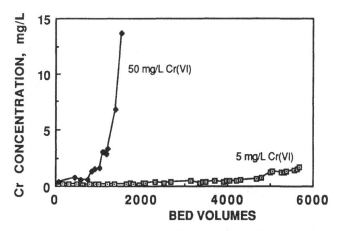

FIGURE 4. Loading curves for a CRO resin (HCl form) with Cr(VI) at 5 or 50 mg/L in 2 g/L chloride solution at pH 4.0.

even higher, while that of the acrylic weak base resin IRA 68 is much lower. Loading curves for Cr(VI)-chloride solutions are shown in Figure 4. The capacity of the resin used, 4.19 meq/g ≈ 1.9 meq/mL, was equivalent to the Cr in 2000 bed volumes (bv) of the 50 mg/L Cr solution and 20 000 bv of the 5 mg/L solution. The usual gradual breakthrough of Cr is observed, but the resin is selective enough to give useful run lengths. Regeneration with 4% NaOH solution (Figure 5) gave a peak Cr concentration greater than 32 g/L = 0.62 M. There was a small amount of tailing but almost all the Cr(VI) loaded was desorbed in 10 bv of regenerant. Regeneration of IRA 900 with 8% NaOH was quite sharp, presumably because of electroselectivity reversal, but the peak Cr concentration was no higher than 16.8 g/L = 0.32 M. The area under the IRA 900 curve is smaller because of that resin's lower capacity.

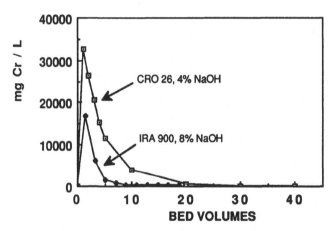

FIGURE 5. Regeneration with NaOH of resins loaded with Cr(VI).

Rates of Cr uptake depend strongly on porosity. Figure 6 shows the uptake rate to be significantly greater for a DETA resin with a surface area of 31 m^2/g than for one with negligible porosity, despite a lower capacity (4.19 vs 6.51 meq/g). Both resins were much slower (and less selective) than the macroporous resin IRA 900.

Resistance to oxidation and hydrolysis was assessed under conditions simulating loading with Cr from a dilute, acidic solution and regeneration at high pH. The table below shows for

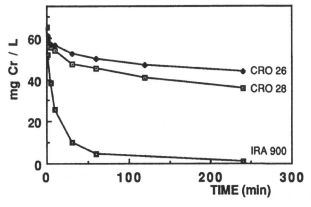

FIGURE 6. Rates of Cr uptake from a solution containing 65 mg/L
Cr and 2 g/L sulfate by DETA resins having surface areas of 2
(CRO 26) and 31 (CRO 28) m^2/g and the macroporous resin IRA 900.

several resins the measured change in capacity. The CRO resins
vary widely in stability, but many showed no capacity change
within experimental error (about 2%). All CRO resins made with
DETA were more stable than IRA 900 or the weak base resin IRA 93
which were readily oxidized at low pH. (A difference between the
proton and chloride capacities implies that some carboxyl groups
have been introduced.) Evidently IRA 900 also suffers partial
degradation of its strong base groups to tertiary amino groups
at high pH, while IRA 68 is relatively stable in both acidic and
alkaline Cr(VI) solutions.

**CHANGE IN PROTON AND CHLORIDE CAPACITY AFTER 7 DAYS IN
DILUTE ACIDIC OR CONCENTRATED ALKALINE Cr(VI) SOLUTION**

RESIN	CAP (meq/g)		%CHANGE (low pH)		%CHANGE (high pH)	
	H$^+$	Cl$^-$	H$^+$	Cl$^-$	H$^+$	Cl$^-$
CRO 7	6.43	6.24	+ 2.3	− 1.8	− 4.2	− 4.6
CRO 8	6.01	6.04	− 0.3	+ 0.8	− 0.2	− 0.5
CRO 24	6.62	6.50	− 2.4	− 3.2	− 5.4	− 6.0
IRA 68	4.93	4.25	+ 0.4	− 4.0	− 3.0	− 4.5
IRA 93	3.69	3.75	−19	−51	− 2.7	− 1.9
IRA 900	1.22	1.43	+35	−14	+24	+15

CONCLUSION

Resins derived from PAN by reaction with DETA are suited to
applications involving Cr(VI) because of their high capacity,

moderate selectivity, outstanding stability and ease of regeneration.

REFERENCES

1. Bolto, B.A. and Pawlowski, L., Wastewater Treatment by Ion Exchange, Spon, London, 1987, pp. 57-62.
2. Kunin, R., New Technology for the Recovery of Chromates from Cooling Tower Blowdown. Amber-hi-lites, No. 151, Rohm and Haas, Philadelphia, 1976.
3. Newman, J. and Reed, L.W., Weak Base Versus Strong Base Anion Exchange Resins for the Recovery of Chromate from Cooling Tower Blowdown. AIChE Symp. Ser., 1980, **76** No. 179, pp. 199-208.
4. Sengupta, A.K. and Clifford, D., Chromate Ion Exchange Mechanism for Cooling Water. Ind. Eng. Chem. Fundam., 1986, **25**, 249-258.
5. Sengupta, A.K. and Clifford, D., Important Process Variables in Chromate Ion Exchange. Environ. Sci. Technol., 1986, **20**, 149-155.
6. Soldatov, V.S., New Fibrous Ion Exchangers for Purification of Liquids and Gases. In Chemistry for Protection of the Environment, ed. L. Pawlowski, A.J. Verdier and W.J. Lacy, Elsevier, Amsterdam, 1984, pp. 353-364.
7. Soldatov, V.S., Shunkevich, A.A. and Sergeev, G.I., Synthesis, Structure and Properties of New Fibrous Ion Exchangers. React. Polym., 1988, **7**, 159-172.
8. Soldatov, V.S., Sergeev, G.I., Martsinkevich, R.V., Pokrovskaya, A.I. and Shunkevich, A.A., Sorption of Cu(II) Ions by Fibrous Materials Containing Substituted 2-Imidazol-ine Groups. Dokl. Akad. Nauk BSSR, 1984, **28**, 1009-1010.
9. Soldatov, V.S., Sergeev, G.I., Martsinkevich, R.V. and Pokrovskaya, A.I., Acid-base and Sorption Properties of a New Fibrous Ion-exchanger Containing Substituted Imidazoline Groups. Zh. Prikl. Khim., 1988, **61**, 46-50.
10. Green, B.R. and Potgeiter, A.H., Unconventional Weak-base Anion Exchange Resins, Useful for the Extraction of Metals, Especially Gold. In Ion Exchange Technology, ed. D. Naden and M. Streat, Ellis Horwood, Chichester, 1984, pp. 626-636.
11. Schwellnus, A.H. and Green, B.R., Structural Factors Influencing the Extraction of Gold Cyanide by Weak-base Resins. In Ion Exchange for Industry, ed. M. Streat, Ellis Horwood, Chichester, 1988, pp. 207-218.
12. Schwellnus, A.H. and Green, B.R., The Chemical Stability, Under Alkaline Conditions, of Substituted Imidazoline Resins and their Model Compounds. React. Polym., 1990, **12**, 167-176.
13. Egawa, H., Nakayama, M., Nonaka, T. and Sugihara, E., Recovery of Uranium from Seawater.IV.Influence of Crosslinking Reagent on the Uranium Adsorption of Macroreticular Chelating Resin Containing Amidoxime Groups. J. Appl. Polym. Sci., 1987, **33**, 1993-2005.
14. Hurwitz, M.J. and Aschkenasy, H., Belgian Patent 637 380 (1964).

COMPARISON OF STRONG BASE RESIN TYPES

JAMES A. DALE AND JAMES IRVING
Purolite International Ltd.
Cowbridge Road, Pontyclun, Mid-Glam., Wales, CF7 8YL.

ABSTRACT

Strong base anion exchange resins have been prepared using amines which are homologues of N,N-dimethylethanolamine (Type-II), namely N,N-dimethylpropanolamine (Type-III) and N,N dimethyl-isopropanolamine (Type-IIIiso). These lesser known resins have been compared with the established resins. Type-II resins are often preferred to the original Type-I resins, because of higher regeneration efficiency. Their main drawbacks are poorer silica removal and thermal stability . Comparing the latter has helped in the understanding of thermal degradation. Investigation of regeneration efficiency and silica removal shows the Type-III resin to be worth complete evaluation.

SYNTHESIS

The anion exchange resins were prepared from one intermediate (chloromethylated poly(styrene-co-divinylbenzene), to ensure comparability(1). A twofold excess of each amine was added at 3-5°C, held overnight, and heated to 40°C, for 2 h., then washed with excess hydrochloric acid and rinsed. Calculations using the initial weight capacity values given in Table 2 shows identical activation. The moles of water per functional group were 14.8, 14.1, 14.3, and 14.2 for the Types-I, -II, -III and -IIIiso respectively.

RESIN REGENERABILITY

The performance of these resins is limited by the extent of regeneration. Measurements on self diffusion with radioactive tracers, first by Soldano and Boyd[2], later by Diamond *et al.*[3], suggested that selectivity is determined by changes in water structure inside and outside the resin beads. Hence the hydrated hydroxide ion prefers the aqueous phase. This mechanism was supported by further work of Diamond *et al.*[4], and the presumed high hydration of the hydroxide ion was later confirmed by its extraction into organic phases, with three moles of bound alcohol, using quaternary ammonium phase-transfer catalysts[5]. The ease of regenerating Type-II resins, relative to Type-I, was evaluated by Wheaton and Bauman[6]. The fixed hydroxyl group, aids the accomodation of hydroxide ions. Higher homologues of the hydroxyalkyl series are investigated here. Table 1 gives regeneration capacities, using sodium hydroxide solution at 4% (w/v) for regeneration levels of 65g and 150g /L of each resin, chloride and sulphate forms. For the chloride form, tests were validated by chloride release and hydroxide uptake. Surprisingly, chloride removal was greater than sulphate [7,8]. The capacities of both Type-III's are higher than for the Type-I resin.

TABLE 1   REGENERATION CAPACITY eq/L (% OF TOTAL)

RESIN	REGENERATION LEVEL g/L NaOH	REGENERATION CAPACITY			
		Cl		SO_4	
		eq/L	(%)	eq/L	(%)
TYPE-I	65	0.58	(40.7)	0.54	(38.6)
	150	0.81	(57.9)	0.70	(50.0)
TYPE-II	65	1.12	(88.4)	1.00	(72.5)
	150	1.31	(94.6)	1.10	(79.7)
TYPE-III	65	0.79	(58.4)	0.78	(57.4)
	150	1.02	(75.3)	0.96	(70.6)
TYPE -IIIiso	65	0.92	(68.4)	0.78	(58.6)
	150	1.14	(85.6)	0.90	(67.7)

Type-II resins have two main drawbacks; i) heat stability--the maximum temperature is 35-40°C (95-105°F)--and ii) silica leakage--higher from a Type-II resin than a Type-I. Therefore, hot sodium hydroxide cannot be used to improve silica removal.

THERMAL STABILITY

Experimental and Results

The resins were converted to the hydroxide form using 300g of NaOH/L. Higher conversion to hydroxide increases degradation beyond working limits, hence the choice of this regeneration level. Resin samples, under water, in individual sealed plastic bottles, were held at 80°C for up to 16 weeks. Tables 2,3 give dry weight and volume capacities at specified intervals. The figures in brackets at zero weeks give the degree of conversion of the starting resin to hydroxide.

TABLE 2   DRY WEIGHT CAPACITY eq/kg

WEEKS, 80°C	TYPE-I	TYPE-II	TYPE-III	TYPE-IIIiso
0	4.23 (58%)	3.81 (94%)	3.58 (75%)	3.57 (83%)
2	3.97	3.42	3.38	3.86
4	3.86	3.35	3.30	3.84
8	3.77	3.15	3.16	3.79
16	3.58	3.19	3.10	3.67
% STABLE 0-16 WEEKS	84.60	83.70	86.60	102.00

TABLE 3   VOLUME CAPACITY eq/L

WEEKS, 80°C	TYPE-I	TYPE-II	TYPE-III	TYPE-IIIiso
0	1.40 (58%)	1.38 (94%)	1.36 (75%)	1.33 (83%)
2	1.28	1.19	1.19	1.38
4	1.21	1.15	1.15	1.42
8	1.19	1.14	1.01	1.45
16	1.08	1.20	1.14	1.45
% STABLE 0-16 WEEKS	77.10 (82.6 wk 8)	87.00 (74.2 wk8)	83.80	109.00

Type-III resins are interesting, showing good apparent stability. The Type-III(normal) is close to the Type-I and Type-II. Type-IIIiso shows an increased total volume capacity. The retention of strong base capacity (%SBC/total volume capacity) is a major advantage of Type-I resins; see Table 4.

The Type-I resin differs markedly from the others. However, the Type-III resin appears more stable than the Type-II. The Type-IIIiso resin is least stable, although the SBC is steady after 4 weeks; this is because the residual strong base groups are not regenerated: 83% conversion of the resin for test, leaving 17% residual, virtually equivalent to the stable SBC.

TABLE 4　　　PERCENTAGE STRONG BASE CAPACITY eq/L

WEEKS, 80°C	TYPE-I	TYPE-II	TYPE-III	TYPE-IIIiso
0	100.0	100.0	100.0	100.0
2	100.0	73.5	80.0	22.7
4	100.0	56.0	63.3	16.8
8	100.0	36.4	51.0	16.7
16	94.7	36.4	51.0	16.7

The % SBC quoted, refers to the heat treated resin. Table 5 gives values in eq/L based on the reference volume (as new). The overall % of stable resin sites after the 16 week treatment is given in the last row. Thus, the advantage of the Type-I resin is not so dramatic. The Type-III resin sits midway between the Type-I and Type-II.

TABLE 5　　　STRONG BASE CAPACITY eq/L
BASED ON RESIN AT START, HYDROXIDE FORM, REFERENCE VOLUME

WEEKS, 80°C	TYPE-I	TYPE-II	TYPE-III	TYPE-IIIiso
0	1.02	1.23	1.15	1.17
2	1.02	0.82	0.82	0.36
4	0.96	0.65	0.70	0.30
8	0.86	0.48	0.54	0.32
16	0.80	0.23	0.43	0.28
% STABLE 0-16 WEEKS	78.00	18.70	37.40	23.40

Finally, table 6 gives the moisture retention values. Resins prepared on the same polymer matrix predictably show moisture differencies, and higher moisture clearly associates with improved heat stability(9).

Changes in moisture retention with heat treatment occur. Type-I
resin shows an increase. Type-III also increases over the
first 8 weeks. All the hydroxyalkyl resins peak in moisture
retention. The earlier the peak the more unstable the resin.

TABLE 6   PERCENTAGE MOISTURE RETENTION

WEEKS, 80°C	TYPE-I	TYPE-II	TYPE-III	TYPE-IIIiso
0	52.9	49.2	47.9	47.7
2	54.2	51.0	50.4	49.9
4	55.3	51.4	50.7	48.2
8	55.0	49.3	54.6	46.5
16	56.7	47.4	48.3	46.5

Comments

There are at least three decomposition modes available to
hydroxyalkyl resins. In addition to direct nucleophilic
displacement by hydroxide ion (Hofmann degradation) possible in
Type-I resins(10,11), the Type-II resins also have the Hofmann
elimination pathway available(12). This reaction for the
synthesis of olefins, historically famous in traditional low-
molecular-weight chemistry, is detrimental to anion exchangers,
and also to phase-transfer catalysts. Furthermore, quaternary
ammonium alcohols form epoxides, five and six membered cyclic
ethers(13). In fact, there may be no way to categorically
increase quaternary ammonium ion stability. Obviating one
decomposition pathway -- even the predominant one -- may only
allow others to take over, as Hatch and LLoyd found with a
neophyl structure(9). Anion exchange resins have been made from
hydroxypropyl containing amines before(2,14). However, so far
as we know, both operating performance and thermal stability
were not evaluated. In the literature cited, a rather high 6%
divinylbenzene copolymer was chosen, and it appears that the
copolymers were only partially aminated.

The above results for thermal stability and regenerability
suggest the Type-III resins have a use, because they offer a
useful compromise with better capacity than Type-I, and better

stability than Type-II. The latter arises because the
unregenerated sites are also the source of the apparent heat
stability of the Type-IIIiso resin. To obtain a true picture,
it is necessary to deduct the residual unregenerated sites from
the total SBC. For Type-III, 0.43eq/L minus 0.34 eq/L, leaves
0.09eq/L, hydroxide form. Similarly one finds 0.21eq./L for the
Type-I and 0.15eq/L for the Type-II. Hence the Type-III resin
is not fundamentally more stable than the Type-II resin.
Nevertheless, Type-III resins may offer practical advantages
under controlled operating conditions. Type-III resin would be
particularly useful if its silica removal performance were
superior to that of Type-II resin. A first investigation is
presented.

OPERATING PERFORMANCE

Local, decationised, feed water contained the natural silica,
necessary for the evaluation, but at a low level. In some cases
the mineral acid was deliberately increased to increase the
load. This had the practical benefit of shortening the run.
Table 7 gives operating capacity obtained for Type-III. This is
compared with published data for the Type-I, Type-II, and
acrylic Type-I resins under the same operating conditions. This
is shown graphically in figures 1 and 2.

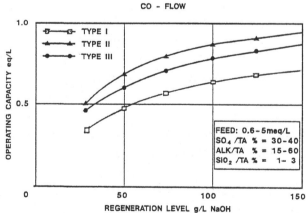

FIGURE 1   OPERATING CAPACITY VERSUS REGENERATION LEVEL
CO - FLOW

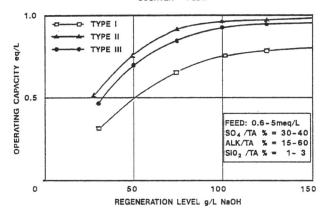

FIGURE 2 OPERATING CAPACITY VERSUS REGENERATION LEVEL
COUNTER - FLOW

TABLE 7 OPERATING CAPACITY OF TYPE-III
SODIUM HYDROXIDE REGENERATION
TO 100mg/m³ SILICA LEAKAGE END-POINT

REGENERATION LEVEL, MODE	OPERATING CAPACITY	PUBLISHED DATA, eq/L		
		TYPE-I	TYPE-II	SBA ACRYLIC
CO-FLOW	eq/L			
60g/L	0.62	0.50	0.66	0.69
150g/L	0.87	0.66	0.95	0.86
COUNTER-FLOW	eq/L			
60g/L	0.74	0.60	0.80	0.72
150g/L	0.95	0.80	1.00	1.00

The capacities obtained are higher than for Type-I resins and
lower than for Type-II or acrylic Type-I resins. Values at
150g/L sodium hydroxide counter-flow were obtained from
published data, by extrapolation. Table 8 gives base silica
leakage. Values are between those for Type-I and-II or acrylic.

TABLE 8 SILICA LEAKAGE OF TYPE-III

REGENERATION LEVEL, MODE	% SILICA/ TOTAL ANIONS	TYPE-III SiO₂ mg/m³	PUBLISHED DATA, SiO₂ mg/m³	
			TYPE-I	TYPE-II
CO-FLOW				
60g/L	1%	18	15	40
150g/L	3%	14	10	20
COUNTER-FLOW				
60g/L	1%	12	8	20
150g/L	3%	10	6	15

SUMMARY

The differences between the resins lie in three areas: operating capacity, silica removal, and heat stability. These can be summarised:

CHARACTERISTIC	TYPE-I	TYPE-II	TYPE-III	TYPE-I ACRYLIC
OPERATING CAPACITY	-	+	+	+
SILICA REMOVAL	+	-	1/2	+
HEAT STABILITY	+	-	1/2	-
TOTAL	2+	1+	2+	2+

CODE: + = GOOD ; - = POOR ; 1/2 = INTERMEDIATE

By this assessment, the Type-III resin may well have a role where moderate heat stability and silica removal are combined.

NOMENCLATURE

Capacity	--Volume, regeneration	eq/L
	--Weight	eq/kg
	Feed (solution)	meq/L
ppb	Concentration	mg/m^3
Reg.	Regeneration	g/L
SBC	Strong base capacity	eq/L
TA	Total Anions	

REFERENCES

(1) Roberts, G.O.& Millar, J.R., Ion Exch.Proc. Ind.1970,42-46.
(2) Soldano, B.A.& Boyd, G.E., J. Am. Chem. Soc.,1953,75,6099.
(3) Bucher, J., Diamond, R.M.& Chu B., J.Phys.Chem.,1972,76,2459.
(4) Chu, B., Whitney, D.C.& Diamond, R.M., J.Inorg.Nucl.Chem., 1962, 24, 1405-1415.
(5) Agarwal, B.R.& Diamond, R.M., J.Phys. Chem.,1963,67,2785.
(6) Wheaton, R.M.& Bauman, V.C., Ind. and Eng. Chem,1951, 43, 1088.
(7) Boari, G., Liberti L., Merli, C.& Passino, R. Desalination,1974 15,145-166.
(8) Nolan, J.D.& Irving, J., Ion Exchange Technology,1984,160-168.
(9) Hatch, M.J.& Lloyd, W.D., J. App. Poly. Sci., 1964, 8, 1659.
(10) Baumann, E.W., J. Chem. Eng. Data. 1960,5,376.
(11) Vasicek, Z., Stamberg, J.,& Dufka, O., Coll. Czech. Chem. Comm., 1971,36,1817.
(12) Cope, A.C.& Trumball, E.R., Org. Reactions, 1960, II,317.
(13) Ref.12:352,389,390.
(14) Bauman, W.C.& McKellar, R., U.S.Pat. 2614099 (1952).

NITRATE REMOVAL FROM WATER : THE EFFECT OF STRUCTURE ON THE PERFORMANCE OF ANION EXCHANGE RESINS

P.TEBIBEL and E.ZAGANIARIS

Rohm and Haas Co., Chauny Research Laboratories,
02301 Chauny, France

ABSTRACT

NO_3^-/Cl^-, $SO_4^=/Cl^-$ and HCO_3^-/Cl^- selectivity coefficients of strong base anion exchange resins having trimethyl ammonium or triethyl ammonium functional groups and varying in total exchange capacity, moisture and porosity, were determined.

Both the NO_3^-/Cl^- and the $SO_4^=/Cl^-$ selectivity coefficients were found to depend on the type of functional groups and the moisture of the gel polymer phase (not the overall moisture content of the resin). The $SO_4^=/Cl^-$ coefficient also depended on the total weight capacity.

A computer program was developed which calculates the effluent concentration histories as a function of the equilibrium separation factors. Good agreement was found between experimental data and calculated values when separation factors were used as functions of the resin composition.

By using this computer program, NO_3^- removing capacities were calculated as a function of resin parameters such as moisture, porosity and total volume exchange capacity. The parameters giving maximum removal of NO_3^- are indicated.

INTRODUCTION

The use of ion exchange resins to remove NO_3^- from potable water is a well established technology. Strong base anion exchange resins (SBR) in the Cl^- form are used. The NO_3^-, along with the other anions present in the water, are exchanged for Cl^-. Since only the NO_3^- are the undesired anions, the useful capacity of the resins is determined by the NO_3^- concentration in the effluent up to a chosen end-point. Consequently, it is the selectivity of the resins for the anions present in the water that determines the NO_3^- removing capacity. In the past years, the effect of the functional groups and the resin matrix on the NO_3^- / $SO_4^=$ selectivity has been studied (1, 2) and anion exchangers have been developed having increased NO_3^- / $SO_4^=$ selectivity in comparison with the conventional SBR. These NO_3^- selective resins have triethyl ammonium (TEA) functional groups instead of trimethyl ammonium (TMA) groups of the conventional, Type I, SBR.

In the present work, conventional and NO_3^--selective resins varying in total exchange capacity, moisture content and porosity have been studied for their selectivity for NO_3^-/Cl^-, $SO_4^=/Cl^-$ and HCO_3^-/Cl^-. The performance in NO_3^- removal was analyzed by using a computer program for the calculation of the effluent concentration histories and the concentration profiles.

EXPERIMENTAL

The selectivity coefficients of NO_3^-/Cl^-, $SO_4^=/Cl^-$ and HCO_3^-/Cl^- were determined by placing a given quantity of resin in Cl^- form in a solution containing the counter-ion as Na^+ salt. Column capacities were determined using three different water compositions, indicated in Table 1.

TABLE 1
Influent Analysis (meq/L)

	Water N° 1	Water N° 2	Water N° 3
$NO3^-$	1.2	1.2	1.2
$SO4^=$	1.2	2.4	3.6
Cl^-	1	1	1
$HCO3^-$	4	4	4

The ion exchange resin columns were of 2 cm diameter and the resin bed 80 cm high. The loading flow rate was 35 BV/h at ambient temperature. Regeneration used 90 g NaCl/liter of resin using a 6 % NaCl solution in a counter flow mode. Regeneration flow rate was 2 BV/h. Following the NaCl solution, 3 BV of deionized water were used at the same flow rate to displace the regenerant.

The ion exchange resins used in this work are described in Table 2. The porosities indicated in Table 2 were calculated from the apparent and skeletal densities of the resins.

The computer model developed uses a set of equilibrium equations, represented by the separation factors, as well as a set of differential

TABLE 2

	Physical structure	Functional groups	Porosity (mm3/g)	moisture (%)	Total Exchange capacity (meq/g)	(eq/L R)	HMS (mm)	Gel moisture (%)
Resin A	gel	TMA	NA	46.0	3.57	1.33	0.56	46
Resin B	gel	TMA	NA	44.8	3.82	1.46	0.60	45
Resin C	gel	TMA	NA	56.0	4.31	1.30	0.69	56
Resin D	MR	TEA	400	55.6	3.45	1.05	0.70	45
Resin E	MR	TEA	580	56.0	2.66	0.80	0.61	39
Resin F	MR	TEA	400	65.0	3.51	0.83	0.67	57
Resin G	MR	TEA	540	44.0	2.43	0.95	0.65	20
Resin H	MR	TMA	1250	65.0	0.55	0.14	0.60	38
Resin I	Gel	TEA	NA	47.9	3.02	1.19	0.60	48
Resin J	Gel	TMA	NA	51.0	4.10	1.41	0.63	51

NA.: Not Applicable
HMS: Harmonic Mean Size

equations representing the rate at which equilibrium was reached, assuming film diffusion for the loading step and particle diffusion for the regeneration step (3). The program was written in Basic language.

RESULTS AND DISCUSSION

The selectivity coefficients of the resins tested, especially NO_3^-/Cl^- and $SO_4^=/Cl^-$, were found to vary with the resin composition. In general, they decrease as the conversion of the resin to the NO_3^- or $SO_4^=$ form respectively, increases. Table 3 summarizes the average values of the selectivity coefficients, determined at 10 meq/L total solution concentrations.

TABLE 3
Average Selectivity Coefficients

	$K(NO_3/Cl)$	$K(HCO_3/Cl)$	$K(SO_4/Cl)$
Resin A	2.0	0.36	0.084
Resin B	3.7	0.31	0.050
Resin C	2.9	0.49	0.395
Resin D	5.8	0.21	0.019
Resin E	7.4	0.38	0.040
Resin F	5.0	0.32	0.016
Resin G	12.0	0.36	0.044
Resin H	4.4	0.56	0.240
Resin I	5.0	0.35	0.023
Resin J	3.0	0.32	0.100

In order to correlate the above selectivity coefficients with the resin structure, the characteristics of the MR resins were analyzed further, so that they reflect the characteristics of the polymer phase. Thus, instead of the overall moisture content given in Table 2 (which includes the water contained in the macropores), the moisture content of the (gel) polymer phase was calculated from the overall moisture and the porosity of the polymer phase. The results of this calculation are also given in Table 2.

A regression analysis of the average selectivity coefficients of Table 3 as a function of the gel moisture and the weight capacity of the resins gave the following correlations : the NO_3^-/Cl^- selectivity coefficients, K_C^N, were found to depend upon the gel moisture and upon the type of the functional groups of the resin (TEA or TMA), and to be independent of the total weight exchange capacity. The $SO_4^=/Cl^-$ selectivity coefficients, K_C^S, were also found to depend on the type of the functional groups. For the TEA resins, the K_C^S was found to decrease with increasing total weight capacity. For the TMA resins, K_C^S was found to increase with increasing moisture and to decrease with increasing total weight capacity.

The HCO_3^-/Cl^- selectivity coefficients were found to be below 1 for all resins, and independent of moisture or total capacity.

COLUMN PERFORMANCE

As criterion for the end of the cycle, the NO_3^- concentration in the effluent of 0.8 meq/L was set. In order to better analyze this performance in terms of the selectivity coefficients, a model based upon the Multicomponent Adsorption Theory (3 - 7) was used which gives the effluent composition and the composition of the resin along the resin bed length as a function of the separation factors and the mass transfer coefficients.

For a given total ionic concentration in the water, the $SO_4^=/Cl^-$ separation factor, α_C^S, was calculated from the values of K_C^S, the total exchange capacity of the resin and for a given fraction of $SO_4^=$ in the resin. These values, along with the values for NO_3^-/Cl^- and HCO_3^-/Cl^- separation factors, α_C^N and α_C^B, were used in the calculations. For the regeneration step, the corresponding α_C^S calculated at the regenerant concentration were used, while α_C^N and α_C^B were the same for loading or regeneration.

The mass transfer coefficients were estimated from an adjustment of these coefficients so that the calculated effluent concentration histories as well as the spent regenerant composition fit the experimentally obtained data for one loading and regeneration cycle (usually the first cycle). Figures 1 to 3 illustrate some of these results.

It was found in all cases that good agreement between experimental and calculated values was obtained if separation factors were expressed as functions of the resin composition, in order to reflect the experimentally found results. It was also found that, under the conditions in this work, experimental and calculated results

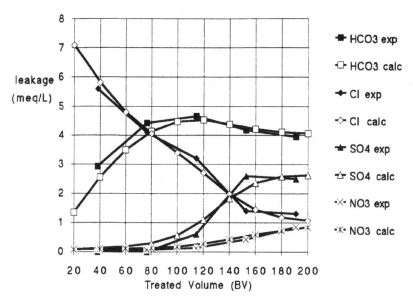

Figure 1. Resin E, Water N° 2, Cycle 3

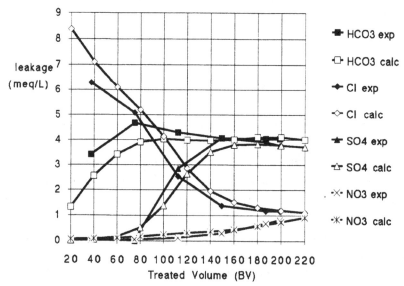

Figure 2. Resin F, Water N° 3, Cycle 3

were in agreement even when only separation factors, without the rate equations, were taken into account, in other words, the effluent concentrations indicated that equilibrium was reached at all plates.

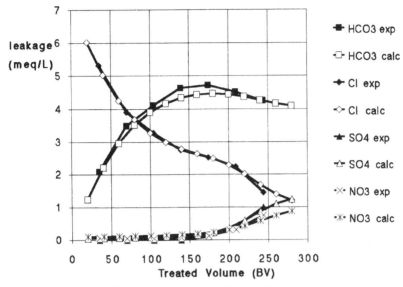

Figure 3. Resin D, Water N° 1, Cycle 3

It follows from the above that the NO_3^--removing capacity of a resin would depend upon the moisture content, the porosity and the total volume capacity of the resin. It is these parameters which determine the values of the separation factors from which, along with the total volume exchange capacity, the NO_3^--removing capacity can be predicted.

Fig. 4 illustrates schematically the calculated NO_3^- removing capacities of TMA or TEA type resins as a function of the overall moisture content and the porosity. As seen in this figure, a curve with a maximum is obtained. This behavior reflects the fact that at low moistures, the total volume exchange capacity increases but at the same time the NO_3^-/Cl^- separation factor increases so that regeneration with NaCl becomes less efficient. At high moistures, the total volume exchange capacity decreases and this results into decreased column capacities. It is interesting to note that for MR type resins, a different curve is obtained at different porosities, since at a given overall moisture content, porosity affects the gel phase moisture and consequently the separation factors.

It is pointed out finally that figure 4 illustrates only trends, since absolute values depend upon factors such as regeneration level and mode, water composition and manufacturing process of the resins.

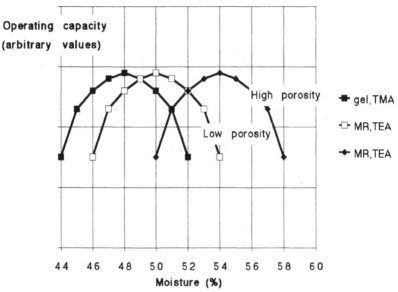

Figure 4. Calculated NO3-Removing Capacity (schematic)

REFERENCES

1. Jackson M.B. and Bolto B.A., Reactive Polymers, 12 (1990), 277-290.

2. Clifford D. and Weber W.J., Jr., Reactive Polymers, 1 (1983), 77-89.

3. Vermeulen T., Separations by Adsorption Methods, In "Advances in Chemical Engineering", T.B. Drew and J.W. Hopes, Jr., Eds., Vol. II, Academic Press, New York, p. 147.

4. Klein G., Tondeur D. and Vermeulen T., I & EC Fundamentals, 6 (1967), 339.

5. Helfferich F.G., idem, p. 362.

6. Clifford D., Ind. Eng. Chem. Fundam. 1982, 21, 141-153.

7. Solt G.S., Nowosielski A.W. and Feron P., Chem. Eng. Res. Des. 66 (1988), 524-530.

DEVELOPMENT OF EXXCHANGE, A CONTINUOUS ION EXCHANGE PROCESS USING POWDERED RESINS AND CROSS-FLOW FILTRATION.

C J COWAN AND MICHAEL COX
Division of Chemical Sciences,
Hatfield Polytechnic
Hatfield, Herts. AL10 9AB, UK

B T CROLL AND P HOLDEN
Anglia Water Services Ltd
Chivers Way
Histon, Cambs.UK

J B JOSEPH
Shanks and McEwan (Waste Services) Ltd
Church Road
Woburn Sands, Milton Keynes, Bucks. UK

A J REES
Crossflow Microfiltration Ltd
11, Charles Street
London WC1, UK

R C SQUIRES
Exxchange Technology
11, Charles Street
London WC1, UK

ABSTRACT

The development of a new type of continuous ion exchange contactor combining the use of powdered ion exchange resins with cross-flow microfiltration is described. This equipment, developed in the laboratory, has also been experimentally evaluated at pilot scale to remove nitrate ions from potable water and the results of these experiments on extraction and regeneration cycles are presented.

INTRODUCTION

Continuous ion exchange offers a number of advantages over fixed bed contactors operating in semi-batch mode. One of particular interest to the water treatment industry is the ability to produce a constant quality product without the need for large product mixing tanks, (in some cases problems may arise with corrosion of pipework and fittings when high peaks of chloride ions arise during the operational cycle of conventional beds). As the majority of the resin is being actively used in the continuous process at any time a lower resin inventory is required. A number of designs of continuous ion exchange plants have been proposed using standard size resins,i.e.ca.0.5-1.2mm diameter beads, but these continuous ion exchange systems can suffer from problems of bead attrition during operation. This problem could be overcome by using powdered resins e.g. the Powdex process,where they are used in thin layers on a filter as a static pre-coat which is removed on exhaustion by backwashing, and discarded.

Cross-flow microfiltration offers an alternative means of removal of fine particles by filtration without the large pressure drops normally found with barrier filtration. The velocity of the fluid flow restricts the deposition of the filtered solids thus maintaining a constant thickness of deposit and pressure drop thereby producing a reasonably constant filtrate flow.

The combination of cross-flow microfiltration and powdered resins seemed to offer a new approach to continuous ion exchange combining the benefits of rapid kinetics and high exchange capacity with the simple design and low civil engineering costs of the cross-flow filtration equipment.Accordingly the process was developed further initially at a laboratory scale and later with a pilot plant.

LABORATORY STUDIES

Figure 1 shows a typical rig design used for these studies.The filter tubes consisted of a woven fabric (UK patent 2191107) which could retain particles down to 0.2µm. The resin used in these studies was MicroIonex OH (Rohm and Haas) with a size range 5-90µm; although this range is outside that associated with microfiltration attrition of these particles could occur to give material below 1µm. In all of the studies the resin was air-dried before use and weighed as such. All chemicals used were reagent grade and the water was a normal potable supply.

The aim of this preliminary study was to investigate the
feasibility of the process and so runs on extraction of nitrate
from contaminated water and regeneration of the resin were
carried out. Because of problems with adequate supplies of water
the process could only be operated in a batch mode and no
attempts were made to try to integrate the extraction and
recovery operations.

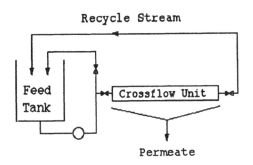

Figure 1   Schematic Diagram of Laboratory Rig

CROSS-FLOW   FILTRATION

Before carrying out any experiments on extraction it was
necessary to discover whether long-term operation of the
equipment with powdered resin in cross-flow mode would cause
problems of blockage of the filter, and to obtain the product
permeate rate at steady-state. The filter was run under full
recycle with the product water being returned direct to the feed
tank thereby not increasing the slurry concentration with time.
The results, table 1, show that the permeate flow rate reaches
steady-state rapidly and was maintained over 24 hours.

Time: (s)	0	30	75	200	270	320	380	440	24 (h)
Flow: (cm^3/min)	102	108	116	132	127	130	127	130	138

Table 1   Permeate Flow from 2 m tube 25 mm diameter
with a 0.1% w/v Resin Slurry on Full Recycle

EXTRACTION EXPERIMENTS

The aim of these experiments was to test the performance of the powdered resin for nitrate extraction. Again these could only be carried out batchwise and the experimental rig was designed with the feed tank fitted with a recycle loop prior to the filter and the latter was operated in a dead-end mode. This allowed the test solution, water/nitrate, to be recycled prior to addition of the powdered resin which was added as a slurry. After recycling for a set time the valve to the filter was opened and the contents of the tank filtered through the unit. The nitrate content of the feed and permeate were determined and the total resin contact time noted. This was measured from the time resin was added to the system to the end of filtration. Table 2 shows that a satisfactory reduction in nitrate levels was obtained in all cases.

Slurry	Contact Time	Feed (NO_3 conc)	Permeate (NO_3 conc)	%Extraction
(% w/v)		(mg/dm^3)	(mg/dm^3)	
0.40	120	44.4	7.2	84
0.30	120	113.2	27.2	76
0.20	200	98.4	20.0	80
0.20	264	108.0	26.0	76
0.10	170	115.8	48.0	58
0.05	170	152.0	94.6	38

Table 2 Nitrate Extraction as a Function of Slurry Concentration, Single Tube in Dead-end Mode.

REGENERATION EXPERIMENTS

These were carried out using the same experimental rig as extraction with the addition of a second tank for the brine regenerant solution. In these initial experiments regeneration took place on a sample withdrawn from the recycle resin stream using dead-end filtration. Thus once the extraction had been carried out brine was passed through the filter cake to regenerate the resin. The efficiency of such regeneration was determined by measuring the amount of nitrate removed from the feed and comparing this figure with the amount of nitrate which appeared in the spent regenerant filtrate. Again the results showed satisfactory regeneration levels (Table 3).

Slurry (%w/v)	Contact Time (s)	Regeneration Time (s)	%Extn	%Regen
0.025	225	30	60	79
0.025	212	30	85	75

Table 3   Extraction of Nitrate from 20L of 85 mg/L Nitrate Solution and Subsequent Regeneration of Resin using 1.6L Sodium Chloride (140 g/L).

Having demonstrated the feasibility of the system it was necessary to investigate the whole process as a continuous operation. The site chosen was the Anglian Water Marham Works in Norfolk where nitrate levels are typically in the range 50-60 mg/L as NO_3^-.

BATCH CONTACTING TESTS

Before commissioning the plant a series of bench tests were performed to determine equilibrium relationships for the Marham raw water and the chosen powdered resin, Purolite A520E. In contrast to conventional column ion exchange, in the Exxchange process there is only a single contact stage between the resin and water and thus the concentration of ions in the final water is determined by the quantity or "dose" of added resin. Two significant observations from the results (Figure 2) are that the nitrate level of treated water cannot rise above that of the feed, and that there are diminishing returns on nitrate removal as the dose is increased so that complete nitrate removal is not possible. The first observation also indicates that low cost non-selective resins could be safely used in the Exxchange process with no risk of high nitrate elution, a key factor in the selection of high cost nitrate selective resins for conventional fixed bed plants.

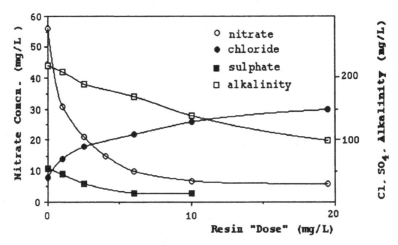

Figure 2   Effect of Resin "Dose" on Ions in Treated
Water

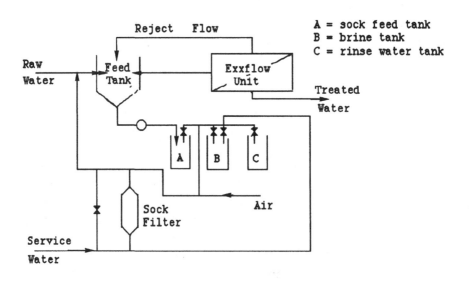

Figure 3   Schematic diagram of Pilot Plant

PERFORMANCE OF PILOT PLANT

The design of the plant is shown in Figure 3 and consisted of a standard Exxflow unit with 8 m curtains of 30 tubes, each 25 mm diameter woven in parallel.

Flow direction manifolds were fitted at each end of the curtains to allow the liquid to travel through six passes of five tubes each, giving a path length of 48 m in each curtain. The process arrangement used was similar to that of the laboratory trials with a side-stream removed for regeneration. Initially the resin for regeneration was removed by a rotary vacuum filter but, because of resin attrition, the filter rapidly blinded and so all subsequent work was carried out using a novel design of sock filter (Patent Appl.PCT/GB 91/01724) with the spent resin slurry pumped direct to the filter and dewatered by air pressure. Rinsing and regeneration are carried out as in a conventional system but with air pressure dewatering between stages to minimise reagent consumption.

The unit has been run continuously with this configuration with typical results of permeate concentrations of 36-40 mg/L nitrate and 87-94 mg/L chloride from feed concentrations of 60 mg/L and 39 mg/L respectively using a resin dose of 5.6 g/L. The flux on the crossflow microfiltration unit initially declined with time to about 0.1 m^3/m^2h, but simply stopping and restarting the feed flow returned the flux to about 0.2 m^3/m^2h. A simple daily stop /start operation has maintained a stable operating regime of 0.15-0.2 m^3/m^2h.

FURTHER WORK

Current work on the plant involves evaluation of different regeneration procedures and resin types. The use of multi-stage contacting during the denitrifying and regeneration cycles will also be investigated to improve the resin dose/ equilibrium relationship and minimise regenerant usage.

CONCLUSIONS

The technical feasibility of a continuous ion exchange process using powdered resin and crossflow microfiltration has been demonstrated in the laboratory and pilot plant for removal of nitrates. Further work is needed to optimise the process.

ACKNOWLEDGEMENT

The authors would like to thank Exxchange Technology, a joint venture between Anglian Water plc and Crossflow Microfiltration Ltd. for permission to publish this paper. Thanks are also due to Purolite International Ltd and Rohm and Haas (UK) Ltd for the supply of powdered resins.The Exxchange process is the subject of a Patent Application (PCT/GB90/01514)

OPERATING EXPERIENCE WITH DIFFERENT WEAK BASE ANION RESINS IN THE FOLLOWING CONFIGURATION: WAC-WBA-SAC-DEGAS-SBA

MIKE E. ROGERS, BOYCE TUCKER
Syncrude Canada Ltd.
Fort McMurray, Alberta
Canada T9H 3L1

and

GUY MOMMAERTS
Bayer Chemical Company

ABSTRACT

There remains a challenge in the treatment of boiler feedwater to adequately remove dissolved organics. This paper deals with some actual in plant experience in using weak base anion [WBA] resins to remove organics. Results are presented for weak base resins supplied by three different manufacturers. Also discussed are the problems and benefits derived from locating the weak base unit in front of the strong acid cation [SAC] unit. The paper draws two main conclusions. Weak base anion resins provide a viable means of removing organics while at the same time assisting in reducing the ionic loading on the strong base anion [SBA] unit at the end of the train configuration. Care has to be exercised in the selection of the type of weak base resin.

INTRODUCTION

In 1978, Syncrude Canada Ltd, a consortium of major oil companies and the government of Alberta, began the operation of the world's second and largest synthetic crude oil mining and processing operation. Located in Northern Alberta, where temperatures go as low as - 40⁰C, the steam/power generation complex, which supports the mining and processing operations, comprises an eight train demineralizer operation. Each train is configured as **WAC-WBA-SAC-DEGAS-SBA.** The degassifier is common to all eight trains. Each train has a design capacity of approximately 4.0 m^3/s. Normal operating configuration is six trains in service and two in regeneration mode. This gives a sustained capacity of approximately 24 m^3/s. The rather unusual location of the weak base unit after the weak acid unit and before the strong acid unit has a history.

HISTORY

Midway during construction of the plant it was realized that problems of organic fouling of the strong base anion resin would be encountered. The reason for this being that the raw water source was from the Athabasca River which contained high levels of humic substances. As a result, organic screen vessels were incorporated into the design. It was expected that there would be sodium leakage and consequently these vessels were placed in front of the strong acid cation units which would take care of any leakage.

RESIN TYPES The organic screen vessels **[OS]** at start-up contained an adsorbent resin which was designed specifically for the removal of organics and had very little ability to remove chlorides and sulphates. Around 1984, resin manufacturers approached Syncrude with advice to switch the adsorbent resin **[Dow S-587]** to true weak base resins and enjoy the benefits of not only organic removal, but also both sulphate and chloride removal. Syncrude was informed that the optimum organic removal would be achieved if the influent water to the weak base units had a high free mineral acidity **[FMA]**. This meant locating the weak base units after the strong acid units. To do this for the existing seven trains [eighth train was added later] required considerable redesign of the control logic and major re-piping. It was decided, therefore, to evaluate the reconfiguration in a pilot plant.

PILOT PLANT

This work was conducted by Dr. Roger Cowles of the Syncrude Research Department and essentially compared the performance of the two arrangements.

Cowles concluded that:

- using a weak base anion exchange resin, instead of an organic screen resin, and changing the configuration gave improved organics removal.
- the ionic loading on the strong base resin was also reduced
- sodium leakage from the weak base resin was minimal.
- weak base anion resin even before the strong acid cation also gave good organics removal and a significant reduction in ionic loading to the strong base anion units.

Based on these results it was then decided to operate the eighth train, which was about to be commissioned, in the new configuration.

TRAIN #8

When Train #8 design started, Syncrude was unsure which configuration would be used. It was decided to obtain maximum flexibility by doing the piping design in such a way that would allow the train to run in either configuration.

PROBLEMS Operating with the configuration where the weak base anion comes before the strong acid cation normally requires the weak base anion to be rinsed for approximately 20 minutes down to a conductivity of 10 mmhos. Any caustic remaining was picked up by the strong acid cation. On the other hand, operating Train #8 with the weak base after the strong acid cation, required a rinse time of two hours as any free caustic affected all trains in service because all streams were combined at the degassifier.

Inability to overcome this problem, without the addition of additional ion exchange facilities like a mixed bed unit, dictated returning to the original configuration of **WAC-WBA-SAC-DEGAS-SBA.**

OPERATING RESULTS

The results presented here represent our experience over the last three years using three different weak base anion resins.

TABLE I

WEAK BASE ANION RESIN PERFORMANCE BY RESIN TYPE AND TRAIN.

PERCENT REMOVAL

TRAIN	RESIN TYPE	RESIN AGE [MONTHS]	TOC	CHLORIDE	SULPHATE
1	WGR2	21	36	0	94
1	WGR2	33	22	0	85
2	AP49	19	61	33	80
2	AP49	31	48	40	80
2	AP49	42	45	44	98
3	WGR2	36	37	1	92
3	AP49	9	65	72	100
4	WGR2	21	30	0	86
4	WGR2	33	25	0	77
5	AP49	10	66	30	99
5	AP49	22	63	10	100
6	AP49	21	56	64	100
6	AP49	33	63	50	80
6	AP49	42	44	31	96
7	MP35	26	65	0	70
7	MP35	38	58	0	97
8	AP49	20	67	92	100
8	AP49	32	63	100	100
8	AP49	45	49	49	88

TABLE II

BAYER AP 49 PERFORMANCE

PERCENT REMOVAL

RESIN AGE [MONTHS]	TOC	CHLORIDE	SULPHATE
9	65	72	100
10	66	30	99
19	61	33	80
20	67	92	100
21	56	64	100
22	63	10	100
31	48	40	80
32	63	100	100
33	63	50	80

TABLE III

DOW WGR2 PERFORMANCE

PERCENT REMOVAL

RESIN AGE [MONTHS]	TOC	CHLORIDE	SULPHATE
21	36	0	94
21	30	0	86
26	65	0	70
33	22	0	85
33	25	0	77
36	37	1	92
38	58	0	97

TABLE IV

BAYER MP35 PERFORMANCE

PERCENT REMOVAL

RESIN AGE [MONTHS]	TOC	CHLORIDE	SULPHATE
26	65	0	70
38	58	0	90

Table I gives the performance of the weak base resins installed in each train. The age of the resins in months is given. The percent removal is determined by taking the difference between the concentration in the inlet water to the weak base unit and the average concentration derived from results taken at specific intervals of the service run. The end of the run is determined not by exhaustion of the weak base unit, but by exhaustion of the strong base as indicated by a rise in silica concentration in the final effluent. Tables II, III and IV are presented for ease in following the performance of each type of resin without reference to which train it might be installed in.

Examination of the results obtained during a run reveals some rather interesting information. To demonstrate this, one train is selected and the performance during the run at different resin ages is analyzed. These results are reported in Table V.

TABLE V
TOC, CHLORIDE AND SULPHATE VS THROUGHPUT

FLOW	TOC			CHLORIDE			SULPHATE		
M^3	21 MTHS	33 MTHS	42 MTHS	21 MTHS	33 MTHS	42 MTHS	21 MTHS	33 MTHS	42 MTHS
757	2.63	2.16	2.6	.145	.07	.08	.359	0	0
2650	2.66	2.43	2.79	.052	.03	.06	.135	0	.08
3758	2.69	2.71	2.96	.053	.71	.17	*	0	0
4750	2.74	2.76	3.13	.215	6.52	1.24	.203	0	0
5700	2.68	3.02	3.19	12.9	14.9	8.09	*	0	0
6650	2.80	6.37	3.48	*	24.5	18.2	.102	1.97	0
7600	2.77	7.69	3.45	*	25.08	24.8	.104	*	0
END	2.94	*	7.38	26.09	28.7	33.6	1.59	0	16.0
INLET	9.47	7.09	6.44	13.18	10.51	15.58	43.9	53.46	51.1

*SAMPLE NOT TAKEN OR RESULTS DISCARDED.

TOC. At the end of the run when the resin is 21 months in service there is still 69% TOC removal efficiency. On the other hand as the resin ages the removal efficiency drops off at the end of the run as indicated by it being only 54% after 33 months and actually falling to zero and starting to throw organics when the resin has been in service for 42 months.

CHLORIDE. For chlorides, the removal efficiency falls to zero about the 5000 m^3 point in the service run when the resin is 21 months old and between 4 and 5 thousand cubic metres when it is 33 months old. The figures for 42 months of resin age for this train actually show continued removal efficiency around 5 to 6 thousand cubic metre mark. This is discussed later in the paper.

SULPHATE. Except for a slight fall off at the end of the run when the resin is 42 months old, sulphate removal remains almost constant at near a 100 % throughout the entire service run for the resin at all other periods in its life cycle so far.

RINSE TIME. The normal rinse time for these units, as stated earlier in the paper, is 20 minutes. After this 20 minute rinse, the conductivity is about 10 mmhos and the train can be put into service. At the present time, with the resin having been in service for 42 months, the rinse time has increased dramatically to 160 minutes.

DISCUSSION Comparing TOC, chlorides and sulphates against throughput during a service run for train #6; the results indicate that when the resin was in service for 21 months or had treated 152 million cubic metres of water, the TOC removal efficiency at the end of the run was just about what it was at the beginning. At the start of the run the removal efficiency was 72 % and at the end it was 69 %. On the other hand, when we look at the performance when the resin was 42 months old and had treated 305 million cubic metres of water, although the removal efficiency started at 70 %, by the end of the run the efficiency had dropped to the point where it was now throwing organics back into the effluent.

With regards to chloride and sulphate removal, there was a consistent break in chloride about midway through the run while sulphate removal remained essentially close to 100 % at all ages of the resin.

It is our belief that we can continue to gain advantage of using the weak base resin by increasing the amount of resin in the vessel. The next problem to be addressed is whether the extended rinse time requirements will require resin replacement or whether a suitable cleaning regime can be adopted either to forestall the degradation of the resin performance or whether even at this stage it can be cleaned and removal efficiency be improved.

CONCLUSION

Weak base anion resins in our system have proven to be able to remove organics from the water at a level sufficient to protect the strong base anion resins. The ability of these resins to remove chlorides and sulphate, even though for only part of the time during the service run, helps relieve the ionic loading on the strong base anion and thus allows longer run lengths. Anyone contemplating using weak base resins in a similar position should be aware that fouling will occur and a suitable cleaning regime must be considered.

THE PROPERTIES AND ADVANTAGES OF UNIFORM PARTICLE SIZE ION EXCHANGE RESINS

STEVE WRIGLEY, ANDRE MEDETE
Separation Systems TS&D Group
Dow Deutschland Inc., 7587 Rheinmünster, Germany

ABSTRACT

This paper describes the properties of ion exchange resins
that have been manufactured with a uniform particle size distribution. The advantages of these resins over conventional
(polydisperse) resins is explained in terms of their kinetic
and physical properties and their improved performance
demonstrated by comparitive tests. Application of these resins
in three major areas is described: (i) water softening, (ii)
water demineralisation and (iii) ultra pure water production.
In each application, the critical resins properties are
discussed and the advantages of uniform resins shown with
respect to chemical efficiency, kinetic performance, rinse
down characteristics, and hydraulic properties.

1. INTRODUCTION

The technology for the manufacture of ion exchange resins has
changed little since the 1940's. The production of resins in a
stirred reactor results in polydispersed beads with the final
resin typically in the -20+50 mesh range. Whilst new resin
product development has been limited by this manufacturing
process, ion exchange application technology continued to
develop with the introduction of mixed beds, layered beds and
packed bed systems. In an attempt to overcome the disadvantages of a wide bead size distribution and to meet the increasingly demanding requirements of these application technologies, these resins had to be screened.

2. UNIFORM PARTICLE SIZE RESINS

The development of uniform particle size DOWEX* MONOSPHERE
resins (*Trademark of The Dow Chemical Company) in the 1980's,
opens the opportunity to tailor resins and optimise their
performance to meet the requirements of each specific application. These resins are manufactured directly as uniform
beads and are not screened from conventional resins. They have
a high degree of bead size uniformity with over 90% of the
beads within ± 50 μm of the mean diameter. The resins have
typical mean diameters are in the range 300 to 1000 μm. The
difference in particle size distribution is illustrated in
Fig. 1 for a conventional -20+50 mesh resin and a 500 μm
uniform particle sized resin.

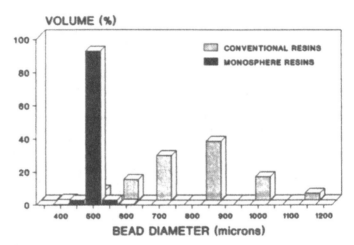

Figure 1. Particle size distribution of conventional and
MONOSPHERE resins.

Screening of conventional resins to produce uniform particle size resins is also possible but has a number of disadvantages, not only for the resin manufacturer but also in
resin quality. To minimise costs, the coarse and fine fractions have to be blended, compromising the quality of other
resin batches. Screened resins also retain the intrinsic
characteristic of the original base product. Screening will
not improve resin mechanical stability and is likely to cause
increased resin damage. The improved mechanical properties are

imparted to MONOSPHERE resins as a result of the manufacturing
process.

3. RESIN CHARACTERISTICS

3.1. Ion Exchange Kinetics

The importance of bead size on ion exchange resin kinetic
performance is widely described (1) - (4) and is a direct
consequence of mass transfer effects. Diffusion at the liquid
boundary layer and within the resin particle itself are both
dependent upon the bead specific surface area which is a
function of the bead diameter. The consequence of reducing
the bead size therefore is to improve the resin kinetics in
both the loading (boundary layer) and regeneration (particle
diffusion) phases. The advantages of the small beads in a
polydispersed distribution are not realised due to the presence of kinetically slower larger beads.

3.2. Mechanical Stability

Standard tests on resin friability (stress), attrition (shear)
and osmotic shock (shrink/swell) show that MONOSPHERE resins
have higher mechanical stability than conventional resins.
This is partly due to the resin matrix structure (degree of
cross-linking) and the manufacturing process (see Table 1).

TABLE 1
Mechanical stability of uniform and conventional resins

	CONVENTIONAL GEL RESINS	MONOSPHERE GEL RESINS	MACROPOROUS RESINS
Friability (crush) (g/bead)	400-700	600-800	400-800
Bead sphericity: - after attrition	75-95	98	60-85
- after osmotic cycling(HCl/NaOH)	50-95(2M)	95-99(2M)	95(8M)

The improved mechanical properties result in longer resin life and fewer fines generation during operation.

3.3 Hydraulic Properties

Head loss in a column of resin beads is described by the Leva equation (6). Important factors relating to the resin are the bead diameter and the bed void fraction. The smaller beads in a polydispersed resin tend to fill the interstitial spaces between the larger beads, thereby reducing the void fraction and increasing head loss. This cancels the advantage of the larger mean diameter (typically 700-750 microns) of the polydispersed resins, so that small uniform particle size resins have similar head loss characteristics to -20+50 mesh resins. This is shown in Fig. 2.

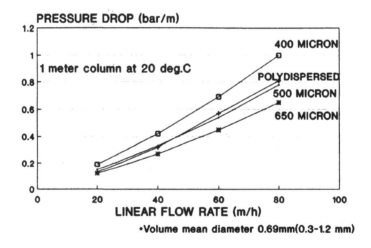

Figure 2. Head loss of uniform and conventional resins.

4. RESIN PERFORMANCE

The advantage of uniform particle size resins over conventional resins have been demonstrated in the laboratory and on working plants in a number of applications using a wide range of operating conditions. Three major application areas are described in the following sections.

4.1 Softening

Traditionally, softener units operate with a large excess of regenerant (typically over 250% of stoichiometry) and at low service flow rates (below 40 BV/h). The fast kinetics of small bead uniform particle size resins challenge this technology by offering the possibility to run smaller units at high flow rates and with greatly improved chemical efficiency. This minimises hardness leakage and results in a substantial reduction of regenerant waste to the environment (3-5 fold at 70 - 200 g/L). In comparative studies between 350 and 400 μm resins against conventional resins of mean diameter 750 μm with uniformity efficient of 1.55, increases in operating capacity of between 30 and 100% have been achieved depending on the particular operating conditions used.

4.2. Water Demineralisation

In water demineralisation, the same advantages have been demonstrated as in water softening in terms of higher chemical efficiency, operating capacity and rinse characteristics together with better effluent quality. This is shown in Fig. 3 for co-flow conditions using a 500 μm type II anion resin compared to a -20+50 mesh conventional resin.

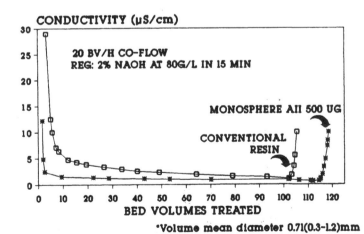

Figure 3. Conductivity profiles of MONOSPHERE & conventional anion resins in demineralisation

Similar advantages can be demonstrated in counter-flow systems such as the UP.CO.RE* packed bed up-flow regeneration process (* Trademark of The Dow Chemical Company). The increased regenerability of MONOSPHERE resins over conventional grades results in regenerant savings of 20-25% on the cation and 15% on the anion at the same plant capacity.

4.3. Ultra Pure Water Production

Ultra pure water for use in the power industry and the manufacture of solid state electronic components requires water of near-theoretical purity (18.2 MOhm.cm). In order to produce such water, ion exchange resins are used in both working and polishing mixed beds. In working mixed beds, good resin separability is a critical factor to ensure high water quality, operating capacity and reduced osmotic damage through cross-contamination.

The degree of resin separation is dependent upon the anion and cation particle size and densities which then determine the terminal settling velocity of the resin beads. Uniform particle size resins have a very narrow range of terminal settling velocities compared to the conventional resins screened for mixed bed operation. Adjustment of the mean bead diameter allows any desired degree of separation of the uniform particle size resins to be achieved. Thus, a 650 micron cation in combination with a 550 micron anion gives an excellent separation for working mixed bed applications. Reducing the cation to 350 microns ensures that the resins remain mixed during operation, an important property of (use once) polishing mixed beds, as any separation would compromise water quality. This is illustrated in Fig. 4.

The 350 μm cation and 550 μm anion combination (MONOSPHERE MR-450 UPE) exhibits substantially better kinetic performance compared to conventional mixed bed grade resins. Using a sulphate leakage test based on Harries (7), a 2:1 volume ration of cation to anion was tested with an influent loaded with 2500 μgkg^{-1} Na$_2$SO$_4$ at flow rates between 50 and 420 BV/H. Even at flow rates of 420 BV/H (corresponding to a linear flow of 60 m/h), the effluent water quality remains excellent compared to the conventional mixed bed (Fig. 5).

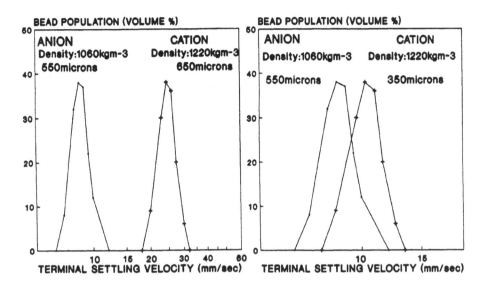

Figure 4. Separability of uniform resins in working & polishing mixed beds: Terminal settling velocity profiles (H/OH forms)

Ionic rinse-down to 18 MOhm.cm conductivity using 0.05 MOhm. cm water in recirculation at 62 m/h in Fig. 6 shows that the conductivity profile of MONOSPHERE MR-450 UPE is shorter than conventional resin mixed beds. This is again a demonstration of the superior kinetics and results in savings in both time and process water.

5. CONCLUSIONS

The characteristics and properties of uniform particle sized resins have been described. The comparative test data given in

this paper show how these resins outperform conventional
resins in all critical areas that are important in ion exchange resin applications.

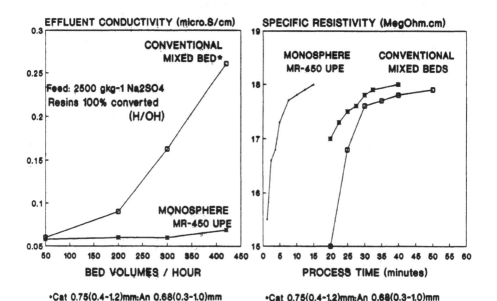

Figure 5. Kinetic evaluation of
MONOSPHERE MR-450 UPE and conventional mixed beds.

Figure 6. Mixed bed rinse
down of MONOSPHERE MR-450
UPE and conventional resins

REFERENCES

(1) Nachod F.; Schubert S. Ion exchange Technology, 1956, 68.
(2) Arden T.V. Water Purification by ion exchange, 1968, 41.
(3) Tremillon Separation par echangeur d'ions, 1965, 64.
(4) Harries R.R. SCI Cambridge, 1988 .
(5) DOWEX analytical method MD 14, 1968.
(6) Leva M., Chem. Eng., 1949, 56(5), 115-117.
(7) Harries R.R., Ray N.J., Effluent & Water Treatment, 1984,
    April, 131

THE SEPARATION OF COMPONENT ION EXCHANGE RESINS OF NARROW SIZE GRADING IN MIXED BED EXCHANGERS, AND EVALUATION OF PERFORMANCE

L.S.GOLDEN AND JAMES IRVING.
Purolite International Ltd.,
Cowbridge Road, Pontyclun, Mid Glam., Wales, CF7 8YL.

ABSTRACT

The operating performance of mixed beds using component resins of narrow size range has been compared with existing systems. To obtain successful separations, it is important to eliminate residual attraction between beads, and to reduce hindered settling. The use of resin density difference between the component resins, rather than particle size difference, can result in superior performance.

INTRODUCTION

Recent work (1,2) has shown that the matching of particle size of components in a mixed bed ion exchanger can improve its performance in virtually all respects. These include; kinetics, regeneration efficiency, ion leakage, throughput between regenerations, and rinse rates. This paper discusses ways in which separation of component resins of similar fine particle size, superior to that from current techniques can be achieved.

OPERATING PERFORMANCE -GEL RESINS

The figure 1 gives operating curves for mixed beds run on a feed of 585 mg/m³ sodium chloride at 50 bed volumes per hour using 60g/L of the appropriate regenerant (HCl and NaOH). The narrow grade resins have superior operating capacity and lower

leakage. Figure 2 shows treatment of local South Wales Mains
water to remove silica to an end-point of 50 μg/m³. using 80g/L
of regenerants. Here, base silica leakage of the narrow range
resin is marginally higher, and volume to reach this is
surprisingly longer. However the difference between the
capacities is significant. Table 1 gives the operating
capacities of gel-type mixed beds to two leakage end-points.
The advantage could not be achieved without separation of
component resins, at least equivalent to that using existing
tailored resins.

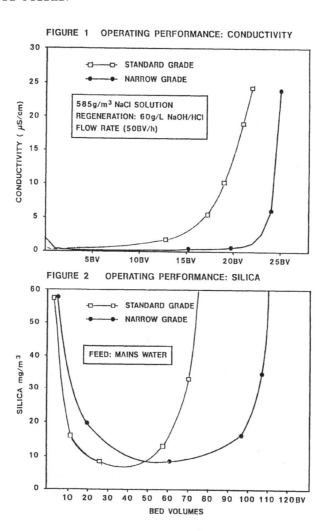

FIGURE 1   OPERATING PERFORMANCE: CONDUCTIVITY

FIGURE 2   OPERATING PERFORMANCE: SILICA

TABLE 1   OPERATING CAPACITY OF GEL-TYPE MIXED BEDS

RESIN	REGENERATION LEVEL	END-POINT	OPERATING (FEED) CAPACITY (SOLN) eq/L
NARROW GRADE	60g/L	$5\,\mu\,Scm^{-1}$	0.24(NaCl)
STANDARD GRADE	60g/L	$5\,\mu\,Scm^{-1}$	0.17(NaCl)
NARROW GRADE	80g/L	$100mg/m^3\ SiO_2$ $50mg/m^3\ SiO_2$	0.195 (LOCAL MAINS) 0.190 (LOCAL MAINS)
STANDARD GRADE	80g/L	$100mg/m^3\ SiO_2$ $50mg/m^3\ SiO_2$	0.170 (LOCAL MAINS) 0.130 (LOCAL MAINS)

RESIN SEPARATION

The separation of fully exhausted component resins of similar
particle size need not be difficult. Table 2 gives data on a
gel cation, and a marginally coarser gel anion resin which, in
the given forms, were surprisingly separated more cleanly than
Trilite components in which the anion resin is approx 0.425-
0.85mm and the cation resin is 0.71-1.2mm. Trilite is the state
of the art technology with specially tailored components to
avoid cross contamination. Unseparated beads from a macroporous
Trilite bed were isolated and size grading measured. No cation
beads were found in the anion resin. Two percent of anion resin
was found in the cation resin. The grading of the contaminant
anion resin was compared with that of the grading of the
component. These data are given in Table 3.

TABLE 2   RESIN CHARACTERISTICS

RESIN TYPE, NARROW GRADE		STRONG ACID CATION	STRONG BASE ANION
IONIC FORM		Na	Cl
BATCH NUMBER		714/89	62/90
PARTICLE SIZE : MICRONS US MESH %			
+ 710	25	NIL	2.8
+ 600	30	TRACE	-
- 425	40	-	TRACE
- 400		1	-
WATER RETENTION: Na, Cl FORMS %		46	48.4
OPTICAL ASPECT %			
PERFECT BEADS		98	95
CRACKED BEADS		2	2
PIECES		TRACE	3
MIS-SHAPES		-	TRACE
TOTAL EXCHANGE CAPACITY eq/L		2.07	1.47
(Na, Cl FORMS) eq/kg		4.58	3.99
RESIN RATIO (MIXED BED) %		40	60

TABLE 3   PARTICLE SIZE OF CROSS-CONTAMINATION; ANION RESIN

PARTICLE SIZE mm	ANION CONTAMINATION %	PUROLITE A-500TL %
< 0.50	0.0	3.2
< 0.60	13.3	20.1
< 0.71	30.0	52.4
< 0.85	86.7	95.1
< 1.00	90.0	97.9
< 1.18	100.0	100.0

Table 3 shows size distribution of the contaminant to be close
to that of the source component. Hence, Stokes law which
predicts faster settling of larger beads is not the major
influence, other factors must be responsible where
contamination is present. During attempts to separate narrow
size range components it was found that traces of residual
hydrogen/hydroxide ion present in resins exhausted to
breakthrough produces significant deterioration in separation
using the resins described in table 2, Na and Cl forms. It may
be concluded, that the cross contamination is attributable to
the interaction either as a result of cation-anion attraction,
by hydrogen bonding, so important in the Powdex Process(3), or
hindered settling between large and small beads(4,5).

Cation-Anion attraction
This is clearly a function of surface area, greater in fine
narrow sized resins. The total surface area of a bed may be
computed from that of individual beads and their number. Table
4 compares the surface area differences of two, simplified,
notionally sized Trilite beds with that of a bed with narrow
range components. More details of the size fractions chosen are
given in Table 5. Furthermore, the surface area effect is
enhanced because a smaller force of attraction is required to
hold beads of smaller mass together. It may therefore be
supposed that the total attraction is a function of surface

area, and the inverse of mass or mass difference of individual beads. Thus total attraction would appear to be a function of bead diameter. In any event it becomes of increased importance to eliminate residual hydrogen and hydroxide ions before component separation is attempted.

TABLE 4   COMPARISON OF SURFACE AREA

PROPERTY \ EXAMPLES OF BEDS	NOTIONAL MIXED BED I			NOTIONAL MIXED BED II			UNI-SIZE BEADS 0.5mm
BEAD DIAMETER d, mm	0.4	0.7	1.0	0.45	0.7	1.0	0.5
SURFACE AREA/BEAD π d^2, mm^2	0.5	1.54	3.14	0.64	1.54	3.14	0.78
GROSS SURFACE AREA/FRACTION	2750	2838	989	1341	2838	1208	7199
TOTAL SURFACE AREA	6577			5387			7199

TABLE 5   CHANGES IN PACKING WITH CHANGE IN GRADING

PROPERTY \ EXAMPLES OF BEDS	SIMPLIFIED MIXED BED I			SIMPLIFIED MIXED BED II			UNI-SIZE BEADS	
							0.7mm	0.4mm
SINGLE BEAD DIAMETER d, mm	0.4	0.7	1.0	0.45	0.7	1.0	0.7	0.4
SINGLE BEAD VOLUME mm^3	0.03	0.179	0.52	0.048	0.179	0.52	0.179	0.033
PORTION OF 1000 ml BED OF SIZE d	250	500	250	250	450	300	100	100
SPHERE VOLUME IN 1000 ml	660			660			600	600
PORTION OF VOLUME, ml OF DIAMETER d	165	330	165	165	297	198	600	600
NUMBER OF BEADS, N IN EACH PORTION	5500	1843	315	2096	1659	381	3351	17905
N$^{1/3}$ d, CONTRIBUTION TO ONE DIMENSION	7.1	8.6	6.8	5.8	8.3	7.2	10.47	10.46
TOTAL, T	22.5			21.3			10.47	10.46
LENGTH, l AVAILABLE 75% EXPANSION	12.1			12.1			12.1	12.1
PACKING OVERLAP T/l	1.85			1.76			0.86	0.86

Hindered Settling

This is clearly of importance when packing is dense. Table 5
gives the calculation of bead numbers in fractions which form
the basis of simplified notional mixed beds. It shows just how
dense the packing is for standard mixed beds during backwash.
Notionally sized Trilite mixed resins are compared with narrow
size range components in 75% backwash expansion. The final row
shows that the available space, expressed in one dimension, is
much larger when beads are of uniform size. Clearly any
hydraulic interference which occurs as a result of differing
particle sizes(1) is avoided. The use of narrow sized ranges
eliminates the advantage of Stokes' Law, and the disadvantage
caused by hindered settling, which can be responsible for the
trace cross contamination in component beds of differing
particle size. Any bed containing uniform beads, regardless of
size gives constant wide spacing.
A further observation is that gel and macroporous resins appear
to have different attraction properties. Although it has been
reported that macroporous resins give fewer and less severe
clumping problems than their gel counterparts, separation tests
using narrow size macroporous resins were less successful,
regardless of ionic form. Unused macroporous resins have to be
antistat treated and fully exhausted, before good separation is
possible. However, perfectly clean separation (less than 1
bead/1000), was achieved by backwashing, after passing 1.5 bed
volumes of 5% ammonium sulphate to fully exhaust the bed.

OPERATING PERFORMANCE-MACROPOROUS RESINS

The use of the higher density Purolite C-160 type offers an
attractive solution to the separation problem. The operating
performance of a bed containing Purofine PFC-160 (0.425-
0.710mm) with a specific gravity of 1.30 in the sodium form,
and Purofine PFA-500, (similar narrow grading) specific gravity
of 1.08 in the chloride form, gives perfect separation after
exhaustion, then followed by full exhaustion with ammonium
sulphate. Only 1 bead in 2000 was found in the separated

components. The use of the same operating procedure also markedly improved the separation of standard Trilite resin components to sub-trace levels, though these were higher than for the Purofine system. The operating profiles are given in the following figures 3 and 4.

FIGURE 3   OPERATING PERFORMANCE: CONDUCTIVITY

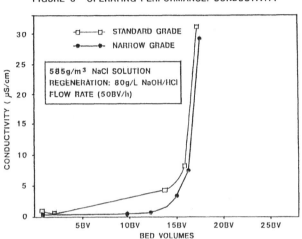

FIGURE 4   OPERATING PERFORMANCE: SILICA

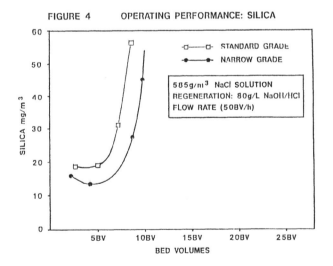

SUMMARY

Near perfect separation of mixed bed components of similar narrow size ranges is possible, provided that attention is paid to the elimination of resin bead attraction, particularly between residual hydrogen form cation sites and hydroxide form anion sites which may be present at breakthrough. The phenomenon of entrainment of small and large particles, which can be caused by hindered settling is completely avoided. The choice of matched components in terms of fine narrow particle size range improves regenerability, kinetics in the treatment cycle, and therefore both operating capacity and treatment quality. Furthermore, the use of resins with moisture contents which would normally be unacceptably low, could be a superior option. Such resins have high volume capacity. Hence overall operation capacities of such mixed beds will be further improved.

ACKNOWLEDGEMENTS

Thanks go to Laurent Dugas, Stuart Price, and Elsa Davies for contributions to the practical work; to John Millar for discussions and helping with references. Finally to the Purolite Company for permission to publish this paper.

NOMENCLATURE

BV	bed volumes	
d	particle diameter;	mm

REFERENCES

(1)   Irving J., Int. Water Conf. Pittsburgh. 1991 52 IWC-91-53.
(2)   Wrigley S.P., Medete A., S.C.I.-Ultra Pure Water 4, Oct. 1991.
(3)   Duff J.H., & Levandusky J.A., Proc. Amer. Power Conf. 1962 24, 739-750
(4)   Rosato A., Strandurg K.J., Prinz F., & Swendsen R.H., Phys. Rev. Let. 1987, 58, No. 10, 1038-1040.
(5)   Scott K.J., CSIR, CENG 497, 1984, ISBN 0798832487..

IMPROVEMENTS IN REGENERATION, RESIN SEPARATIONS,
RESIN RESIDUALS REMOVAL IN CONDENSATE POLISHING

G. J. CRITS
AQUA-ZEOLITE SCIENCE, INC.
268 Glendale Rd., Havertown, Pa., USA

ABSTRACT

There are many condensate polishing, mixed bed regeneration
systems operating with various degrees of satisfaction. The
many problems in resin separations, resin and chemical cross
contamination, resin residuals removals, and excessive use of
regenerants will be reviewed. Utilities and semi-conductor
manufacturers require more stringent performance in respect
to obtaining lower residuals or leakages below 0.1 µg/L.
An economical improved Four Tank Regeneration System is
presented that provides for a superior and more precise
control of the resin separations and regeneration.

1. INTRODUCTION

Mixed bed (Deep Beds) condensate polishing and ultra pure
semi-conductor water treatment have required more stringent
performance in respect of obtaining lower residuals or
leakages. After periodic cleaning and chemical regeneration,
the cation and anion leakages such as sodium, chloride,
bisulfate, bicarbonates now are required to be below 0.1 µg/L

The critical requirements in the regeneration are:
  . Thorough separation of the resins
  . Perfect transfers of each type of resin without
    cross resin contamination
  . Proper chemical injections without cross contamination

. Proper rinses to remove chemical residuals
. Complete removal of residual resins in the operating,
  regeneration, or storage tanks

2. REVIEW OF VARIOUS CPP SYSTEMS

There are many condensate polishing, mixed bed regeneration
systems operating with various degrees of satisfaction. But
many have problems in resin separations, resin and chemical
cross contamination, resin residuals removal, and excessive
use of chemical regenerants. These include the following
condensate polishing plants (CPP):
. Single tank mixed beds with external or internal regenerations.
. External two tank regeneration system where the oversized
  anion tank serves both for regeneration of the anion resin
  and for storage of the mixed resins.
. External three tank regeneration system being the most
  prevalent used, however the cation regen. tank is large.
. The Tri-ammonex (1) and the Belco "Trouble Layer" (2)
  systems which utilize a small fourth tank for storage of
  the inert or interface resins. The cation tank is large.
. The "Conesep system" (3) with and without "Seprex" (4)
  utilizes an oversize anion regeneration tank.
. The "RSTS system" (5) utilizes a multi compartment
  separation vessel for separation of the resins by
  floatation and screening. Resin cross contamination was
  reported as much as 0.40%.

Most of the above systems have problems with anion resin
cross contamination of the cation layer, with values above
0.2%.

The Auerswald (6) San Onfre CPP plant, similar to this
author's system, has shown superior results both in leakages
and resin separations; but uses an extremely expensive five
tank system for resin separations and regeneration. Anion
resin cross contamination was less than 0.15%.

3. IONIC LEAKAGES FROM RESIN CROSS CONTAMINATION

Resin cross contamination (RCC) from incomplete resin separation (or from chemical cross contamination by regenerants transgressing into the wrong resin layer) results in ionic leakage during the service or exhaustion run. Much of the leakages are due to hydrolysis or "equilibrium leakage", for instance:

3.1 Cation resin left in the anion layer - where sodium from the caustic soda is imparted to the cation RCC (or trans- gression of caustic soda into the cation resin layer). Ammonex (7) replaces the sodium with ammonium ions which are compatible with condensate treatment in fossil utility plants. But with some neutral/zero solids or nuclear utility plants or with ultra high pure water plants, Ammonex is not applicable. Therefore, resin separation is most important. Caustic soda residuals left on the anion resin if not completely rinsed also imparts sodium to the cation resin after mixing. Sodium leakage may hydrolyze from the cation resin or from the anion resin itself. Ammonex has helped the anion rinse but again would be applicable to ammoniated condensates.

3.2 Anion resin left in the cation layer- where sulfuric or hydrochloric acid is imparted to the anion RCC (or trans- gression of acid into the anion resin layer). Bisulfate or sulfate ions from the sulfuric acid or chloride ions from the hydrochloric acid on the anion resin will cause equilibrium leakage during service runs. When the resins are relatively new, a certain amount of chemical or anion resin cross contamination may be tolerated but as the resins age or foul, leakages are greater due to kinetic resin problems. Particularly with the leakage of bisulfate or sulfate ions which have large ionic sizes. Since there is no special process (such as Ammonex for sodium) to relieve this cross contamination, therefore the resin separation is most im- portant. The acid cation rinse also needs special attention too to remove the residual acids.

4. IMPROVED FOUR TANK SYSTEM

To avoid some of the problems mentioned above and provide for a more effective regeneration system, the Improved Four Tank System (IFTS) has been developed as shown in Figure 1. This system comprises:

Tank #1- A narrow, tall tank for receiving the exhausted resins, for scrubbing the resins, for backwash-separation, for transferring both the anion and cation resins to their respective regeneration tanks, and finally retaining the "trouble interface/cross contaminated resins". The anion resin is removed at a collector well above the resin interface (150 mm minimum). The cation resin is removed from the bottom, leaving 150 mm or more of the cation fines in the trouble interface layer. This 300 mm trouble interface layer remains in this first tank to be mixed with the next batch of exhausted resins. With very large tank diameters, the interface trouble layer may be selected even deeper than 300 mm for proper separations. This tank has a white rubber lining to help with the light/photo cell controller that controls the amount of trouble interface resin left behind. No regeneration is performed in tank #1.

Tank #2- Serves as the anion regeneration tank. It also serves as the "anion collection trap" during the backwash-separation step which allows for higher than normal backwash-separation flows in tank #1. Rates higher than 12 m/h may be used in the separation to ensure complete resin separations. Any anion resin carried over during the backwash-separation will be deposited into the anion tank where eventually all the rest of the anion will be transferred. In this way, tank #1 need not have any extra freeboard beyond that normally used. Often, in the conventional systems, it is this limited freeboard and expansion in the cation/anion separation step that hinders the proper resin separation.

TANK #3- Serves as the cation regeneration tank. Both this tank and the anion tank are sized for a deep bed in each, conducive to a good and efficient regeneration and rinse.

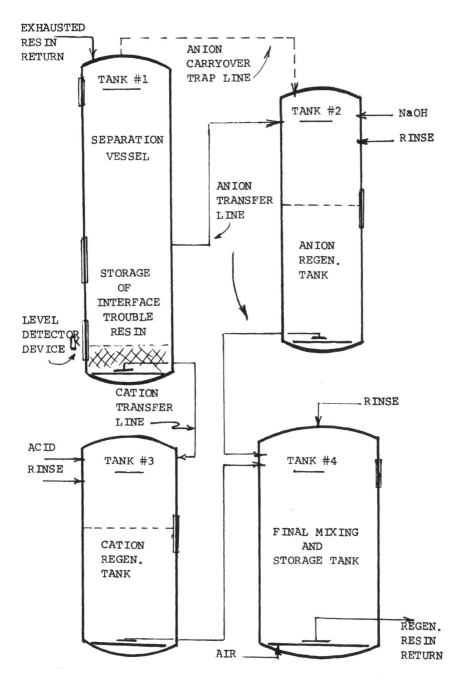

Figure 1. CPP Four Tank External Regeneration System

TANK #4- is the regenerated mixed resins storage tank. Here the regenerated resins are received, air mixed, and the final rinse is performed.

The residual resins in tanks #2, #3, & #4 may amount to 1 to 5% by volume after the simple hydraulic water resin transfer. If these residuals represent the "equilibrium" amount, repeating after each regeneration, this should not pose a problem since there is no cross contamination of any kind. However, should one require to remove all traces of the resin each time, then the "AIR ASSIST RESIN TRANSFER" scheme will have to be practiced. This involves four steps:
. Step 1- Removing most of the resin hydraulically as normally done. 3 to 5% resin may remain.
. Step 2- Drain 300 to 600 mm of water from the tank.
. Step 3- Apply air at 850 L/Sq.Meter or more into the bottom with the vent closed and resin transfer line open. The residual resin is fluidized and the tank dome is simultaneously pressurized which drives out the water-resin mixture. This is continued until the water level reaches 100 to 200 mm above the bottom.
. Step 4- Refill with water to within 600 mm of top of tank and repeat steps 3 and 4 as required.
After the first cycle of the air assist transfer, residual resin may be about- 1%; after the second cycle- 0.1%; and after the third- less than 0.05%.

Additional Features of this Four Tank System:
. Separation vessel tank #1, is designed hydraulically for the scrubbing and resin separation: Not to be used for any regeneration.
. Anion Tank #2 and Tank #3 are designed for the resin regenerations utilizing deep beds for an efficient regeneration and rinse. No resin cross contamination should occur here.
After regeneration and rinse, both anion and cation resins are transferred to Tank #4 for mixing and final rinsing.

. Resin residuals in tanks #2, #3, and #4 should be of no
  concern: there is no resin or chemical cross contamination.
. Ammonex, ammoniation of the anion resin is performed by
  circulating the 0.3% ammonia through tanks #2 & #3 if
  elected.
. Acid and caustic regenerations may be performed
  simultaneously if ammoniation is not required.
. Separated resins should contain substantially less than
  0.1% cross contamination.
. Variations in the cation resin volume may be tolerated as
  much as 150 mm and the anion resin as much as 300 mm in
  each batch processed without affecting the separation
  efficiency.
. Control of the cation resin transfer (and retention of the
  trouble resin interface) by the light/photo cell controller
  is a simple arrangement, fully and easily adjustable at the
  bottom sight glass of tank #1. Complete automatic control
  without any manual interactions should be expected for the
  norm.
. Retrofit- Any three tank system may be retrofitted by
  addition of Tank #1, controls, a few extra valves and pipes
  Acid and alkali regenerant lines are not relocated.

CONCLUSION

The improved four tank regeneration system is applicable to
mixed bed condensate polishing or to the preparation of
ultra pure water for the semi-conductor industry. The
superior and more precise but straight forward design enables
not only efficient resin separations but also efficient
chemical utilizations. Leakages below 0.1 µg/L and 18
megohm water should be easily obtained from this system.

REFERENCES

1) Crits, G.J., U.S.Patent # 4,472,282, Tri-Ammonex

2) Belco, U.S. Patent # 4,375,252

3) O'Sullivan, J., Experience with Conesep Polishing Plant At Aghada, Workshop on Condensate Polishing, EPRI, Little Rock, Arkansas, 1989

4) O'Brien, M.J., Recent Improvements in Deep Bed Separation Technology, 52nd Annual Meeting, International Water Conference, Pittsburgh, Pa., 1991

5) Strauss, S. D., Resin Regeneration Key to Improved Condensate Polishing, Power, 1991, p.47-48

6) Auerswald, D. C., Effects of Cation Resin Leachables on Condensate Polisher System performance, ION EXCHANGE FOR INDUSTRY, Streat, M., Ellis Harwood Publ., 1988, p.11-21

7) Crits, G.J., Condensate Polishing with the Ammonex Procedure- 1984, ION EXCHANGE TECHNOLOGY, Naden & Streat, Ellis Harwood Publ., 1984, p.119-126

CHALLENGES IN THE MANUFACTURE OF ULTRA-PURE WATER

YAIR EGOZY, GARY A. O'NEILL and KITTY K. SIU

Millipore Corporation
80, Ashby Road, Bedford, MA 02173 USA

ABSTRACT

Factors affecting the performance of laboratory water purification systems are discussed. Strong emphasis is placed on methods for the elimination of organic matter originating from either the feed, or as extractables and degradation products from system components. Details on the performance of properly designed systems are given. The current status of analytical instrumentation for characterization of ultra pure water is also discussed.

INTRODUCTION

Requirements for high purity water have become increasingly demanding in recent years. In all frontiers of science, it has been found that high purity water is critical to the success of many operations. While semiconductor manufacturers have been the leaders in requirements for ultrapure water, the power and pharmaceutical industries follow the same trends.

In the laboratory, ultrapure water is needed for the preparation of standards, blanks and eluents for analytical instruments (HPLC, GC, IC, ICP etc) and as a solvent for preparations. It is needed by all disciplines: life sciences, chemistry and physics. Usually the most advanced, pioneering work is performed first in a laboratory environment. Therefore, laboratory water purification systems need to have the capability to exceed the specifications set forth by industry.

In recent years, there has been considerable interest in water containing less than 50 $\mu g/m^3$ (part per trillion - ppt) of inorganic ions (eg. metal ions), and less than 1000 ppt total organic matter (TOC). Ion chromatography is commonly used for analysis of ions of concentrations below 1 mg/m^3 (parts per billion - ppb). For example, using ion chromatography with trace enrichment, it has been shown (1) that the concentration of sodium and chloride in Milli-Q water were approximately 10 ppt and 17 ppt, respectively.

These values correspond to 0.8×10^{-9} and 0.5×10^{-9} equivalents per liter. For reference, note that the concentration of (H^+) and (OH^-) in pure water is 1×10^{-7}. This means that a good water purification system can remove simple ions *to levels approximately 100 times lower than that of (H^+) and (OH^-)*. Moreover, based on our calculations (1), with properly cleaned ion exchange resin, it is even possible to achieve purity **1000 times lower than that of (H^+) and (OH^-)**.

This paper describes our experience with the development, operation and characterization of the Milli-Q® water purification systems. Emphasis is placed on the pretreatment and other means of removal of organic matter as this is the key to the overall success of the water purification system.

REMOVAL OF ORGANIC MATTER

Organic matter should be removed from water for two reasons:
- **To prevent organic fouling of the resin, and allow it to function properly.**
- **To reduce the organic matter in the product water.**

Usually those goals are coupled: when most of the organic foulants are eliminated in the pretreatment stage, the resin can exchange the inorganic ions <u>and</u> remove a significant fraction of the residual organics much more efficiently than in the presence of high of organic load (2,3).

> *A Milli-Q® water system was challenged with a surface water source (containing ≈ 2.4 ppm TOC organic, with very high Silt Density Index) in Bedford, MA. In each test, the system was protected by either Service Deionization (SDI) or by Reverse Osmosis (RO). HPLC chromatographs (4) were taken of the feed water, downstream of the pretreatment and the final product.*

Figure 1 shows the results: Both RO and SDI removed a substantial fraction of the organic matter. RO provided much better protection of the polishing water system, which resulted in much flatter chromatograms. SDI was not cabable of removing some of the organic contaminants from the water. As a result, the final product water contained significantly more organic matter.

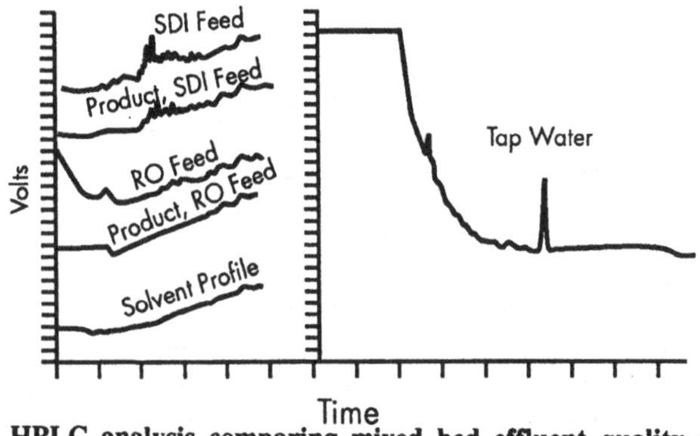

Figure 1: HPLC analysis comparing mixed bed effluent quality as affected by pretreatmemt.

ORGANIC CONTAMINANTS; EXTRACTABLES FROM SYSTEM COMPONENTS

Even with the best protection against organic fouling from feed water contaminants, the organic quality of the product water may be limited by the water treatment system itself. Essentially all of the components of the system; tubing, fittings, housings, ion exchange resin, carbon; contain organic matter. For this reason, extractables and degradation products from these components can be expected to contribute to the TOC level of the product water.

A simple calculation can be made to demonstrate the effect of a single source of extractables, the natural breakdown of ion exchange resins, on the TOC level of product water. It is generally accepted that anion exchange resins lose capacity over time. This loss of capacity can be translated into a total organic carbon value to show that, if this source of organic contamination is not removed in some fashion, the levels of organics produced from this single source are nearly equivalent to organic levels now achievable by the best laboratory water systems.

For the sake of the example, assume that an anion exchange resin loses 0.2 equivalents of exchange capacity per year. As a conservative assumption, assume that only one carbon atom is associated with this loss of capacity and released to the water. If one liter of anion exchange resin is being used to treat water at a flow rate of 1L/min, the level of organics released to the water from this source is:

$$\frac{0.2 \text{ eq/year} \times 12 \text{ g C/eq}}{1 \text{ L/min} \times 365 \text{ day/year} \times 24 \text{ h/day} \times 60 \text{ min/h}} = 4.6 \times 10^{-6} \text{ gC/L}$$

Thus, in this example, the natural breakdown of the anion exchange resin would contribute 4.6 ppb TOC to the product water under continuous use conditions. If the resin is not in continuous use, the TOC level at system start-up could be considerably higher. However, in a well-designed water purification system, other system components such as cation exchange resin, activated carbon and, perhaps, UV oxidation, can remove considerable portions of these extractables. Therefore, it is still possible for a carefully designed water system to produce water with less than 5 ppb TOC and to minimize start-up difficulties.

As can be seen from the calculation above, extractables from system components can have a considerable impact on the quality of the product water. For this reason, it is important to understand the types and levels of extractables from all system components. Moody (5) has made an important contribution to the understanding of inorganic impurities from fluorinated containers. However, in this work, we will focus on organic impurities in water.

Removal Of Extractables and Degradation Products

The previous section shows how a single component in a water system can be a significant contributor to the organic level in the product water. Indeed, extractable testing (6) of individual ion exchange resins has found substantial, widely varying, levels of organic extractables. However, at the polishing stage of an ultrapure water system, mixtures of individual components, not single components, are typically employed. In order to achieve the lowest level of ionic impurities (highest resistivity) mixed beds of strong base and strong acid ion exchange resins are employed. In a similar fashion, Hegde and Ganzi (7) found that a mixture of mixed bed ion exchange resin and activated carbon in the final stage of an ultrapure water system resulted in the lowest levels of organic impurities in the product water. Therefore, to assess the affect of system components on final water quality, the water resulting from treatment by the entire system, not by a single stage of the system, must be analyzed.

To study the impact of component quality, various replacement cartridges designed to be used in Millipore Milli-Q water systems were tested. The cartridges were employed in a 4-bowl purification scheme consisting of carbon, two cartridges of mixed bed deionization resin, and an Organex-Q® type polishing cartridge (7). The polishing system used to evaluate the cartridges was fed with water treated by reverse osmosis. The materials used in the various cartridges varied significantly. Millipore uses specialty grade ion exchange resin, thermally-bonded polypropylene cartridges and a double O-ring design as an added safety factor to prevent water by-pass. A number of the other cartridges tested used ABS plastic that was solvent bonded and had a single O-ring design. In spite of these differences, all cartridge sets were able to achieve 18 Mohm-cm resistivity in the product water. However, there were significant differences between the cartridge sets in the organic content of the product water.

Figure 2 shows the TOC content of product water produced by the various cartridge systems as a function of water throughput. Millipore Technical Brief no. TB035 provides full details of the experimental procedure. As Figure 2 shows, the TOC performance of the various cartridge sets is quite different. The TOC specification of the Milli-Q system used in the testing is 15 ppb. Only the Millipore replacement cartridges achieved this level on the first day of operation. Even on the second day, it took nearly 5 times more water throughput before any of the other cartridge sets achieved the TOC specification. One set of cartridges did not achieve the specification even after over 500 L of water throughput.

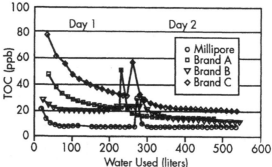

Figure 2: **TOC flush-out of IX cartridges**

The product water from the various cartridge sets was also analyzed by reverse phase HPLC (Figure 3). The same types of differences between cartridge sets detected by TOC analysis are seen rather vividly in the HPLC chromatograms. Obviously, the quality of the resins, carbon, cartridge material and cartridge bonding technique have a significant impact on the organic quality of water from an ultrapure water system.

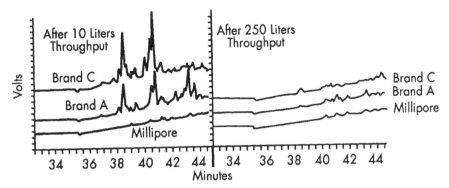

Figure 3: HPLC analysis of effluents from polishing systems

UV OXIDATION OF ORGANICS

When high intensity UV treatment is introduced into a water system loop, it can enhance the oxidation of organics in water including extractables and degradation products. Careful design of a mercury lamp and fused quartz sleeve allow for emission of a full spectrum of UV light including the high-energy 185 nm band which is beyond the emission range of conventional, sterilizing UV lamps. This band begins the oxidation process by catalyzing the reaction of dissolved oxygen in water to ozone.

In the presence of 254 nm energy from the UV lamp,ozone is converted to hydroxyl radicals which oxidize organics in water. In some cases, the oxidation may proceed completely to carbon dioxide. However, even if the oxidation is incomplete, removal of the more highly charged fragments (such as carboxylic acids) can be achieved using Organex-Q type polishing cartridges. Figure 4 shows an example of oxidation of an alcohol (methanol) involving hydroxyl radicals.

By including the ultraviolet light and polishing cartridge within the recirculation loop of an ultrapure water system, the advantages of UV oxidation can be achieved each time the system recirculates. This can reduce the organic load on the system and lead to enhanced overall performance. Employment of this technology has allowed consistent production of ultrapure water with less than 5 ppb TOC. In addition to quantitative differences in water quality as measured by TOC analysis, water from a Milli-

$$CH_3OH + 2OH \bullet \xrightarrow[UV]{} HCHO + 2H_2O$$
$$formaldehyde$$

$$HCHO + 2OH \bullet \xrightarrow[UV]{} HCOOH + 2H_2O$$
$$formic\ acid$$

$$HCOOH + 2OH \bullet \xrightarrow[UV]{} CO_2 + 2H_2O$$

Figure 4 : UV Oxidation of methanol.

Q UV Plus water system has provided significantly enhanced performance when used in a wide range of analytical techniques including HPLC, GC, and ion chromatography. The following is an example of the improved quality of UV treated water in a less conventional, but highly sensitive technique.

Physicists at a major university are studying the nature of organic monolayers at the air/water interface using high-energy X-ray diffraction. Prior to that analysis, they use more conventional interfacial measurements to determine the level of organic impurities which might interfere with their analyses. In general, the impurities of interest are long-chain organics, such as surfactants, which accumulate at the water surface. Fortunately, the presence of these impurities can be easily detected using surface tension measurement techniques.

By measuring changes in surface tension as a function of surface area, the researchers can calculate the number of insoluble organic impurities in water. To do this, the water is placed in a Langmuir trough and a Teflon barrier is swept across the water surface. Because the impurities are essentially insoluble in the bulk water, the movement of the barrier concentrates the impurities into a progressively smaller surface area. As the impurities become more concentrated, the surface tension of the water at the air/water interface drops. This change in surface tension is monitored using a Wilhelmy plate force balance technique. The change in surface tension as a function of surface area data are then curve-fitted to a two-dimensional ideal gas law equation which allows for the calculation of the number of insoluble organic impurities present in the water.

Figures 5 (a and b) show the results of these analyses for Milli-Q UV Plus water versus HPLC-grade bottled water. While the Milli-Q UV Plus water contained 5.0×10^{13} molecules of insoluble surface impurities, the HPLCgrade bottled water contained 1.5×10^{16} molecules of insoluble organic impurities. Thus, the UV treated water contained 300 times less organic impurities than the bottled water. The influence of this lower level of organics on the interfacial research was far more significant than a simple quantitative measure of TOC might have indicated.

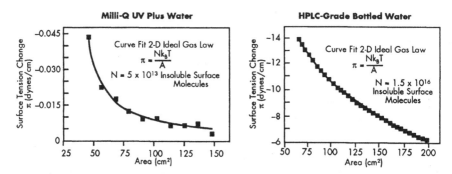

Figure 5: Calculation of Insoluble Impurities
(a) Milli Q water   **(b) Bottled water**

CHALLENGES FOR THE FUTURE

In the future, we can expect to see major advances in two areas:
- **Improved analytical methods for determination of trace level contaminants.**
- **Further advances in purification, with even greater emphasis on the removal of trace organics.**

Improvement In Analytical Capability:
The fundamental process for the removal of ions from water is ion exchange, mostly in the form of a mixed bed of cation and anion exchange resins. When used properly, ion exchange resins are highly effective in removing inorganic ions to extremely low levels, so that even measuring the concentration of the remaining ions becomes difficult.

The most common tool for the estimation of concentration of inorganic ions in water is measurement of the resistivity of the purified water. For NaCl in water, resistivity declines roughly by 1 Megohm-cm with the addition of 1 ppb of NaCl. Therefore, resistivity cannot be used as a guide for water purity when the contamination level is below the 1 ppb range.

Advances in water purification go hand-in-hand with progress in analytical instrumentation. Highly sensitive analytical instruments require ultra pure water for calibration, or as an eluent. A new development of special interest is Capilary Ion Analysis (11) which allows for rapid, low level detection of numerous ionic species in single samples.

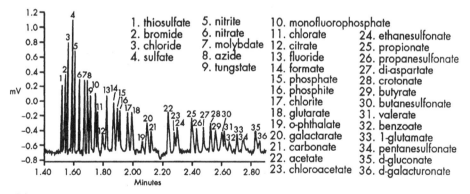

1. thiosulfate	5. nitrite	10. monofluorophosphate
2. bromide	6. nitrate	11. chlorate 24. ethanesulfonate
3. chloride	7. molybdate	12. citrate 25. propionate
4. sulfate	8. azide	13. fluoride 26. propanesulfonate
	9. tungstate	14. formate 27. di-aspartate
		15. phosphate 28. crotonate
		16. phosphite 29. butyrate
		17. chlorite 30. butanesulfonate
		18. glutarate 31. valerate
		19. o-phthalate 32. benzoate
		20. galactarate 33. 1-glutamate
		21. carbonate 34. pentanesulfonate
		22. acetate 35. d-gluconate
		23. chloroacetate 36. d-galacturonate

Figure 6: Capillary Ion Analysis of 36 ions, separated over 83 seconds.

Removal of Trace Organics by Continuous Deionization (CDI)
A significant fraction of extractables can be eliminated from the system using the Continuous Deionization (CDI) process (9) rather than conventional mixed bed resin. THe CDI process uses electric energy to transfer ions from the product stream into the waste.

Wilkins and McConnelee (10), using an industrial scale CDI module, found that when such a device is fed with RO water, the TOC of the product decreased while that of the waste stream increased, as shown in the following table:

Stream	TOC, ppb
RO feed	75
CDI waste	500
CDI product	<3 (non detectable)

The CDI approach is valid for ionizable matter only. For nonionic matter, UV oxidation would be the most beneficial for removing organics released by the system. The combination of both UV and CDI could help to keep the system in the best possible condition, leading to optimal performance with minimal maintenance over an extended period of time.

It should be noted that this process is still under development. Additional optimization of module size and components will be required before a device based on this principle can be offered as a part of laboratory water purification system.

REFERENCES:

1. Denoncourt, J.P. and Egozy, Y. "Trace level analysis of high purity water. Part I: Inorganic ions" Ultrapure water 3 (4), July 1986, p 40.
2. Egozy Y., Denoncourt J.P., Wilkins F., Jha A., Ganzi G.C., and Hamelin D., "Evaluations of Pretreatment Alternatives for Ion Exchange Demineralizers", The 21th Annual Liberty Bell Corrosion Course, Philadelphia, PA, 1983.
3. Egozy Y., Denoncourt J.P., Ganzi G.C., "The Role of Reverse Osmosis in the Production of Laboratory Water" in "Reverse Osmosis Technology - Application for High Purity Water Production", Parekh, B.S. , Ed, Marcel Decker 1988.
4. Denoncourt J.P., Egozy Y. and Jandik P., "Trace level analysis of high purity water. Part III: High performance liquid chromatography" Ultrapure water 4 (1), January 1987, p 44.
5. Paulsen P. J., Beary E. S., Bushee D. S. and Moody J. R., "Analysis of Ultrapure Reagents from a Large Sub-Boiling Still made of Teflon PFA" Anal. Chem. 61,(8)827 (1989).
6. Egozy Y. and Denoncourt J.P., "Trace level analysis of high purity water. Part II: Total Organic Carbon" Ultrapure water 3 (6), November 1986, p 52.
7 Hegde R. S. and Ganzi G.C., "Method and Apparatus for the Production of Ultra Pure Water", U.S. Patent #4,430,226 (1984).
8 Jandik, P. and Jones, W.R., "Analysis of Anions in low ppb range using Capilary Electrophoresis" paper no.163, presented at the 1991 Pittsburgh Conference & Exhibition on Analytical Chemistry, March 3-8, 1991, Chicago, Illinois, USA.
9. Ganzi, G.C., Egozy, Y., Giuffrida, A.J. and Jha, A.D., "High purity water by Electrodeionization, the performance of Ionpure Continuous Deionization System" Ultrapure Water 4(3), April 1987.
10. McConnelee, P.A. and Wilkins, F.C., "Continuous Deionization in the preparation of microelectronics water". Solid state Technology, August 1988.

ION EXCHANGE STUDIES ON STRONGLY BASIC ANION EXCHANGE RESINS PREPARED WITH TERTIARY AMINES OF VARYING MOLECULAR WEIGHT

FRANCIS X. McGARVEY and REGINA GONZALEZ

Sybron Chemicals Inc., Birmingham, New Jersey 08011, USA

ABSTRACT

Anion exchange resins based on triethyl, tripropyl and tributyl amine functionality have been prepared and given a systematic evaluation using column and batch methods. These resins were compared with the standard Type I strong base resins which had been prepared using trimethylamine.

Equilibrium values were determined for the ion pairs:

NO_3^-/SO_4^{2-} NO_3^-/Cl^- NO_3^-/HCO_3^- NO_2^-/Cl^- NO_2^-/SO_4^{2-}
NO_3^-/SO_2^- HPO_4^{2-}/Cl^- $H_2PO_4^-/Cl^-$ SO_4^{2-}/Cl^-

and the results used to develop column studies where these resins were used to remove nitrate from a natural water supply. A theory of interaction was also developed and other resins were considered in the nitrate removal application.

INTRODUCTION

For many years the commercially available strongly basic anion exchange resins have been prepared by the reaction of chloro-methylated styrene-divinylbenzene copolymers with a tertiary amine to yield a quaternary ammonium functional group. Naturally the amine having the lowest molecular branches, namely the methyl group, would have the highest weight capacity and was selected for commercial development. These resins were prepared on a gel copolymer which had been chloromethylated using a zinc chloride. Amination was achieved using a solvent.

APPLICATION

In recent years the need to remove nitrates from drinking
waters has resulted in the preparation of other strong bases
using triethyl, tripropyl and tributyl amines. These have been
shown to have unusual selectivity for nitrate over sulfate at
the concentrations usually encountered in water treatment (1,
2). The exchange properties of these resins have been determined and their affinity for various ions found in waters have
been measured. In addition, a series of application tests have
been performed and the results illustrated in a series of
Figures 1 through 8.

CHEMICAL DETAILS

In Table 1 the ion exchange capacity values are reported.

TABLE 1
Ion Exchange Characteristics of Strong Base
Resin Prepared with a Tertiary Amine Series

Resin	Amine	Strong Base Capacity meq/g dry		Moisture Retention %	% DVB
		Structure	Synthesis		
A	Trimethyl	3.70	3.25	45	6
B	Triethyl	3.01	2.83	42	6
C	Tripropyl	2.54	2.04	33	5.8
D	Tributyl	2.19	1.90	44	6.0

AFFINITY STUDIES

Resins were converted to various forms and these were contacted
with the appropriate anion. The amount of counter ion in the
test was selected to equal the amount of the anion on the resin
(3). The resin sample was held to 10-11 mL and the counter ion
was contacted at about 10-15 meq depending on the capacity.
The samples were allowed to equilibrate for 16 h and the fractions were analyzed and the affinity number was calculated
using the mass action equations (4). Where possible the anions
in solution were determined by ion chromatography.

EXPERIMENTAL RESULTS

The results of these tests are summarized in Table 2 where the
values are listed as affinity constants for a particular

experimental condition. These values can be used to give a displacement series for each resin. Generally the affinity values increase with the complexity of exchange site. The nitrate-sulfate values are most interesting from an application standpoint.

TABLE 2
Affinity Values for Various Quaternary
Ammonium Strongly Basic Anion Exchangers

Resin	---------------Affinity Values-------------			
	A	B	C	D
NO_3^-/SO_4^{2-}	0.8	21	71	461
NO_3^-/Cl^-	2.2	2.5	2.4	6.9
NO_3^-/HCO_3^-	1.3	3.2	2.4	7.3
NO_3^-/NO_2^-	1.5	1.7	2.0	6.0
NO_2^-/Cl^-	1.2	3.2	3.7	3.9
NO_2^-/SO_4^{2-}	2.3	7.2	12	45
SO_4^{2-}/Cl^-	1.5	1.7	2.3	15
HPO_4^{2-}/Cl^-	2	5	20	65
$H_2PO_4^-/Cl^-$	2	2	3	7

Several papers have been written on the selectivity of the anion exchange resins. Clifford and Webber (5) studied several resins having amine sites although the more complex quaternaries were not available to them. They found sulfate selectivity was greater than unity for all the commercial resins available in 1983. They reported that matrix structure and spacing were important factors. Jackson and Bolto (6) had access to the complex amines and developed data indicating that the complex amine sites are indeed increasingly selective for nitrate over sulfate. The crosslinkage of the polymer structures was not indicated in Jackson's study.

COLUMN STUDIES

The nitrate-sulfate selectivity was examined by column tests where the chloride forms of the resins were used to remove nitrate from a water which contained 1.3 meq/L nitrate, 2.6 meq/L sulfate and 2.6 meq/L bicarbonate.

The columns were 25 mm in diameter and the beds were 1 m in height containing 490 mL of the resins. The resins were regenerated with various amounts of 10% sodium chloride. The beds were exhausted at 16 bed volumes per hour and the effluent analyzed for chlorides, alkalinity, pH, sulfates and nitrate. Several cycles were run on each resin. Typical runs are shown in Figure 1 for a run with Resin A, the trimethylamine quaternary resin. Figure 2 illustrates a typical run using Resin B,

the triethylamine quaternary amine resin. Figure 3 shows
Resin C, the tripropylamine quaternary resin, and finally
Figure 4 gives the results of Resin D, the tributylamine
quaternary resin. Resins B, C and D all show selectivity as
expected from the affinity study and there is a relationship
between breakthrough points for the various resins as shown in
Figure 5, a plot of affinity coefficients as a function of
breakthrough points for sulfate and nitrate.

The important differences in these resins are related to
sulfate-nitrate interchange. Figure 6 summarizes the previous
displacement curves based on percentage of influent concentration. Regeneration of these resins was typical from a displacement standpoint when sodium chloride was used. A typical
resin displacement curve for Resin D is shown in Figure 7 for
240 grams NaCl/L regeneration level. Figure 8 gives y vs. x
diagrams for the various resins.

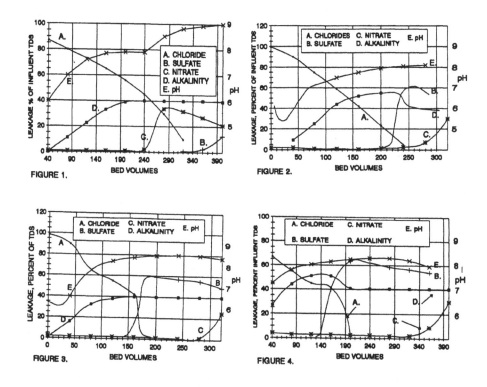

FIGURE 1.

FIGURE 2.

FIGURE 3.

FIGURE 4.

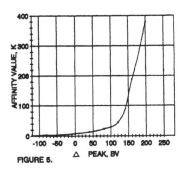

FIGURE 5.

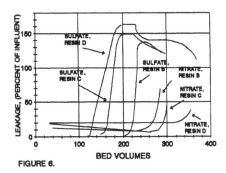

FIGURE 6.

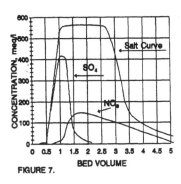

FIGURE 7.

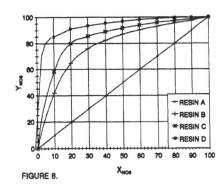

FIGURE 8.

Figure 1. Log of run made with Resin A (trimethylamine).
Regeneration level 240 grams NaCl/L.

Figure 2. Log of run made with Resin B (triethylamine).
Regeneration level 240 grams NaCl/L.

Figure 3. Log of run made with Resin C (tripropylamine
quaternary). Regeneration level 240 grams NaCl/L.

Figure 4. Run made with Resin D (tributylamine). Regeneration level, 240 grams NaCl/L.

Figure 5. Importance of affinity coefficient on separation of
sulfate-nitrate values.

Figure 6. Comparison of sulfate displacement curves. Regeneration level 240 g NaCl/L (33% NO_3/SO_4).

Figure 7. Regeneration Profile - Resin D. 240 grams NaCl/L
(33% Na_3/SO_4).

Figure 8. Y vs. X curves for the quaternary resins.

DISCUSSION

These studies show that there is a particular relationship

102

between nitrate and sulfate affinity for these resins. The
affinity of nitrate over sulfate increases with length of
branch chains around the exchange site. The affinity values
for mono-mono-valent exchange were not pronounced and this is
also evident from the chloride-alkalinity relationship. In
the test runs there is a gradual increase in pH during the
run. These changes follow the alkalinity in the water.
Noticeable changes are also found in alkalinity when sulfate
breaks.

The affinity values are a good indication of the ability
of these resins to separate sulfate from nitrate. Additional
studies are needed to fully explain this phenomenon, but site
spacing appears to be a likely explanation from a qualitative
standpoint. Why the nitrite affinity is not as high as the
phosphate should be viewed as very preliminary due to a
limitation in the analytical methods. The ion chromatographic
system really measures HPO_4^{2-} and it is doubtful if the tri-
valent phosphate salt of a strong base anion exchanger can
actually be prepared by the methods used here.

CONCLUSION

The affinity coefficients support the conclusion that the
tributylamine quaternary resin is the most effective resin for
nitrate removal in the presence of high sulfate concentration.
Since it is also the least efficient for regeneration, it is
likely that counter operation would be preferred.

ACKNOWLEDGMENT

The authors wish to acknowledge the help of Mrs. E. Irons and
Mr. D. Dally in preparation of the manuscript.

REFERENCES

1. Guter, G. A., Removal of nitrate from contaminated water
   supplies for public use, Report 600/2-82-042. U. S.
   Environmental Protection Agency, Cincinnati, Ohio, March
   1982.

2. Guter, G. A., Nitrate removal from Contaminated Water
   Supplies, Report 600/52/87/034. U. S. Environmental
   Protection Agency, Cincinnati, Ohio, August 1987.

3. Ambrus, P., Grantham, J. and Gresser, D., nitrate removal
   by selective ion exchange - Presented at the 5th International Ion Exchange Conference, Cambridge, England, 1988.

4. Helfferich, F., Ion Exchange, McGraw Hill Book Company, New

York, 1962.

5. Clifford, D. and Weber, W., The determination of divalent/-monovalent selectivity in anion exchangers. Reactive Polymers, 1983, 77, 1.

6. Jackson, M. and Bolto, B., Effect of ion-exchange resin structure on nitrate selectivity, Reactive Polymers, 1990, 12, 277.

ALTERNATIVE ROUTES TO CHLOROMETHYLATION FOR PREPARATION OF ANION EXCHANGERS

V.C. MALSHE, S.E. MICHAEL, V.V. PANDYA, P.G. JOSHI
Ion Exchange (India) Limited,
Plot C-4, M.I.D.C. Chemical Zone, Ambarnath 421 501, India.

ABSTRACT

Chloromethylated anion exchange resins based on styrene have always posed a potential health hazard in production. As a safe alternative, three new resins were prepared based on the chemistry of polyamine epoxy, acrylic epoxy and phenolic epoxy with epichlorohydrin. Samples of strong, weak and mixed basicity with varying characteristics were made for water treatment applications. Pilot plant samples show promising results in laboratory operating data and performance in field trials.

INTRODUCTION

The manufacture of anion exchange resins, if not monitored carefully, can be hazardous. The basic steps of polymerization - chloromethylation - amination are followed for the manufacture of all polystyrenic based anion exchangers. Chloromethyl methyl ether (CME) is used in the chloromethylation step. CME, even if prepared 'in situ', is a potential health hazard and a carcinogen; bis-chloromethyl ether, a byproduct, is even more hazardous (1-3). Intensive research work has been undertaken to find a substitute for CME or a substitute process. In the case of polystyrenic based anion exchangers, the known alternatives are:

i) chloromethylation with chloromethyl long chain alkyl ethers(4-5),

ii) copolymerization of chloromethyl styrene and divinyl
    benzene,
iii)halogenation of methyl substituted polystyrenes(7-9).

The long chain alkyl ethers, although less hazardous, are expensive and the formation of bis-chloromethyl ether cannot be ruled out. The processes based on these ethers are ruled out due to poor availability and higher prices of chloromethyl styrene and paramethyl styrene monomers.

Weak base anion exchangers (equivalent to Amberlite IRA68, Duolite A375) as well as strong base anion exchangers (equivalent to Amberlite IRA458, Duolite A132) based on acrylic matrix structure are commercially available. It is known that the synthetic route for preparing this kind of resin is safe and involves aminolysis of crosslinked acrylic esters with polyfunctional amines containing at least one primary amino group and one secondary or tertiary amino group(10). The primary amine reacts with polyacrylic ester to form an amide. Secondary or tertiary amine groups form the active group of a weak base anion exchanger.

This weak base anion exchanger can be further reacted with methyl chloride or dimethyl sulphate(11) to give a strongly basic anion resin (SBA). Although these types of resins have superior organic fouling resistance, they are more expensive and have a severe temperature limitation (for SBA 37°C) and hence in tropical countries the performance becomes suspect.

Another route tried and tested by many researchers(12-14) is the synthesis of epoxyamine-based anion exchangers. Our aim was to avoid chloromethyl ether and materials which would pose economic constraints. The use of formaldehyde and hydrochloric acid in combination was ruled out. The versatile molecule epichlorohydrin is reasonably priced and widely used. TLV value TWA for epichlorohydrin is 2 cm^3/m^3

(ppm) compared to 0.001 cm^3/m^3 for chloromethyl ether[16].

EXPLORING ALTERNATIVE PROCESSES

Many ideas were considered and the three investigated were:

1. Phenol-formaldehyde polymers to be epoxidised and further functionalised.

2. Reaction of epichlorohydrin with acrylic/methacrylic acid based ion exchangers and then functionalisation with suitable amine.

These routes were studied in greater detail for the following reasons:
i)   fewer than three steps for synthesis,
ii) safer, cheaper raw materials,
iii) the chemistry of cationic polymers becomes the basis of synthesis, hence well-known and predictable.

EXPERIMENTAL

A. **Epichlorohydrin-amine based anion exchangers**

A-WBA - (Weak Base Epoxy-Polyamine Anion Exchangers)

Triethylene tetramine (TETA) (1.52 moles) was stirred as a solution in xylene (9.5 moles). Epichlorohydrin (4.55 moles) was added to the stirred solution of TETA/xylene. The exotherm was controlled by external cooling and controlled addition of epichlorohydrin. With the build-up in viscosity, discrete polymer beads separated out and were cured by heating to the final product. Xylene was then separated and the polymer beads washed with surfactants to remove excess xylene.

A-SBA - Intermediate Base Anion Exchanger

Epichlorohydrin (10 moles) was reacted with dimethylamine (3.60 moles). The reaction was carried out at temperatures below 35°C. TETA (2.13 moles) was added in a controlled manner to control the exotherm. The polymer was then isolated and suspended to form beads and cured. The rest of the process is similar to A-WBA.

B. **Addition condensation products of epichlorohydrin with weak acid cation exchanger**

B-WBA and B-SBA

Weak acid methacrylic or acrylic acid based cation exchangers (1 mole) which are the products of the existing range of Ion Exchange (India) were reacted with epichlorohydrin (9 moles) in the presence of a phase transfer catalyst. After removal of the excess epichlorohydrin, these polymers were converted into strong base or weak base anion exchanger by reacting with trimethylamine and dimethylamine respectively.

C. **Phenol-formaldehyde based Anion Exchanger**

C-WBA and C-SBA

Commercially available EPN have not been considered as starting materials due to high cost. The phenol-formaldehyde polymer was prepared using a mole ratio of 1:3. The PF prepolymer was suspended in water and cured at higher temperatures to form spherical beads. The phenol-formaldehyde polymer was reacted with excess epichlorohydrin (15 moles) in alkaline medium. The product on treatment with dimethylamine gave a weak base anion exchanger and with trimethylamine a strong base anion exchanger.

All the resins were tested for their operating capacity. The feed water composition used is given in Table 1 and operating data in Tables 2 and 3.

Table 1 - The feed water composition for the resins

Feed	Composition	SBA	WBA
EMA	as mg/L $CaCO_3$	100	300
Cl^-	as mg/L $CaCO_3$	75	225
SO_4^{-2}	as mg/L $CaCO_3$	25	75
CO_2^{-2}	as mg/L $CaCO_3$	5	5
SiO_2	as mg/L $CaCO_3$	25	Nil

Table 2 - Operating capacity data

Resin	Operating Capacity	Regeneration
A-WBA	57.5	60% of 3% NaOH
A-SBA	32.2	60% of 5% NaOH
B-WBA	46.5	120% of work done
B-SBA	22.0	60% of 5% NaOH
C-WBA	45.0	60% of 1% NaOH
C-SBA	18.0	60% of 4% NaOH
Conventional WBA	35.6	60% of 3% NaOH
Conventional SBA	27.2	60% of 5% NaOH

Table 3 - Typical analysis

Resin	W.R.	W.C. g/g	T.C.C. meq/dry g	% S.B.C. meq/L
A-WBA	1.36	4.84	1610	28.2
A-SBA	1.87	5.34	1410	53.8
B-WBA	1.10	---	1600	Nil
B-SBA	1.08	3.45	1300	99.0
C-WBA	1.44	2.60	1180	11.2
C-SBA	0.77	2.13	630	98.0
Conventional WBA	1.05	3.90	1100	24.8
Conventional SBA	1.10	4.05	1200	97.0

RESULTS AND DISCUSSION

The results indicate that these alternative routes give anion exchangers comparable to those made by the chloromethylation route to get a commercially viable product. All the products synthesized were critically examined from a commercial point of view.

Epichlorohydrin - amine based anion exchangers

Tables 2 and 3 show that the weak base anion exchanger (A-WBA) has comparable capacity to the conventional styrenic resins with advantages of single step synthetic process, a non-hazardous route and no acidic effluent.

The intermediate anion exchanger (A-SBA) has around 50% of weak base and strong base anion exchanger. This resin has higher capacity than the conventional strong base type I anion exchanger. The molecular weight per functional group is much lower than for conventional resins, but this resin is suitable for operation up to 40°C only. Hence, it is best suited for non-water applications.

Addition condensation products of epichlorohydrin with weak cation exchanger

Existing acrylic based anion exchangers have the disadvantage of temperature limitations. The ester linkage introduced should be more stable than the usual amide linkage. The resins have lower capacities than the conventional acrylic resin (due to difference in molecular weight per functional group). Similar work has been reported(15). The stability of this resin was reported satisfactory and no hydrolysis of ester linkage was observed. The stability in hydroxyl form is still suspect and the resin has been recommended for use in special applications only.

Phenol-formaldehyde based anion exchangers - The most probable alternative

The phenol-formaldehyde based products have the potential to pass the final test of commercialisation. Early ion exchangers were based on phenol-formaldehyde polymers, but these were not stable towards alkalies. The mode of synthesis of these early resins was based on the Mannich reaction.

The weakly acidic hydroxy group present is converted to - O*Na during regeneration and gives sodium leakage during service. During service, decrosslinking also takes place due to oxidation. The smaller fraction so formed has more solubility due to the - O*Na group and leads to faster degradation, especially in alkaline conditions. These problems have been overcome by blocking and functionalising the phenolic hydroxy group.

The ether linkage gives stability to the resin in both acidic and basic conditions and the problem of sodium slip never arises. During the synthesis uneven shape and size of the polymer beads, as well as loss of epichlorohydrin due to self polymerization, were observed. The beading problem was overcome by the use of an appropriate solvent/non-solvent system. However, work is still in progress to find a solution to the homopolymerization of epichlorohydrin.

The resins have lower capacity than the conventional resins (Tables 2 and 3) and there are problems regarding degree of crosslinking. Future work will concentrate on overcoming these two disadvantages. Polyhydroxy phenols and aminophenols can be used instead of phenol to overcome this problem of lower capacity.

CONCLUSION

These studies show that the phenol-formaldehyde based anion exchangers have the potential to replace the conventional styrenic anion exchangers in commercial production with the added advantage of providing a safe manufacturing environment. However, these resins have to be further tested for osmotic, thermal and life cycle tests before commercial viability is fully established.

REFERENCES

1. Van Duuren, B.L., Sivak, A., Goldschmidt, B.M., Katz, C. and Melchionne, S., J. Nat. Cancer Inst., 1969, 43, 481-486.
2. Van Duuren, B.L., Katz, C., Goldschmidt, B.M., Frenkal, K. and Sivak, A., J. Nat. Cancer Inst., 1972, 48, 1431-1439.
3. Figueroa, W.G., Raszkonski, R. and Weiss, W., New England Journal of Medicine, 1973, 288, 1096-1097.
4. Olah, G.A., Beal, D.A. and Olah, J.A., Synthesis, 1974, 560.
5. Warshawsky, A. and Deshe, A., J. Polym. Sci., Polym. Chem., 1985, 23, 1839.
6. Balakrishnan, T. and Ford, W.T., J. Appl. Polym. Sci., 1982, 27, 133.
7. Barrett, J.H., U.S. Patent 3,812,061 (to Rohm & Haas), May 21, 1974.
8. Mohanraj, S. and Fort, W.T., Macromolecules, 1986, 19, 1470.
9. Qureshi, A.E. and Fort, W.T., Reactive Polymers, 1989, 10, 279-285.
10. Krauss, Sabrowski, Schwachula, Plasti Kautsch, 1982, 29 (8), 449-451.
11. Mocanu, Carpov, Rom. RO, 28th February 1983, 81, 410.
12. Meteyer, T.E. and Fries, W., U.S. Patent 4,184,019, Jan. 15, 1980.
13. Ward, E.L., U.S. Patent 4,189,539, Feb. 19, 1980.
14. Vakulenko, V.A., Izotov, V.M., Keelus, L.S., Razbacv, V.N., Slobstov, I.E. and Khlyupin, A.A., USSR, SU 1,512,985, Oct. 7, 1989.
15. Itradil, J. and Svec, F., Reactive Polymers, 1990, 13, 43-53.
16. US.EPA, Noyes Date Corporation, Vol.1, 1988.

WEAK BASE ANION EXCHANGE RESIN:
SIMPLIFICATION OF AMINATION PROCESS AND CONTROL ON SBC

S.V. VAIDYA, S.S. BAPAT, A.S. KALE AND S.V. MOKASHI
Thermax Limited, Chemical Division, 97E General Block,
Bhosari Industrial Estate, Pune, India 411 026

ABSTRACT

In the manufacture of weak base anion (WBA) resin, it is important to control strong base capacity (SBC). In the conventional process, it is achieved by adding moist or semi-dried chloromethylated beads to amine.

At Thermax, commercial production of macroporous WBA resin with simplified unit operations, improved productivity, safe handling and cleaner environment is regularised by transfer of aqueous slurry of chloromethylated beads to amine.

By aqueous slurry transfer, SBC is controlled by detailed study of the parameters; a) amount of water, b) type and amount of solvent, c) pH and d) temperature.

INTRODUCTION

Macroporous weak base anion (WBA) resin is characterised by tertiary amine groups attached to a cross-linked polystyrene matrix.

The matrix is prepared using divinyl benzene as cross-linking agent, in the presence of a linear polymer such as polystyrene to introduce porosity and form a macroporous matrix.

The macroporous matrix is further chloromethylated by reaction with paraformaldehyde, methanol, hydrochloric acid or chlorosulfonic acid in the presence of Friedel-Craft catalyst such as aluminium chloride, zinc chloride or ferric chloride.

The chloromethylated (CM) beads are washed with water and aminated with dimethyl amine (DMA).

The macroporous (WBA) resin thus prepared has a unique structure which offers superior kinetics, resistance to osmotic shock and excellent adsorption and desorption characteristics for organic matter.

Furthermore, macroporous WBA resin possesses high operating capacity on caustic regeneration. The resin can be regenerated with other alkalies such as sodium carbonate and ammonium hydroxide. For regeneration, stoichiometric quantities of regenerant are adequate. This results in better economics of operation.

Due to these advantages the macroporous WBA resins find wide variety of applications; a) removal of mineral acids, b) removal of organic matter, c) removal of colour, d) demineralisation of glucose liquors, e) removal of formic acid, f) removal of chromates, etc.

Thus, macroporous WBA resin assumes a high importance amongst commercially available anion exchange resins due to its versatile performance. It is, therefore, essential to focus attention on controlling its properties - especially the strong base capacity (SBC).

In order to control SBC, which results from bridging tertiary ammonium groups, in a conventional method, CM beads are washed with water. Water is filtered out and semi-dried or moist beads are charged into the amine and amination is completed.

The conventional method has the following drawbacks:

a)   additional work force for washing of CM beads
     and charging into amine;

b)   washing and charging of CM beads leads to exposure
     to chemical fumes and loss of solvent;

c)   only smaller lots can be aminated since
     amination is highly exothermic.

With a view to eliminating these drawbacks, commercial production of macroporous WBA resin is simplified at THERMAX by transferring aqueous slurry of CM beads into previously cooled DMA and completing amination at low temperature.

As mentioned earlier, macroporous styrene-divinyl benzene copolymer is chloromethylated and forms the starting material for manufacturing WBA resin.

GENERAL PROCEDURE FOR MANUFACTURE OF MACROPOROUS WBA RESIN

Chloromethylated beads are washed in a closed-kettle with water. Any free CMME is quenched during washing. The pH of CM beads is made highly alkaline with caustic soda solution. The reaction mass is uniformly mixed and cooled at $15^{\circ}C$.

In a clean aminator, equipped with an agitator and a condensor, DMA is charged and cooled to $10^{\circ}C$. The solution is made highly alkaline by addition of caustic flakes under mechanical agitation without allowing temperature to rise beyond $10^{\circ}C$.

The alkaline slurry of CM beads is transferred to the aminator. The amination is completed at low temperature. After amination, hydrochloric acid is added and the solution is maintained distinctly acidic. The reaction mixture is slowly heated and the solvent is recovered by distillation. The batch is cooled and washed until acid free. The resin is then converted to the freebase form.

PARAMETERS

This paper covers details of various parameters which were studied in simplifying the amination process at Thermax.

Amount of Water:

For transfer of alkaline CM beads into amine, varying amounts of water were added. The resultant pH of the reaction mixture was around 8. Figure 1 shows changes in Total Chloride Capacity (TCC) and %SBC with varying amounts of water.

Type and amount of solvent:

The organic solvent is useful in improving the kinetics of the amination reaction and also the physical properties of the finished product. Chlorinated hydrocarbons such as methylene chloride, 1-2 dichloro ethane (EDC) etc. may be used. For our studies, EDC was used as solvent.

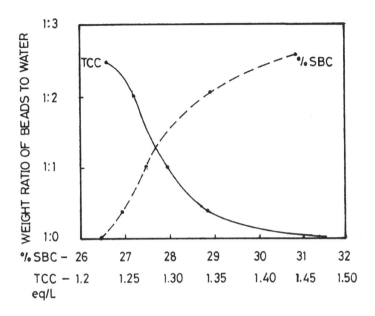

<u>Figure 1.</u> Change in %SBC with variation in water quantity.

WEIGHT RATIO OF BEADS TO SOLVENT	TCC eq/L.	% SBC
1:0.3	1.55	29.2
1:1.0	1.63	26.4
1:1.5	1.65	23.5

<u>Table 1.</u> Change in %SBC with variation in solvent quantity.

pH variation:

Resultant pH of reaction mixture of alkaline CM beads and Amine:

The CM beads were washed and pH was adjusted to 8. Varying amounts of caustic flakes were added to DMA. After transfer of CM beads into the amine, the resultant pH was adjusted from 8 to 11.

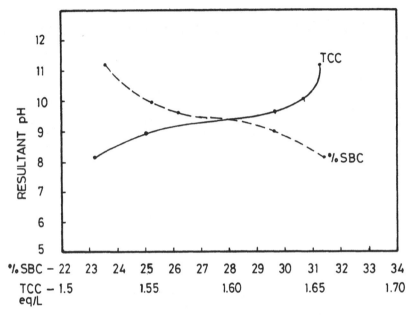

Figure 2. Change in %SBC with variation in pH.

pH during recovery of solvent:

After amination, EDC is recovered by distillation. Figure 3 shows changes in %SBC at varying pH before and after neutralization.

Temperature rise during Amination:

In these experiments, addition of CM beads into amine was done at 10°C. After the addition, the temperature was allowed to rise from 30 to 70°C. Figure 4 shows the changes in % SBC with varying temperature rise during amination.

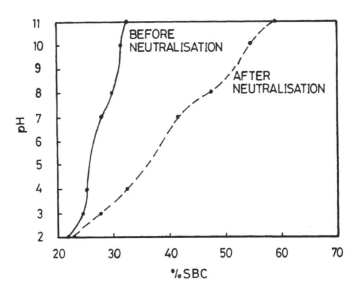

Figure 3. Change in % SBC with variation in pH during recovery of solvent.

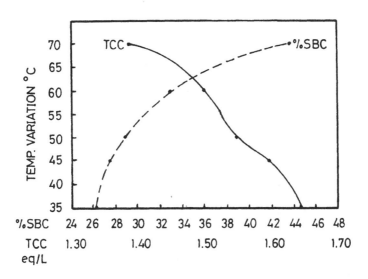

Figure 4. Changes in %SBC with variation of amination temperature.

DISCUSSION

Amount of Water:

For transfer of CM beads to the aminator, the presence of water to form a slurry is essential. If the amount of water is excessive, dilution of amine occurs, resulting in low TCC and high SBC.

Amount of EDC:

The solvent EDC swells the CM beads and facilitates efficient amination. Hence, an appropriate amount of solvent is essential. In the absence of solvent, amination occurs superficially and low TCC and high %SBC are obtained.

pH:

Resultant pH of Amination:

The resultant pH during amination is maintained towards the alkaline side. For this purpose, both CM beads and amine are made distinctly alkaline. Such an alkaline condition during amination facilitates the reaction of amine with CM beads, resulting in high TCC and low SBC.

pH during recovery of solvent:

In alkaline medium, EDC forms a complex with a tertiary amine group [1]. The formation of the complex is enhanced with increase in temperature. The resultant resin is dark-brown in colour and has a high %SBC. Hence, before distillation the reaction mass is acidified by adding hydrochloric acid and pH is maintained at 2. Under these acidic conditions, formation of the EDC-tertiary amine complex is controlled, resulting in high TCC and low %SBC.

Temperature control during Amination:

When CM beads are added to the amine, the reaction is exothermic. If the temperature rises above 40° C., bridging of tertiary amine groups takes place, giving rise to high strong base capacity.

Operational Safety:

Since CM bead handling is totally avoided and amination is conducted under closed conditions, exposure to chemical fumes is prevented. This leads to a cleaner and safer environment.

Economic advantage:

Due to simplified material handling and closed plant operations, loss of solvent and CM beads is avoided. Since amination is completed in a closed system and at low temperature, a larger batch size is possible. The net effect is increased productivity at lower material cost.

CONCLUSION

At THERMAX, refinement of the commercial production of WBA resin has been successfully accomplished. The process facilities have improved productivity, safe handling, cleaner environment and consistent quality.

ACKNOWLEDGEMENT

We would like to express our gratitude to Mr. R.D. Aga, our Chairman & Managing Director and Mr. A.M. Nalawade, Corporate Director for their constant encouragement.

We wish to acknowledge the co-operation extended by our Manufacturing Department and thank Mr. N. Raghuraman who took keen interest in regularising this process.

Finally we sincerely thank Mr. S. Krishnamurthy for preparing the draft and figures for this presentation.

REFERENCES

[1] Malovik V.V. Demchenko, Ya.P.; Semenii V.Ya.; Lozinskii M.O.; Golik G.A.; Zavatskii V.N.; Reaction of EDC with Dimethyl Amine; Tetramethyl ethylene diamine; Institute of Organic Chemistry, Academy of Sciences, Ukranian SSR; U.S.S.R.; 546, 605 (Cl CO7 C87/16); 15 February, 1977; Appl.2,303,326; 29th December, 1975.

MULTI-COMPONENT MIXED-BED ION EXCHANGE MODELING
IN AMINATED WATERS

EDWARD J. ZECCHINI, GARY L. FOUTCH, TAEKYUNG YOON
School of Chemical Engineering, Oklahoma State University
Stillwater, Oklahoma 74078-0537

ABSTRACT

A model for multi-component mixed-bed ion exchange in pH adjusted water is developed. Film diffusion controlled mass transfer is combined with bulk-phase reaction to determine mixed-bed effluent-concentration profiles. The cationic resin is initially in the hydrogen form. As exchange progresses, the hydrogen cycle is replaced by the amine cycle which produces a characteristic of amine form operation, sodium throw. The model predicts the transient sodium outlet concentration surge.

INTRODUCTION

Many nuclear and coal-fired power plants operate with a pH adjusted water cycle. A pH above 9.0 has reduced corrosion and iron transport and gives longer equipment life. However, weak base (amine) addition creates problems for the removal of dissolved ions which contribute to corrosion. The removal of these ions is accomplished effectively by ion exchange. There are two mixed-bed ion exchange (MBIE) operating modes with pH adjustment; a) convert the cationic resin to the weak base form and regenerate the bed as needed, and b) operate in the hydrogen cycle in the presence of a weak base. Hydrogen form resin removes both ionic contaminants and dissociated base, as a result, water must be redosed with amine to maintain operating pH. Redosing leads to additional chemicals and man-power expenses. This disadvantage of the hydrogen cycle can be overcome by operating the bed past amine breakthrough.

Amine cycle operation has been addressed earlier (1,2). A model that can address operation with pH control in either the amine or hydrogen cycle will result in improved operation and lower costs for water purification systems currently in use.

MODEL DEVELOPMENT

The model developed here describes MBIE using the hydrogen cycle with pH control. The weak bases used predominantly in industry, ammonia and morpholine, each have unique influences on cycle performance, primarily through different selectivity coefficients and dissociation constants. Operating with a weak base requires multiple reaction equilibria with bulk-phase neutralization as a correction for bulk-phase concentrations.

The assumptions used in model development are presented elsewhere(1,2). Most important of these is that neutralization occurs in the bulk phase. Allowing neutralization within the film surrounding the cationic resin would be more accurate, however, the required information cannot be obtained for this system. The film neutralization model for binary exchange requires that concentrations be specified at the reaction plane (3). The location of this plane must be known for multi-component exchange, but the static film model does not allow this point to be specified since the film thickness, δ, around the bead is not known. Using a different film model may be appropriate for ion exchange systems, but the flux expressions will now be based on the finite film thickness that these other models predict. Either approach requires estimated values or assumptions that reduce the accuracy of the intended method.

The exchange process is described by determining the flux of each of the exchanging species within the film surrounding the resin (4). Film diffusion control is assumed and the Nernst-Planck equation is used to express the fluxes, as:

$$J_i = D_i \left(\frac{dC_i}{dr} + z_i \frac{FC_i}{RT} \right) \frac{d\phi}{dr}$$

Solving each species flux equation in terms of concentration determines the effective diffusivity. The development of the binary flux equations for the anionic resin and the amine cycle

are available (1), as are ternary univalent flux expressions
(2). These equations can be used here with appropriate changes.
The flux expressions are solved to obtain $J_i \delta$ as a function of
species concentrations. With the product of the film thickness
and the flux, the static film model can be used to determine the
effective diffusivities. Ternary exchange requires that two
effective diffusivities be determined. The third component
fractional concentration is found by applying a material balance
on the resin and bulk phase. The binary effective diffusivity
for chloride exchange is:

$$D_e = \frac{2 D_o D_c}{(D_o - D_c)} \left(\frac{c_o^o + c_c^o - c_c^* - c_o^*}{c_c^o - c_c^*} \right)$$

The effective diffusivity for species i in the ternary exchange
on the cationic resin is:

$$D_{ei} = \frac{2 D_i}{(c_c^o - c_o^*)} \left(\frac{1 - (c_p^*/c_p^o)(c_i^*/c_i^o)}{1 + c_p^*/c_p^o} \right)$$

The ternary effective diffusivity is determined by using a
pseudo-single co-ion within the cation film. The effective
diffusivities are used in the rate expressions to describe the
exchange process.

The static film model is used to determine the rate of
exchange in terms of an overall mass-transfer coefficient. The
effective diffusivity to the 2/3 power correlates well with the
packed bed overall mass-transfer coefficient (5). Re-defining
the relation for the overall mass-transfer coefficient (6) as:

$$R_i = \left(\frac{D_e}{D_i} \right)^{2/3} = K_i'/K_i$$

re-combining the value of R_i with the static film model gives
the rate of exchange in terms of resin phase fraction as:

$$\frac{\partial y_i}{\partial t} = K_i R_i \frac{a_s}{Q} (c_i^o - c_i^*)$$

The rate of exchange together with the material balance
determines the effluent-concentration profile for MBIE. A

change of variables is necessary so that the material balances lend themselves to an appropriate method of solution. The new dimensionless distance, ξ, and time-distance, τ, variables are of the general form proposed by Kataoka et al. (7). The material balances for species i for resin and mobil phases are:

$$\frac{\partial x_i}{\partial \xi_x} + FC_j \frac{\partial y_i}{\partial \tau} = 0$$

$$\frac{\partial y_i}{\partial \tau} = 6R_i\left(x_i^o - x_i^*\right)$$

These equations can be solved by the method of characteristics.

The complex equilibria of the bulk phase must also be considered. The two competing reactions are:

$$H^+ + OH^- \longleftrightarrow H_2O$$

$$Amine^+ + OH^- \longleftrightarrow Amine + H_2O$$

Since neutralization is restricted to the bulk phase, two equilibria equations must be solved at each grid point by accounting for the species released from the resin. New bulk concentrations are then determined, based on equilibrium, by solving a system of non-linear algebraic equations. The equations that describe the exchange process can now be applied to estimate column performance.

DISCUSSION

The model is used to consider ammonia, the traditional weak base used for pH control, and morpholine, an alternative to ammonia with certain advantages. The morpholine distribution coefficient gives a higher film pH than ammonia, which is highly desirable (8). Morpholine also has a selectivity coefficient less than 1.0 for sodium exchange. Morpholine draw-backs are degradation and a lower dissociation constant. Comparing MBIE performance for these additives will provide insights into selection criteria.

Figure 1 shows an effluent-concentration profile for hydrogen cycle MBIE with ammonia present. One feature involving ammonia in the hydrogen cycle is the sodium "blip," a peak in the effluent concentration following shortly after ammonia

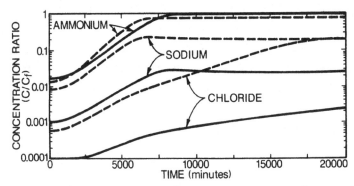

Figure 1.   Ammonia Cycle Operation Past the Ammonia Break for pH
9.6 and C/A of 1/1. $C_f = 1.0 \times 10^{-6}$ M.

break.   The shape of the sodium curve agrees with Emmett (9) and
Salem (10).   The fact that the sodium surge occurs after the
ammonia break, even though ammonium is preferred by the resin,
is related to the significantly higher bulk phase concentration
of ammonium and the presence of undissociated ammonia.   Figure 1
also demonstrates that for many cycle choices, the breakthrough
of chloride is negligible compared to sodium.   Lower inlet
concentration does correspond to a longer time until sodium
breakthrough, but the earlier than expected curve is a result of
the ammonium acting as an eluant, much as a carrier does in gas
chromatography, displacing the sodium down the bed.   Since pH is
the same at both concentrations, the ability of the ammonium to
displace the sodium in less time was expected.

Since chloride break trails sodium considerably, a higher
cation-to-anion resin ratio (C/A) may give longer run time.
Figure 2 shows runs for different C/A with conditions as in
Figure 1.   With a higher C/A, the initial leakage is
considerably lower, and the time at which breakthrough occurs is
later.   An interesting feature is the height of the sodium blip.
As C/A increases, the transient sodium peak height increases due
to more effective removal of sodium.   The resin phase loading of
sodium is higher and, therefore, the ammonium has a greater
amount of sodium to displace.   Increasing C/A leads to earlier
leakage of chloride, however, even at a ratio of 2/1, chloride
leakage lags significantly behind sodium leakage.

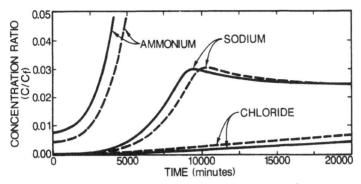

Figure 2. Amine Cycle Operation Past the Ammonia Break for pH 9.6 and C/A of 1.5/1 (solid) and 2/1 (dash). C_f = 1.0 x 10^{-6} M.

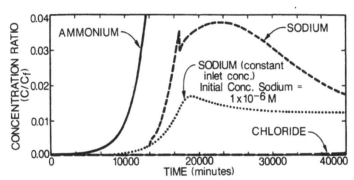

Figure 3. Amine Cycle Operation Past the Amine Break for pH 9.6 and C/A of 1/1 with a simulated condenser tube leak.

A simulated condenser leak is shown in Figure 3. The inlet concentration was stepped from 1x10^{-6} M to 1x10^{-5} M for a short time. The exaggerated peak height, followed by a decline to an outlet concentration higher than the original inlet concentration is caused by the intermittent leak. The ability of the bed to handle leaks of this nature when operating in the amine cycle is extremely important.

Figure 4 shows morpholine effluent concentrations for the conditions of Figure 1. A selectivity coefficient for sodium over morpholine greater than 1.0 removes the transient "bump" that is evident with ammonia. The lower dissociation constant also contributes to the sharpness of morpholine break and the

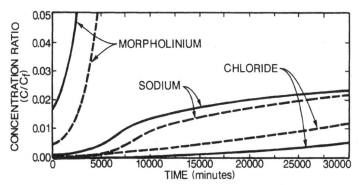

Figure 4. Amine Cycle Operation Past the Morpholine Break for pH 9.6 and C/A of 1/1 (solid) and 2/1 (dash). $C_f = 1.0 \times 10^{-6}$ M.

initial leakage. Morpholine shows lower initial sodium leakage with a slower rise to equilibrium. At a higher C/A, initial leakage levels are lower, since there is greater cation exchange capacity. The higher capacity equalizes the tremendous difference in the selectivity coefficients for the cation and anion resins. The drawback of a higher C/A is that morpholinium removal is also enhanced, requiring more make-up morpholine.

Figure 5 shows the effect of a step-change in sodium concentration at the conditions of Figure 3. The results are a slight transient bump due to the change in inlet concentration and that the long-term removal capacity for the bed is reduced. The cycle is able to reduce the impact of the concentration change, as does ammonia, which is the desired result.

Industrially, inlet concentrations could be 100-fold lower than those presented here. Lower sodium concentrations mean

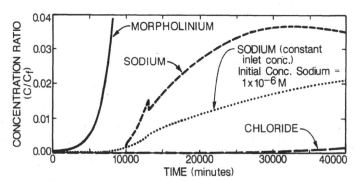

Figure 5. Amine Cycle Operation Past the Morpholine Break for pH 9.6 and C/A of 1/1 with a simulated condenser leak.

127

that most of the bed capacity is consumed by morpholinium which thereby reduces the sodium removal efficiency (1).

CONCLUSIONS

Hydrogen cycle exchange past the amine break requires that complex equilibria and ternary exchange be addressed. The model was evaluated for two different pH control additives, ammonia and moroholine. The transient sodium "blip" is predicted and the ability to operate past the ammonia break demonstrated.

NOMENCLATURE

a_s	interfacial surface area (m^2/m^3)
C_i	concentration of species i (meq/m^3)
D_{ei}	effective diffusivity of species i (m^2/s)
D_i	diffusivity of species i (m^2/s)
F	Faraday's constant (C/mol)
FC_j	resin volume fraction
J_i	flux of species i in the film $(meq/s\ m^3)$
K_i	mass transfer coefficient (including flow)(m/s)
K_i'	mass transfer coefficient (excluding flow)(m/s)
Q	capacity of the resin (meq/m^3)
r	radius from particle center (m)
R	universal gas constant
R_i	ratio of mass transfer coefficients
t	time (s)
T	temperature (K)
x_i	bulk phase concentration fraction of species i
y_i	faction of species i on the resin
z_i	charge of species i
ξ	dimensionless distance variable
ϕ	electrical potential $(mol\ m^2/s\ C)$
τ	dimensionless time-distance variable

REFERENCES

1) Zecchini, E.J. and G.L. Foutch, Ind. Eng. Chem. Res., 1991, 30, 1886.
2) Zecchini, E.J., Ph.D. Dissertation, Okla. St. Univ., 1990.
3) Haub, C.E. and G.L. Foutch, Ind. Eng. Chem. Fund., 1986, 25, 373.
4) Wildhagen, G.R.S., R.Y. Qassim, K. Rajagopal, and K. Rahman, Ind. Eng. Chem. Fund., 1985, 24, 423.
5) Kataoka, T., H. Yoshida, and T. Yamada, J. Chem. Eng. Jpn, 1972, 5, 132, .
6) Pan, S.H. and M.M David, AIChE Symp. Ser., 1978, 74, 74.
7) Kataoka, T., H. Yoshida, and Y. Shibahara, J. Chem. Eng. Jpn., 1976, 9, 130
8) Sawochka, S.G., Power, 1988, April, 67.
9) Emmett, J. R., Effluent and Water Trmt. J., 1983, 507.
10) Salem, E., Proc. Amer. Power Conf., 1969, 31, 669.

LEACHABLES FROM CATION RESINS IN DEMINERALIZATION
AND POLISHING PLANTS AND THEIR SIGNIFICANCE
IN WATER/STEAM CIRCUITS

KATE WIECK-HANSEN
The Chemical Engineering Group of the ELSAM Corporation
I/S NORDKRAFT, Postbox 238, Østerbro 7, 9100 Aalborg, Denmark

ABSTRACT

All the resins in our water treatment plants release organic
matter to some degree. This can be divided into two groups:
that present initially and that due to ageing. The former can
be largely removed by thoroughly cleaning of new resins before use. This paper focuses on cation resins. Even though
the anion resins remove some of the leachables, the remainder
enters the water/steam circuit of the boiler, where due to
the high pressure and temperatures it decomposes into acetate, formate, carbonate and sulphate. These products can accumulate in the steam system leading to initiation of corrosion during outages.

BACKGROUND

In Danish power stations ground water is purified by ion exchange and used for boiler make-up water. The demineralization plants consist typically of weakly and strongly acid
cation resin beds, degasser weakly and strongly basic anion
resin beds and finally mixed bed units. Full flow condensate
purification, consisting of cation and mixed bed units is usually provided.

In some of our plants we have encountered operations problems. In one case it was so severe that it caused salt-initiated fatigue fracture in a turbine blade. There were two main reasons for the poor water quality: poor kinetics of the resin caused by organic fouling and release of organic matter from the ion exchange resin matrix.

We described earlier (1) the malfunctions of our condensate purification plants and how we have improved the kinetics of the resin by regular hydrochloric acid and alkaline brine treatments. At that time we thought that we were dealing with humic acid held on the anion resin, which could be removed by the cleaning procedure. It is now clear that the fouling matter was leached from the cation resin. Humic acid only caused a reduction in capacity and poor rinsing after regeneration.

It has been stated (2) that all types of resins release organic matter which can block active groups. Normally the amounts are so small that they can not be measured directly and are only detected because they increase conductivity after cation, but they accumulate in the water/steam circuit. Therefore we have concentrated on identifying leachables, their origin, how they damage the resin and how to avoid them.

LEACHABLE MATTER

Leachable matter consists of water soluble organic material, derived from the ion exchanger matrix. This arises from manufacturing and partly from the ageing, i.e. decay of the matrix. To a large extent, this material is sulphonated polystyrene derivates. Leachables are released from all kinds of resin, but in our initial work, we have concentrated on leachable matter from cation resins, as they contain sulphonated groups that cause the worst problems in the water/steam circuit. We discriminate initial leachables from permanent leachables, though in practice they occur together in plant. Initial leachable matter arises from the manufacturing process. Leakage will occur for the lifetime of the resin. Permanent leachables are break-down products from the resin matrix. The rate of decay is determined by the operating conditions, e.g. temperature, redox potential, catalysts and osmotic changes. The structure of the matrix also plays a significant role, especially the extent of cross linking.

SURVEY PROCEDURES

For an assessment of the release of leachables, we have introduced an analytical procedure which is a modification of the method employed by the Southern California Edison (3 and 4). In the laboratory the ion exchanger is treated initially with acid and salt to regenerate it and convert it to the sodium form. This results in maximum leakage for cation exchangers. The leachables from a 100 g sample of ion exchanger are added to 100 mL of ultra pure water and heated to 60°C for 24 h. Organic Carbon (TOC) is measured, and the UV spectrum is scanned.

We have started to determine the amount of leachable matter passing a mixed bed filter unit, as a measure of the amount that will enter the water/steam circuit.

WATER TREATMENT PLANTS

In water treatment plants, it is the cation resin in the cation and mixed bed units that releases leachables, largely in the form of sulphonated polystyrene compounds. Most of the released material is removed by basic anion resins and by the mixed bed resins. Unfortunately, a significant amount in the make-up water reaches the condensate purification plant and adds to those released here. The amount retained depends on the type of leachable matter and type of anion resins.

What remains in the water passes into the water/steam circuit. Exposed to the high temperatures and pressures, the organics matter decomposes resulting in increased acid conductivity of the feed water, high pressure steam and reheated steam.

Our initial experience resulted from the commissioning of new resin in a make-up demineralization plant. We observed an increased conductivity in the water/steam circuit despite a low conductivity from the condensate purification plant. For about 24 h the feedwater conductivity was between 0,1 and 0,2 µS/cm (usually 0,08 µS/cm). In the high pressure steam, the conductivity reached 0,5 µS/cm for a short while. The fluctuations in conductivity depended on the quantity of make up water. The increased conductivity could be explained by the organic content of the water. The TOC after mixed beds was 0,216 mg/L instead of the normal value 0,04 mg/L. High TOC were seen in feedwater and high pressure steam.

CASE HISTORIES

1. Make-Up-Plant

a) On one plant, problems occurred after regeneration due to increased conductivity. For about 30 min the conductivity was greater than 0,1 µS/cm measured after the mixed bed. The increased conductivity was caused by the release of leachables from the cation resin.

In terms of TOC and following normal regeneration, the cation resin released 23,6 mg C/L. After 1-2 bedvolumes, the value decreased to approx 0,5 mg/L, the same level as in the raw water. This high level of TOC moved through anion and mixed bed units (Fig. 1)

TOC-leakage, measured after a demin plant.

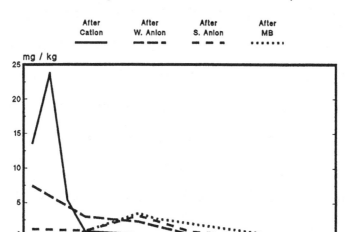

In the laboratory test the gel resin gave 971 mg/kg TOC on the dry resin basis. When the resin was replaced with a new material of a similar type, the problem was solved. The release of leachables from the new resin measured in the laboratory was only 41,8 mg/kg dry resin.

2. Condensate purification plant

a) In an 11 year old plant, the anion resin had to be removed after 7 years and after another 3 years of operation, due to poor kinetics. Both resins in the mixed bed were replaced after 10 years whilst the resin of the cation units were retained. All the resins were macroporous. For the first 11 years we had focused our attention on humic acid but following the change of mixed bed we noted that leachables had become a problem. In laboratory tests we found that the release of TOC was 153 mg/kg dry resin for cation resin and 31 mg/kg dry resin for the anion resin. Macroporous cation resins typically release leachables of only 25-40 mg/kg dry resin. Commissioning procedure was designed to eluminate the initial leachable matter, and was as follows:

Alkaline treatment - Rinse - Acid treatment - Regeneration - Rinse to a conductivity point < 0,1 µS/cm - Batch
rinsing with twelve 2 BV of water and allowed to stand
for 2 hours.

During the batch rinsings, the TOC was measured. It was
found that the release rate was low at first but later increased and finally reached a plateau. At the end of the
12 rinsings the concentration returned to low levels. The
initial performance of the new resins was followed closely. Two 50% streams of the plant were in operation all
the time, one with old resin and the other with new, so
that only half of the flow contained the initial release
of leachables. TOC, chloride, acetate and formate were
measured on samples taken when the conductivity peaked
(Fig. 2).

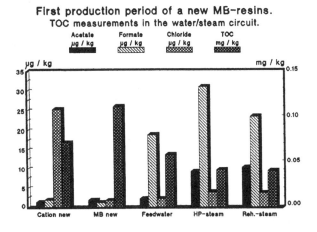

First production period of a new MB-resins.
TOC measurements in the water/steam circuit.

It appears that the organic leakage stems both from the
old cation resin and the new cation in mixed bed. The old
cation resin gave in laboratory tests the equivalent of
102 mg/kg TOC on the dry resin basis. Part of this passed
through the mixed bed units. It was not possible to identify the contribution of leachables from the mixed bed
resin directly, because of the large contribution from
the cation units. The leakage could be detected as TOC in

the water/steam circuit, and decomposition into acetate and formate was observed.

During the start up of the power plant where steam systems get wet we found acetate, formate and high acid conductivity. We found values of up to 160 mg/kg formate and 400 mg/kg acetate.

Leachables from lead cation resin are most probably the cause of the earlier mentioned problem of fouling and poor kinetics of anion resins. A great deal of these leachables pass the mixed-bed filter, enter the boiler and after the decomposition cause salt deposits released at certain conditions, e.g. as in start ups (5).

b) Study of a condensate purification plant with 10 year old resins showed that a significant amount of leachables are released just after a regeneration of lead cation resin. In the laboratory evaluation it released 11,1 mg/kg dry resin as TOC, which is a low value.. Nevertheless increases in conductivity in the water/steam circuit were observed. During the final rinsing and the return to service commissioning, TOC was measured. A significant rinse of sulfate, acetate, formate and even chloride was observed in high pressure steam.

CONCLUSIONS

All ion exchange resins release leachable matter to some extent. Leachable matter can lead to severe water chemistry problems in the water/steam circuit. The problem arises because the leachables foul the anion exchangers. Resins will lose kinetic performance and the risk of chloride and sulphate breakthrough during condenser leaks will increase.

The leachables pass through the rest of the water treatment plants and decompose into acetates, formates, carbon dioxide and sulphates at the high pressures and temperatures. In this way, leachables cause contamination of the water/steam circuit with an increased risk of corrosion during outage due to accumulation of salts and acids in various places.

Leachables can be divided into two types, those released initially which stem from the manufacturing process, and those resulting from decomposition of the resin themselves. Releases of leachable matter will increase immediately after a regeneration. There is a difference between the leachables released from gel and macroporous resins. In laboratory tests, gel resins typically gave between 100 and 400 mg/kg dry resin, whereas macroporous resins gave only 30-40 mg/kg dry resin. These values were measured on cation resins converted to the sodium form or anion resin converted to the chloride form. The following precautions can be taken to minimise problems. Resins are selected according to the lowest amount of leachables. All new resins are assessed for leachate release. All new resins are subject to a rinsing procedure involving several acid and alkaline treatments, rinsing and regeneration.

REFERENCES

1. Success with a condensate polisher starts at the demineralization plant. K. Daucik, I/S Skærbækværket. IEX 88, Cambridge, Soc.chem.Ind.

2. Are ion exchange resins soluble. F. X. Mc Garvey, Sybron Chemicals Inc. Birmingham.

3. Comparative studies of cation resin extractables. F. M. Cutler, Southern California Edison. EPRI Condensate Polishing Workshop, May 1987.

4. Determination of leachable organics in ion exchange resins. Southern California Edison. F. M. Cutler 1987.

5. Chemical changes in water/steam cycle of Fossil Power Plants under Cycling Conditions. K. Daucik. IWC 90-62.

MUNICIPAL DRINKING WATER TREATMENT
BY THE CARIX ION EXCHANGE PROCESS

WOLFGANG H. HÖLL[1] and KLAUS HAGEN[2]
[1]Kernforschungszentrum Karlsruhe, Institut für Radiochemie,
Abteilung Wassertechnologie, P.O.Box 3640, D-7500 Karlsruhe, Germany
WABAG WASSERTECHNISCHE ANLAGEN GmbH&Co. KG,
Lichtenfelser Str. 53, D-8650, Kulmbach, Germany

ABSTRACT

The CARIX ion exchange process has been applied in municipal water treatment for the combined reduction of hardness, nitrate, and sulfate concentrations. The performance of two plants demonstrates that a considerable improvement of water quality is easily achieved. CARIX has proved to be a reliable process which allows a non-polluting partial demineralization of drinking water. Results and costs are discussed in detail.

INTRODUCTION

Analyses of many drinking waters over the last twenty years reveal an important increase of hardness and nitrate/sulfate concentrations [1]. Due to both health risks and economic reasons a partial demineralization with reduction of the content of neutral salts has become necessary for many water supplies. This problem can be solved by the CARIX ion exchange process. CARIX applies a mixed bed consisting of a weak-acid resin in the free acid form and a strong-base resin in the bicarbonate form [2, 3]. Dissolved neutral salts are thus replaced by carbonic acid which is split into CO_2 and H_2O:

$$R_c - (COOH)_2 \qquad\qquad R_c - (COO^-)_2\ Ca^{2+}$$
$$+ Ca^{2+} + SO_4^{2-} \quad \Leftrightarrow \qquad\qquad\qquad + 2\ H_2CO_3$$
$$R_a - (HCO_3^-)_2 \qquad\qquad R_a - SO_4^{2-}$$

Regeneration develops as the direct reversal of the service cycle: CO_2 is dissolved in water under pressure to generate carbonic acid, which is conducted across the mixed bed filter. Thus, the waste water contains the same amount of salt that was eliminated during the service cycle. Carbon dioxide is required in excess, but these excesses can be recovered. CARIX is therefore a non-polluting process. Carbonic acid is a weakly-effective regenerant which does not completelyconversion of the resins convert the resins to the free acid and the bicarbonate forms. As a consequence, only a partial demineralization of water is achieved which, however, is sufficient for drinking water production. One further important advantage stems from the fact that there is no restriction with respect to the ratio of exchanger volumes. As has been shown in previous publications the process can therefore be adapted either to raw water quality or to the desired kind of partial demineralization [4 - 9].

The fundamental properties of the CARIX process have been investigated at the Karlsruhe Nuclear Research Center, whereas the development in the technical scale was made by the WABAG WASSERTECHNISCHE ANLAGEN Company which is the licencee of the process. Between 1982 and 1984 CARIX was demonstrated in several water works at pilot scale. The first full scale plant for municipal water treatment went into service in 1986, the second one in 1991. In 1991 a third water authority in South West Germany decided to purchase a CARIX plant.

PERFORMANCE OF FULL-SCALE CARIX PLANTS

Plant at Bad Rappenau: Bad Rappenau is a spa in south west Germany. The city and the surrounding region are are supplied by the water supply group "Mühlbach". There are about 28,000 inhabitants in 17 smaller towns and villages. The annual consumption of water amounts to about $1.7 \cdot 10^6$ m^3. The major part is local ground water but about 300,000 m^3/a are bought from the Lake of Constance water supply. The local groundwater contains calcium, magnesium and sulfate at undesirable concentrations. Due to the high hardness there were many home softeners which discharged considerable amounts of brine solution during regeneration [8].

Table 1: Raw water at Bad Rappenau and Kilchberg

Parameter	Bad Rappe-nau	Kilchberg
$Ca^{2+} + Mg^{2+}$, mmol/L	5.40	4.75
HCO_3^-, mmol/L	6.50	4.90
SO_4^-, mmol/L	1.67	2.05
NO_3^-, mmol/L	0.65	0.42
Cl^-, mmol/L	1.69	1.75
pH	7.30	7.46
κ, μS/cm	910	908

In 1984 the authority decided in favour of a CARIX plant for partial demineralization after comparison of several treatment processes. Objectives of the design were the reduction of hardness to 2.70 mmol/L and of nitrate to about 25 mg/L as recommended by the European Community. The maximum throughput was to be 170 m^3/h.

The Bad Rappenau CARIX plant consists of three filters from which always two are operating in the service cycle while the third one is regenerated or waits for service (Figure 1).

After half of the throughput between two regenerations the next filter starts its service cycle. Thus there is a cyclic switch of the filters. Product water passes a degasifier in which the carbonate balance is adjusted by means of air. Excess CO_2 from the product water is not recovered. Finally the product water is chlorinated with 0.05 mg/L Cl_2 before being pumped to a 1,200 m^3 reservoir.

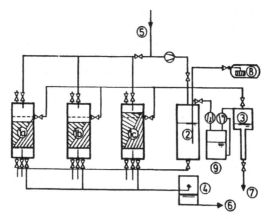

Figure 1: Scheme of the Bad Rappenau plant: 1 = resin filters, 2 = regenerant vessel, 3 = vacuum degasifier, 4 = product water degasifier for adjusting the carbonate balance, 5 = raw water, 6 = product water, 7 = waste water, 8 = carbon dioxide reservoir, 9 = intermediate tank

A separate vessel contains carbonic acid for regeneration which is carried out in counterflow with 4.5 bed volumes. Regeneration is followed by a rinsing step with one bed volume. The spent regenerant passes a vacuum degasifier in which about 90 % of the unspent CO_2 is recovered and pumped back into the regenerant vessel. Rinsing water and part of the spent regenerant are re-used. As a consequence the waste water volume amounts to about 10 % of the throughput of raw water. The effluent is directly discharged into the river Neckar. Operating conditions were chosen according to results with the pilot plant. Data are given in table 2.

Average values of main parameters in the product water between February 1986 and December 1990 are listed in table 3. The results demonstrate that hardness is reduced by about 60 %, sulfate by 75 % and nitrate is approximately decreased to 25 mg/L. Chloride reduction is negligible. The anion exchanger also reduces the concentration of humic acids in the water. Due to the acrylic matrix of the anion exchanger the sorption is completely reversible and there is no fouling [8]. During the first months of service the filters became contaminated with bacteria. However, without taking any measures the contamination had disappeared after a few regeneration steps which might be due to the high CO_2 concentrations during regeneration. Since 1989 the raw water has been disinfected by UV irradiation.

Table 2: Design and operating conditions of CARIX plants

	Bad Rappenau	Kilchberg
Data of filters		
Total resin volumes WAC/SBA, m³/m³	54/42	60/35
Column diameter, m	4	3.2
Column height, m	5	5.5
Bed depth, m	2.55	4.0
Service cycle		
Rate of filtration, m/h	6.8	16.2
Throughput per run, m³/BV	1300 / 41	1200 / 37.5
CO_2 recovery pressure, mbar	---	40
Regeneration		
Regenerant volume per filter, m³	165	135
CO_2 pressure, bar	5.5	5.0
Rate of filtration, m/h	4	5.6
CO_2 recovery pressure, mbar	250	250

Table 3: Average performance of CARIX plants

Parameter	Bad Rappenau		Kilchberg	
	Conc.	% Removal	Conc.	% Removal
$Ca^{2+} + Mg^{2+}$, mmol/L	2.40	55.6	2.40	49.5
HCO_3^-, mmol/L	3.20	50.8	2.70	44.9
SO_4^{2-}, mmol/L	0.42	75.0	0.94	55.0
NO_3^-, mmol/L	0.42	35.0	0.37	11.5
Cl^-, mmol/L	1.55	8.3	1.13	35.5
pH	7.65	-	7.50	-
κ μS/cm	470	48.4	500	44.9

Plant at Kilchberg: The Kilchberg water work belongs to the water supply authority "Steinlach". It supplies about 17,000 inhabitants. The annual consumption of water amounts to 1,700,000 m³. Between 1980 and 1985 the quality of the local ground water has deteriorated considerably: hardness has increased from 3.77 to 4.75 mmol/L, sulfate concentration from 135 to 205 mg/L and nitrate content from 12.5 to 26.5 mg/L. Considering the recommendations of the European Community nitrate therefore does not present a problem. Alkalinity has increased only slightly from 4.77 to 4.90 mmol/L. Data are summarized in Table 1 [9]. In 1988 the authority decided for a partial demineralization by a CARIX plant in order to adjust drinking water quality approximately to that from the Lake of Constance.

The maximum throughput was to be 260 m³/h. Objectives with respect to water quality were the reduction of hardness to about 2.4 mmol/l and of sulfate to 90 mg/l. Since cost estimations had shown that at high alkalinities of the raw water the recovery of CO_2 from the product water would be economic an important objective was to reduce the consumption of carbon dioxide by recovery from both the spent regenerant and from the product water.

As in Bad Rappenau the plant consists of three filters which are operating in the same way as in the first plant. Data for the plant and its operation are given in Table 2. The important difference from the Bad Rappenau plant consists of the recovery of carbon dioxide also from the product water which requires a second set of vacuum pumps/compressors (see Figure 2).

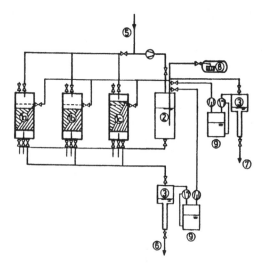

Figure 2: Scheme of the Kilchberg plant. 1 = resin filters, 2 = regenerant vessel, 3 = vacuum degasifiers, 5 = raw water, 6 = product water, 7 = waste water, 8 = carbon dioxide tank, 9 = intermediate tanks

Product water concentrations of alkaline earth ions, sulfate, and hydrogen carbonate are in the expected range. Average values from the first three months of service are given in Table 3.

COST OF THE TREATMENT

Plant at Bad Rappenau: The plant at Bad Rappenau went into service in February 1986. Therefore, detailed information about the cost of the treatment is available. Operating costs stem from the consumption of electric energy, carbon dioxide, chlorine, and manpower. The plant operates completely automatically. Thus, work is restricted to routine analyses, normal control and minor repairs. So far no severe troubles have occurred. Table 4 gives a detailed list of the consumption of energy and chemicals and the required manpower. The resulting operating costs per m³ of product water are listed in Table 5.

Table 4: Consumption of energy, chemicals, manpower at Bad Rappenau

Parameter	1986	1987	1988	1989	1990	average
Energy, kW/m³	0.34	0.27	0.33	0.32	0.34	0.32
CO_2, kg/m³	0.400	0.430	0.390	0.424	0.447	0.423
Cl_2, g/m³	0.05	0.05	0.05	0.05	0.05	0.05
Manpower, h/a	543	438	547	516	712	553

The capital costs resulting from building, CARIX plant, reservoir, pipelines, access to the area, etc. amounted to DM 6,551,782 (= US-Dollar 3,801,018 at a ratio of 1.75 DM / 1.00 US-Dollar). For calculation of specific capital costs it has to be taken into account that construction of this first plant was strongly supported by the Federal Ministry of Research and Technology and by the State Government of Baden-Württemberg by a total of 80 % of the above sum. The resulting cost which had to be covered by the water authority thus was only DM 1,323,096.

Table 5: Operating costs at Bad Rappenau

Parameter	1986	1987	1988	1989	1990	average
Energy, Dpf/m³	7.50	9.12	8.11	6.98	6.82	7.76
CO_2, Dpf/m³	12.84	13.57	12.61	13.63	14.36	13.54
Cl_2, Dpf/m³	0.20	0.20	0.22	0.22	0.22	0.22
Manpower, Dpf/m³	1.68	1.36	1.66	1.58	2.17	1.69
Repairs, Dpf/m³	1.53	1.14	1.69	5.67	2.99	2.87
Inspection, Dpf/m³	-	4.11	-	-	-	-
Total, Dpf/m³	23.75	29.49	24.29	28.08	26.56	27.11

(1 Dpf = 1 German Pfennig = 0.01 DM)

Plant at Kilchberg The Kilchberg CARIX plant went into service in June 1991. Thus, only a litte information about operating and capital costs can actually be given (october 1991). A further difficulty is that at Kilchberg a complete water works has been constructed including a treatment by oxidation, multilayer filtration, and activated carbon filtration prior to partial demineralization by CARIX. The costs of the CARIX plant without building, pipelines, etc. amounted to approximately DM 5,500,000. Construction of the extended CO_2 recovery was funded by the Federal Ministry of Research and Technology to the extent of DM 1,500,000.

Detailed operating costs for electric energy, carbon dioxide, chlorine, and manpower cannot be given yet. However, due to the recovery of CO_2 from the product water the consumption is in the range between 50 and 80 g/m^3. Thus one of the main objectives of the operation has been achieved.

Table 6: Capital costs at Bad Rappenau

Total cost, DM	
Building	1,100,659
Access, etc.	146,397
CARIX plant	4,227,940
Product water reservoir	941,853
Pipelines at plant	131,933
Total	6,551,782
Depreciation[1]	
Charge per year, DM	59,157
Average charge, Dpf/m^3	4.50
Interests at 6.5 % interest rate[1]	
Charge per year, DM	43,000
Average charge, Dpf/m^3	3.27

[1]Based on DM 1,323,096.

COST SAVINGS DUE TO CARIX PLANTS

For the customers the use of partially demineralized drinking water has led to important cost savings.

• As a consequence of the decreased water hardness most of the home softeners have been shut down. The service of such devices caused additional costs in the range of DM 2.00 - 3.00 per m^3 of water which are now saved [8].

• Due to the decreased hardness about 25 % of detergents, soap, etc. could be saved in private households. Detailed analysis of the various parts yields cost savings of about 1 DM/m^3 [8].

- Life time of water heaters is increased due to less scaling.

- Water softening or full demineralization for industrial purposes requires less brine or acid/caustic for regeneration of the resins.

REFERENCES

[1] H. Sontheimer, Entsalzung und Enthärtung wann, wie und warum, DVGW Schriftenreihe Wasser 1986, *106*, 17 - 32.

[2] W. Höll, B. Kiehling, Verfahren zur Teilentsalzung von Wässern mit einer Kombination von schwach saurem und basischem Austauschermaterial und anschließender Regeneration des Ionenaustauschermaterial, EP 81109498.6 (1984).

[3] W. Höll, B. Kiehling, Partial demineralisation of water by ion exchange using carbon dioxide as regenerant, in: Ion Exchange Technology (D. Naden and M. Streat eds.) SCI, Ellis Horwood, Chichester, 1984.

[4] W. Höll, W. Feuerstein, Partial demineralization of water by ion exchange using carbon dioxide as regenerant. Part III: Field tests for drinking water treatment, Reactive Polymers, 1986, *4*, 147 - 153.

[5] W. Höll, treatment of drinking water by the CARIX ion exchange process - Experiences from two years of operation of the first full scale CARIX plant Proc. Int. Water Conf. Pittsburgh, 1988, *49*, 80 - 88.

[6] K. Hagen, Betriebsergebnisse mit der CARIX - Teilentsalzungsanlage, Vom Wasser, 1987, *69*, 259 - 267.

[7] W. H. Höll, W. Kretzschmar, Combined nitrate and hardness elimination by the CARIX ion exchange process, Water Supply, 1988, *6*, 51 - 55.

[8] B. Steeb, Teilentsalzung nach dem CARIX-Verfahren, DVGW-Schriftenreihe Wasser, 1988, *107*, 167 - 195.

[9] Steinlach-Wasserversorgung, Mössingen, private communication.

[10] W. H. Höll, Umweltfreundliche Teilentsalzung bei der Trinkwasseraufbereitung mit Ionenaustauschern, in: Neuere Technologien in der Trinkwasseraufbereitung, Berichte aus Wassergütewirtschaft und Gesundheitsingenieurwesen, Technische Universität München, 1990, *95*, 255 - 281.

CATALYTIC REMOVAL OF DISSOLVED OXYGEN FROM WATER

R. WAGNER, A. MITSCHKER, T. AUGUSTIN
Lewatit Research and Applications Department
Bayer AG, 5090 Leverkusen, Germany

ABSTRACT

It is often necessary to remove dissolved oxygen from water. The following discussion describes how this may be accomplished in a catalytic process, which is increasingly becoming the method of choice in a variety of processes ranging from ultrapure water applications to offshore oilfield injection water.

INTRODUCTION

In applications requiring water with low residual oxygen concentrations, thermal or vacuum degassing systems are normally utilized. Their major disadvantage however, lies in high operating costs and and bulky construction.

An alternative method can be found with the catalytic removal of oxygen, whereby hydrogen is injected into the water to be treated. This is then passed through a bed of palladium doped resin catalyst, resulting in a virtually quantitative removal of dissolved oxygen in accordance with the following equation.

$$2\,H_2 \quad + \quad O_2 \longrightarrow \quad 2\,H_2O$$

With this method it is possible to achieve residual oxygen concentrations below 10 ppb (μg/l) using stoichiometric amounts of hydrogen. In nearly all installations worldwide in which this process is used, the catalyst is supplied by Bayer AG.

ADVANTAGES OF CATALYTIC REDUCTION

Compared with energy intensive vacuum degassing, the operating costs are up to 70% lower. The main reason for this is the relatively low cost of hydrogen, which only needs to be added in small amounts: at a concentration of 8g of oxygen /m³ water, only 1g of hydrogen is required

The process is environmentally clean and does not require additional chemicals. Water is the only reaction product. The system is therefore ideal for removing oxygen from mineral salts-free water.

The process can be used over a wide range of temperatures without any loss of efficiency.

Mineral salts do not interfere with the catalytic reaction.

Apart from feedwater pumps, the system does not contain any moving parts and is virtually maintenance free.

CATALYST STRUCTURE AND REACTION MECHANISM

The catalysts are palladium doped anion exchange resins based on styrene and divinyl benzene. They are supplied as transparent-gelular and opaque-macroporous spherical beads (Fig.1).

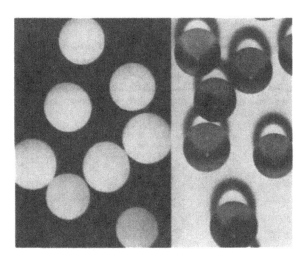

Fig. 1. Macroporous catalysts (left) and gel types (right).

The doping process results in the palladium being dispersed in the outer periphery of the beads as extremely fine particles in metallic form (Fig. 2). Therefore, the reactants hydrogen and oxygen can rapidly access the active sites and the catalyst can attain a very high degree of efficiency.

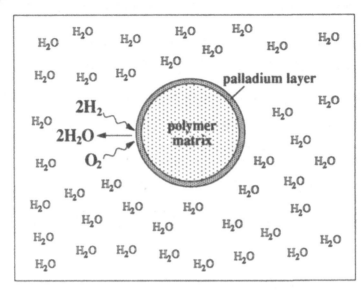

Fig. 2. The reaction between hydrogen and oxygen occurs mainly in the outer periphery of the beads.

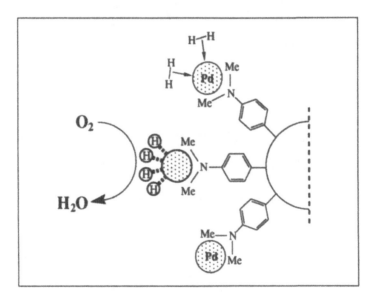

Fig. 3. Schematic presentation of the reaction mechanism: Hydrogen is adsorbed on the palladium surface and can then react with the oxygen.

DEOXYGENATION CATALYSTS

The following catalysts are offered by Bayer AG:

Bayer Catalyst K 6333
A strongly basic gel type in the chloride form for temperatures up to 70°C.

Bayer Catalyst K 7333
A strongly basic gel type in the hydroxide-form for temperatures up to 40°C, when only small amounts of chloride ions can be tolerated in the treated water.

Bayer Catalyst VP OC 1063
A weakly basic macroporous type for temperatures up to 120°C.

PLANT DESIGN AND CONSTRUCTION

A catalytic deoxygenation plant consists of a hydrogen source, a hydrogen gas distributor, a hydrogen gas mixing system and the catalyst bed. In industrial plants a static mixer is positioned vertically above the hydrogen gas distributor in the direction of the feedwater flow, as shown in Fig. 4. Hydrogen is first evenly dispersed in the water stream as small bubbles by the distributor and then completely dissolved by the static mixer. The water downstream from the mixer should not contain any bubbles, in order to enable the catalyst to function optimally.

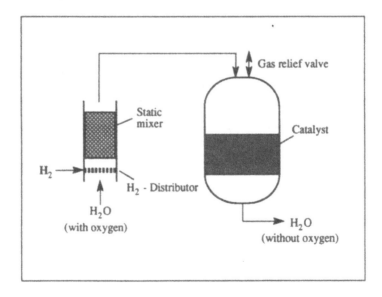

Fig. 4. A commercial catalytic reduction unit

HYDROGEN STORAGE CAPACITY

The catalytic activity of the resins arises from their ability to adsorb hydrogen on the surface of the finely dispersed palladium. Maximum activity is obtained only when the catalyst is saturated with hydrogen. As this takes some time, a delay is observed during start-up, before the residual oxygen reaches ppb (μg)-levels in the treated water.

Conversely, if the hydrogen supply is discontinued, full catalytic activity will be observed for some time before the oxygen concentration in the treated water starts to increase. One cubic meter of catalyst can adsorb 6 g of hydrogen. As 1 g of hydrogen is capable of reducing 8 g of oxygen, 6 m^3 of water at a concentration of 8 g oxygen /m^3 of water (8 ppm or 8 mg/L) would be necessary to desorb the hydrogen from 1m^3 of catalyst.

During start-up it is recommended to use a 50% excess of hydrogen for the first 30 min, in order to accelerate the rate of catalyst saturation.

During shut-down it is recommended to pass an additional 5 - 10 bed volumes of water con-taining dissolved oxygen through the resin, to desorb all of the hydrogen from the catalyst.

OPERATING INFORMATION AND CATALYST DATA

The catalysts are delivered in bead-form with diameters between 0.5 - 1.3 mm.

When the resin bed has a diameter greater than 500 mm the height should be at least 900 mm.

It is essential that the hydrogen-gas is completely dissolved before the water passes through the catalyst bed.

A system pressure of at least 1.7 bar is necessary at 25°C and 2.2 bar at 5°C, to maintain the hydrogen in solution.

The palladium content of the catalysts is 1 g / L. When starting up a plant with new catalyst a minimal loss of palladium (1 - 2 mg / L resin) is observed during the first hour of operation. Further losses cannot be observed at detection limits of 10 μg palladium / L water.

As shown in Fig. 5, the residual oxygen concentration is dependent on the influent concentration and the specific velocities. It is generally recommended to operate at a maximum specific velocity of 80 BV/h when residual oxygen concentrations below 20 ppb (20 μg/L) are required.

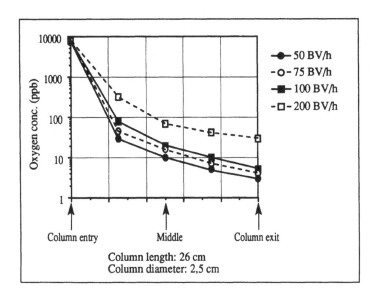

Fig.5. Profiles of oxygen concentrations along a catalyst resin bed at different specific velocities (BV/h). Water temperature: 20°C, [O_2] influent: 8000 ppb (μg/L), [O_2] effluent: 3-30ppb (μg/L).

ALTERNATIVE REDUCING AGENTS

Hydrazine or formic acid, which react very slowly with oxygen at ambient temperatures without catalytic assistance, can be used as alternatives to hydrogen as reducing agents:

$$N_2H_4 \; + \; O_2 \longrightarrow 2\,H_2O \; + \; N_2$$

$$2\,HCO_2H \; + \; O_2 \longrightarrow 2\,H_2O \; + \; 2\,CO_2$$

In contrast to hydrogen they have the advantage of being soluble in water at atmospheric pressure. Their relatively low cost-weight effectiveness however, is often sufficient to preclude their application on an industrial scale.

DEACTIVATION

Irreversible catalyst deactivation can be caused by mercury, cadmium or sulfide contaminants. Deactivation can also be caused by humic acid contaminants or by bacterial growth on the resin.

The best method of ascertaining whether humic acids will pose a problem is to run several thousand bed volumes of the water to be deoxygenated through a 100 ml bed of catalyst. Then a solution of oxygen saturated water containing 15 mg hydrazine / L should be passed through the resin at a specific velocity of 80 bed volumes / h. If the residual oxygen concentration is decreased to below 20 ppb (20 µg/L) within 5 min, no significant contamination of the resin has occurred. If it decreases only very slowly, or not at all, con-tamination has occurred and the water must first be pretreated by passage through a bed of a suitable adsorber, e.g. Lewatit MP 500 A or activated carbon.

APPLICATIONS

The catalytic method of removing dissolved oxygen from water is used for a variety of applications, such as process water in chemical industry, boiler-feed, heating circuits, ultrapure water, drinks manufacture and injection water on offshore oil platforms.

THE WILSON EQUATION APPLIED TO THE NON-IDEALITIES OF THE RESIN PHASE OF MULTICOMPONENT ION EXCHANGE EQUILIBRIA

M.A MEHABLIA, D.C SHALLCROSS and G.W STEVENS
Department of Chemical Engineering
University of Melbourne
Parkville, Victoria, Australia

ABSTRACT

Non-ideal behaviour for liquid and resin phases in an ion exchange system is investigated theoretically and experimentally for the aqueous ternary system K^+, Na^+ and H^+ with Cl^-. The Pitzer model is applied to the solution non-idealities, and the Wilson approach, based on experimental binary data, is applied to the resin phase. The two models interact through the equilibrium constant, which is calculated using the Gaines and Thomas approach. The predicted ternary data based on the three binary exchange results have been found to be consistent with the experimental results.

INTRODUCTION

To predict multicomponent ion exchange equilibrium, activity coefficient models for both the solution phase and the resin phase are required. It is also desired that the parameters of these models are not influenced by the presence of other ions so that parameters obtained from binary equilibrium data can be used to predict multicomponent equilibria.

In this study a model is presented which uses the Pitzer correlation (1,2) to describe activities in the solution phase. In order to improve the description of the solution phase, new Pitzer parameters were obtained from the original data over the range 0 to 2M instead of those optimised for range 0 to 6M. For the resin phase, the equilibrium constant has been calculated using the approach of Gaines and Thomas (3) see also Argersinger (4) and combined with the Wilson equation to regress the binary interaction parameters from the binary equilibrium data.

MODEL DEVELOPMENT

The system studied may be represented by the following three stoichiometric reactions :

$$R-H^+ + Na^+ \rightleftharpoons R-Na^+ + H^+ \tag{1}$$
$$R-H^+ + K^+ \rightleftharpoons R-K^+ + H^+ \tag{2}$$
$$R-Na^+ + K^+ \rightleftharpoons R-K^+ + Na^+ \tag{3}$$

where R denotes the resin.

The equilibrium constants for these equations may be written as:

$$K_{Na,H} = \frac{(f_{Na}y_{Na}Q_{Na})(\gamma_H C_H)}{(f_H y_H Q_H)(\gamma_{Na} C_{Na})} \tag{4}$$

$$K_{H,K} = \frac{(f_H y_H Q_H)(\gamma_K C_K)}{(f_K y_K Q_K)(\gamma_H C_H)} \tag{5}$$

$$K_{K,Na} = \frac{(f_K y_K Q_K)(\gamma_{Na} C_{Na})}{(f_{Na} y_{Na} Q_{Na})(\gamma_K C_K)} \tag{6}$$

where Q_i is the cation exchange capacity in the i form.

A further constraint involving the resin equivalent ionic mole fraction may also be written :

$$y_H + y_{Na} + y_K = 1 \tag{7}$$

Equations 4 to 7 may be used to solve for the composition in the solid phase in equilibrium with a liquid phase at a specified initial composition and concentration. The system is however overspecified because there are four equations and three unknowns. In this case three sets of three equations (4), (5) and (7), (5), (6) and (7), and (4), (6) and (7) are solved to obtain three predicted solid phase compositions, which were then averaged to find a single predicted composition. Variation in the composition predicted by the three sets of equations was within ± 0.01.

LIQUID PHASE ACTIVITY COEFFICIENT : PITZER MODEL

The semi-theoretical thermodynamic model proposed by Pitzer (1) to calculate the activity coefficient of an ion in a multicomponent system is used to consider non-idealities in the liquid phase. It has been reviewed and applied by many workers (5,6). The activity coefficient of cation M present in an aqueous solution of c cations and a anions is given by :

$$\ln\gamma_M = z_M^2 f^\gamma + 2\sum_a m_a (B_{Ma} + (\sum_c m_c z_c) C_{Ma}) + 2\sum_c m_c \theta_{Mc} + \sum_c \sum_a (m_c m_a z_M^2 B_{ca}' + z_M C_{ca} + \psi_{Mca})$$

$$+ \frac{1}{2}\sum_a \sum_{a'} (z_M^2 \theta_{aa'}' + \psi_{Maa'}) + \frac{z_M^2}{2}\sum_c \sum_{c'} m_c m_{c'} \theta_{cc'}' \tag{8}$$

where, $\quad f^\gamma = A_\phi \left[\dfrac{I^{1/2}}{1+1.2I^{1/2}} + \dfrac{\ln(1+1.2I^{1/2})}{0.6} \right]$ (9)

$$B_{MX} = \beta_{MX}^{(0)} + \frac{2\beta_{MX}^{(1)}}{\alpha^2 I^2} [1 - (1+\alpha I^{1/2}) \exp(-\alpha I^{1/2})] \tag{10}$$

$$B_{MX}' = \frac{2\beta_{MX}^{(1)}}{\alpha^2 I^2} \left[-1 + \left(1 + \alpha I^{1/2} + \frac{\alpha^2 I}{2}\right) \exp(-\alpha I^{1/2}) \right] \tag{11}$$

and $\quad C_{MX} = \dfrac{C_{MX}^{(\phi)}}{2|z_M z_X|^{1/2}}$ (12)

The constant A_ϕ is temperature-dependent and has a value of 0.392 at 25°C. $\beta^{(0)}{}_{MX}$, $\beta^{(1)}{}_{MX}$ and $C^{(\phi)}{}_{MX}$ are the pure electrolyte interaction parameters which also depend on temperature. These parameters are regressed from mean activity coefficient data of the pure electrolytes. They have been tabulated for over 250 electrolytes (7). Later Kim and Frederick (8) re-optimised the Pitzer ion interaction parameters for pure salts. A value of 2.0 for α is recommended by Pitzer (1). The values for the ternary mixing parameters, θ and ψ, are known for 69 binary electrolyte mixtures (2). They have been also re-optimised by Kim and Frederick (9).

RESIN PHASE ACTIVITY COEFFICIENTS AND EQUILIBRIUM CONSTANT

The semi-theoretical thermodynamic model of Wilson, developed for vapour-liquid equilibria has been used by several workers to calculate activity coefficients in the solid phase (10). The workers make use of the equilibrium quotient, K', defined for the exchange between cations A and B as:

$$K_{AB}' = \frac{(\gamma_B C_B)^{z_a} (y_A Q_A)^{z_b}}{(\gamma_A C_A)^{z_b} (y_B Q_B)^{z_a}} \tag{13}$$

Values for K' may be determined from equilibrium experiments if the equilibrium compositions of both phases are known and the Pitzer model is applied to calculate the liquid phase activity coefficients. This equilibrium quotient is related to the equilibrium constant by:

$$K'_{AB} = K_{AB} \frac{f_B^{Z_a}}{f_A^{Z_b}} \tag{14}$$

When applied to calculate the resin phase activity coefficients, the Wilson model yields the equation:

$$\ln K'_{AB} = \ln K_{AB} - \sum_{k=1}^{M} \omega_k \left[\left(1 - \ln \sum_{l=1}^{M} y_l \Lambda_{kl} \right) - \sum_{n=1}^{M} \left(\frac{y_n \Lambda_{nk}}{\sum_{l=1}^{M} y_l \Lambda_{nl}} \right) \right] \tag{15}$$

where Λ_{ij} is a binary interaction parameter.

Values for both the binary interaction parameters and the equilibrium constant are then regressed from binary equilibrium data using equations (13) and (15). However, the results obtained are very sensitive to experimental error. In the present study, the equilibrium constant is calculated independently using the Gaines and Thomas approach (3) see alson Argersinger (4):

$$\ln K_H^{Na} = (z_{Na} - z_H) + \int_0^1 \ln K_H'^{Na} \, dy_{Na} \tag{16}$$

This allows the equilibrium constant to be calculated without reference to the Wilson model. Values for the binary interaction parameters are then regressed using equations (13) and (15) but with the equilibrium constant set from equation (16). The interaction parameters are regressed so that the objective function, F, is minimized:

$$F = \frac{\sum\limits_{i=1}^{N} [x_i(1-x_i) y_i(1-y_i)]^2 \left(\frac{K_i'^{exp} - K_i'^{fit}}{K_i'^{exp}} \right)^2}{\sum\limits_{i=1}^{N} [x_i(1-x_i) y_i(1-y_i)]^2} \tag{17}$$

This procedure is repeated for each binary system until all the required equilibrium constants and Wilson binary interaction parameters are known. The equilibrium constant values determined for the three reactions of Na-H-K system may be used directly in eqs (4) to (6). The interaction parameters are substituted into the Wilson equation to calculate the solid phase activity coefficient in the ternary system. Since the three equilibrium constants and all the required activity coefficients in both phases are known, the model equations may be solved to yield the unknown phase composition.

EXPERIMENTAL

For the present study, the ion exchange equilibria for the binary and ternary systems involving Na^+, H^+ and K^+ ions were determined using a batch technique. A mass of resin ranging from 1 to 12 gms is contacted with the 100 ml of solution at a constant temperature of 20°C for three days under intermittent shaking. The binary equilibrium curve is obtained by varying the mass of the resin and by keeping the same volume of the solution. The concentration of hydrogen was determined by NaOH titration and the sodium analysis was performed using atomic absorption.

The ion exchanger used in the experiments was Dowex HCR-S, a macroreticular polystyrene-divinylbenzene resin with active sulphonate groups.

RESULTS AND DISCUSSION

Figure 1 presents the experimental data for one of the three binary systems. The set of curves shown in each diagram represent the binary isotherms predicted by the model described earlier. The figures show good agreement between the model predictions and the experimental observations for the binary systems for liquid phase concentrations of 0.01eq/L, 0.05eq/L and 0.5eq/L.

Table 1 presents the values determined for the three parameters, the K_{ij}, Λ_{ij} and Λ_{ji}. Figure 2 represents in graphical form the equilibrium quotient versus the equivalent ionic mole fraction. An empirical cubic equation was used to fit the experimental data so that the variation of Kc with concentration of K^+ or Na^+ can be obtained analytically.

Table 1 : The equilibrium constants and the interaction parameters of Wilson for each binary system.

System	K_{GT}	Λ_{ij}	Λ_{ji}	F
Na-K	0.7399	2.0846	0.2279	0.001
K-H	2.870	4.1491	0.0487	0.003
Na-H	1.816	0.6968	1.2455	0.002

Ternary equilibrium experimental data are presented in figures 3 and 4. The equivalent fractions of potassium in the resin phase are presented for an initial solution concentration of 0.01eq/L and 0.5eq/L. The model predictions are represented by a set of curves for the equivalent fractions of potassium in the solid phase. The quality of prediction of the ternary system is quantified using a function defined as :

$$F_t = \frac{\sum\limits_{i=1}^{N} \sum\limits_{j=1}^{M} |y_j^{exp} - y_j^{pred}|_i}{MN} \qquad (18)$$

The magnitude of the quality of fit (0.2%) is smaller than the experimental error (1%) for the ternary ion exchange system.

CONCLUSIONS

The model proposed is a modified form of the Smith and Woodburn approach. It calculates the liquid phase activity coeficients using the Pitzer model and the solid phase activity coefficients using the Wilson approach with two parameters regressed. The third parameter which is the equilibrium constant is calculated using Gaines and Thomas' approach. Using this approach multicomponent ion exchange equilibria can be predicted to within experimental accuracy. The model can be easily extended to more complex systems if the Pitzer interaction parameters are provided. This approach allows multicomponent ion exchange equilibria for simple systems to be predicted from binary data.

NOMENCLATURE

A_ϕ Debye-Hückel-Pitzer constant
B_{mx}, B_{mx}' Pitzer correlation variables
C_i Concentration of species i in the liquid (eq/L)
c_{mx}, $c_{mx}^{(\phi)}$ Pitzer Correlation variables
C_0 Total solution concentartion (eq/L)
f   Activity coefficients of species i in the solid phase
f^γ Pitzer correlation variable
F   Objective function
F_t Quality of fit
I   Ionic strength
K   Equilibrium constant
K_{GT} Equilibrium constant calculated from Gaines and Thomas Approach
K'   Equilibrium quotient
m_i Free ion molality of species i in liquid phase   (mol/L)
M   Number of counter-ion species in solid phase
N   Number of experimental points
P   Polynomial degree
n_i Number of moles of species i in the solid phase
Q_i Capacity of the resin for species i (meq/g dry resin in H^+ form)
x_i Equivalent ionic mole fraction of species i in the liquid phase
y_i Equivalent ionic mole fraction of species i in the solid phase
z_i Valence of species i

Greek sysmbols

α Pitzer parameter
$\beta^{(0)}$, $\beta^{(1)}$ Pitzer correlation parameters
γ_i Solid phase activity coefficients of species i
θ, θ' Pitzer correlation parameters

157

Λ_{ij} Wilson binary interaction parameters
ψ Pitzer correlation parameter
ω Stoichiometric coefficient of species k in the resin

REFERENCES

1 Pitzer, K.S., J Phys Chem, 1973, **77**, p268-277
2 Pitzer, K.S., and Kim, J.J., Am Chem Soc J, 1974, **96**, p5701-5707
3 Gaines, G.L., Thomas, H.C., J Chem Phys, 1953, **21**, p714-718
4 Argersinger, W.J., Davidson, A.W., Bonner, O.D., Trans. Kansas. Acad. Sci., 1953, **53**, 404.
5 Horvath, A.L., Handbook of aqueous electrolyte solutions, Ellis Horwood Ltd., England, 1985, p206-232
6 Meijer, J.A.M., Van Rosmalen, G.M., Desalination, 1984, **51** p255-305
7 Pitzer, K.S., Mayorga, G., J Phys Chem, 1973, **77** p2300-2308
8 Kim, H.T., Frederick, W.J., J Chem Eng Data, 1988, **33** p177-184
9 Kim, H.T., Frederick, W.J., J Chem Eng Data, 1988, **51** p278-283
10 Smith, R.P., and Woodburn, E.T., AIChE J, 1978, **24** p577-587

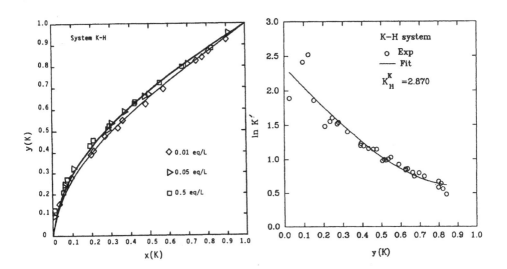

Fig 1: EQUILIBRIUM CURVES FOR K-H SYSTEM.

Fig 2: LN OF THE EQUILIBRIUM QUOTIENT VERSUS EQUIVALENT IONIC MOLE FRACTION OF K IN SOLID PHASE.

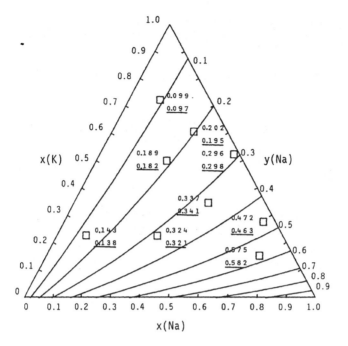

Fig 3 : PREDICTED y(Na) CONTOURS FOR THE K-Na-H EXCHANGE IN CHLORIDE
SOLUTION ON DOWEX HCR-S AT Co=0.5 eq/L. EXPERIMENTAL DATA ARE ALSO
SHOWN WITH CORRESPONDING PREDICTIONS SHOWN UNDERLINED.

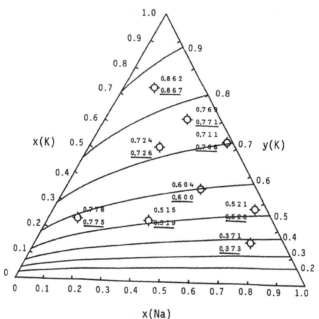

Fig 4 : PREDICTED y(K) CONTOURS FOR THE K-Na-H EXCHANGE IN CHLORIDE
SOLUTION ON DOWEX HCR-S AT Co=0.5 eq/L. EXPERIMENTAL DATA ARE ALSO
SHOWN WITH CORRESPONDING PREDICTIONS SHOWN UNDERLINED.

MATHEMATICAL MODELLING OF ION EXCHANGE
EQUILIBRIA ON POLYMER ION EXCHANGERS

V.S. SOLDATOV
Institute of Physical Organic Chemistry,
Byelorussian Academy of Sciences,
Surganov Str., 13, Minsk 220603,
Republic Belarus

ABSTRACT

A mathematical model for description of dependences of the ion
exchange equilibria parameters on the composition of ion exchangers is suggested. It is assumed that ion exchange sites in
the polymer network are different due to the difference in
their micro-environment characterized by the number of the nearest neighbours and the local ionic composition. A "property-
composition" equation was derived and proved to be adequate to
describe the experimental data for some complicated ion exchange systems.

INTRODUCTION

In 1951 Reichenberg, Pepper and McCauley published results of
their study of ion exchange equilibria of univalent ions on
sulphonated copolymers of styrene and divinylbenzene (1).It
was clearly shown that even for these simple systems the apparent equilibrium constant varied strongly with extent of ion
exchange.(Apparent equilibrium constant is a value calculated
as equilibrium constant with concentrations of the exchangeable ions in the ion exchange phase in place of their activi-
ties, $\tilde{K} = \bar{X}_2 a_1 / (\bar{X}_1 a_2)$). Significant decrease in selectivity
toward preferably sorbed ion with increasing loading of the
resin was explained by irregularity of the resin exchange
sites. Interpretation of this effect since then has become one
of the key points of the theory of ion exchange. This work was

followed by many publications in which dependence of apparent
equilibrium constants or the other parameters of equilibria on
the extent of ion exchange were observed and interpreted in
a similar manner. Later, the increase of the apparent equilibrium constant was reported and explained by "co-operative"
effect (e.g. (2)). The combination of the irregularity of exchange sites with "co-operative" interaction can result in ex-
tremes and other complicated dependences described in the literature. Nevertheless ,physical reasons for non-uniformity of
the exchange sites as well as the "co-operative" effect was
not clear and quantitative interpretation was not offered.

THE NEW MODEL

The model presented here suggests a quantitative interpretation
of parameters of ion exchange equilibria as a function of ionic
composition of ion exchanger (3,4). In this paper the case of
uni-univalent exchange will be considered.
The main statements of the model are as follows. Exchange sites
in an ion exchanger have different local environments characterized by number of neighbouring ions (i) and the local ionic
composition.Two parameters are chosen to describe the local
environment: number of ions type 2 denoted as j, and that of
ions 1 denoted as i-j. Then all possible states of exchangeable ions (I) are expressed as $I(i-j,j)$:

i, number of nearest neighbours	Possible states of ion I
0	$I(0,0)$
1	$I(1,0)$, $I(0,1)$
2	$I(2,0)$, $I(1,1)$, $I(0,2)$
3	$I(3,0)$, $I(2,1)$, $I(1,2)$, $I(0,3)$
....	....
i	$I(i-j,j), j=0-i$

A real ion exchange equilibrium may be regarded as a superpo-
sition of ideal "elementary" equilibria related to each of
the states:

$$\bar{I}_1(i-j,j) + I_2 = \bar{I}_2(i-j,j) + I_1 \qquad (1)$$

$$K(i-j,j) = \frac{\bar{X}_2(i-j,j)\ a_1}{\bar{X}_1(i-j,j)\ a_2} \qquad (2)$$

where $K(i-j,j)$ is the equilibrium constant, $\bar{X}(i-j,j)$ - mole
fraction of the ions with a $(i-j,j)$ micro-environment, a - activities. The barred symbols are related to the ion exchanger
phase. The definitions $y(i-j,j) = \ln K(i-j,j)$ and $Y = \ln \tilde{K}$ are used.
y and Y can stand also for any molar additive property of the
ion exchange system. A property Y can be obtained as a sum of
properties $y(i-j,j)$ for elementary equilibria proportionally
to the contents of the exchangeable ions in the relative states
The result of this summation (for details see (3,4)) gives the
main equations of the model, expressing dependence of the property Y as a function of relative molar fraction of exchangeable ions.

$$Y_i = \sum_{j=0}^{j=i} \frac{i!}{(i-j)!\,j!}\ y(i-j,j)\ (1-\bar{X})^{(i-j)} \bar{X}^j \qquad (3)$$

$$Y = \sum_{i=0}^{i=n} Y_i P_i \qquad (4)$$

where P_i is the probability of the event that i-exchange sites
are present in the micro-environment of an observed ion site,
Y_i is a property Y for the hypothetical ion exchanger in which
all exchange sites have equal numbers of neighbours i. As
shown later in this paper, the number of nearest neighbours is
rarely more than 4. This means that equation (3) represents
the following set of equations(5a - 5e). Since all $y(i-j,j)$
values are constants, these equations are polynomial presented
in a special form for $\bar{X} = [0,1]$. Eq.(3) has some features making
it convenient for consideration of ion exchange equilibria.

$$Y_0 = y(0,0) \tag{5a}$$

$$Y_1 = y(1,0)(1-\bar{X}) + y(0,1)\bar{X} \tag{5b}$$

$$Y_2 = y(2,0)(1-\bar{X})^2 + 2y(1,1)(1-\bar{X})\bar{X} + y(0,2)\bar{X}^2 \tag{5c}$$

$$Y_3 = y(3,0)(1-\bar{X})^3 + 3y(2,1)(1-\bar{X})^2 X + 3y(1,2)(1-\bar{X})\bar{X}^2 + y(0,3)\bar{X}^3 \tag{5d}$$

$$Y_4 = y(4,0)(1-\bar{X})^4 + 4y(3,1)(1-\bar{X})^3\bar{X} + 6y(2,2)(1-\bar{X})^2\bar{X}^2 + 4y(1,3)(1-\bar{X})\bar{X}^3 + y(0,4)\bar{X}^4 \tag{5e}$$

Coefficients of the first and the last terms of equations (3) independently of its power, $y(i,0)$ and $y(0,i)$, are equal to the values of the Y_i at $\bar{X}=0$ and $\bar{X}=1$ correspondingly:

$$y(i,0) = Y_{i(\bar{X}=0)}, \quad y(0,i) = Y_{i(\bar{X}=1)} \tag{6}$$

If all $y(i-j,i)$ have the same value A, then $Y_i = A$ (7)

If all constants $y(i-j,j)$ are changed by a constant value A, then the function Y_i is changed by the same value:

$$Y_i + A = \sum_{j=0}^{j=i} \frac{i!}{(i-j)!j!} (y(i-j,j)+A)(1-\bar{X})^{(i-j)}\bar{X}^j \tag{8}$$

The next important property of equation (3) is that if values $y(i-j,j)$ linearly depend on the local composition, $\bar{X}(i-j,j)$, then $Y = Y_i(\bar{X})$ is also a linear function, i.e. if

$$y(i-j,j) = y(i,0)(1-\bar{X}(i-j,j) + y(0,i)\bar{X}(i-j,i) \tag{9}$$

then

$$Y_i = Y_{i(\bar{X}=0)}(1-\bar{X}) + Y_{i(\bar{X}=1)}\bar{X} \tag{10}$$

It is to be noted that the local compositions are expressed by the sets of discrete numbers $\bar{X}(i-j,j)$ according to the number of neighbouring exchange sites and the combinations of the ionic species in the micro-environment. It is clear that in the case of linear dependence $Y_i = Y_i(\bar{X})$ the values Y_i and $y(i-j,j)$ coincide at the total ionic compositions equal to the local compositions. Thus two "end" constants $y(i,0)$ and $y(0,i)$ control the value of the function Y_i for pure ionic forms while the intermediate constants control its curvature. This is illustrated by Fig.1 for the important particular case i=3.

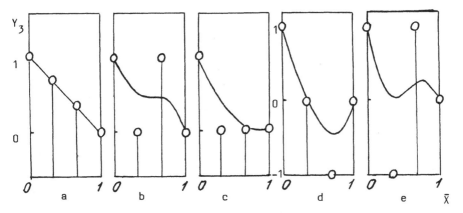

Figure 1. Hypothetical dependences $Y_3=f(\bar{X})$ for the following
sets of constants:

curve	y(3,0)	y(2,1)	y(1,2)	y(0,3)
a	1	1/3	2/3	0
b	1	0	1	0
c	1	0	0	0
d	1	0	-1	0
e	1	-1	1	0

Y_0 is always constant. It corresponds to ideal ion exchange
when ion exchange sites are situated far away from each other.
Y_1 is always a linear function of $\bar{X}$. It corresponds to "regular"
ion exchange systems, as defined in (7). Y_2-Y_4 correspond to
statistically irregular ion exchange systems. A particular
case i=2 has been successfully used for description of many
non-ideal equilibria in (5-7) and many later works. A computer
program was used to fit experimental data to one of these equations according to a given mean square error of interpolation.
  In this way the maximal possible number of nearest neighbours
can be easily found as it is equal to the power of a polynomial. Nevertheless, good fit of one of the equations to experimental data does not mean that the ion exchanger does not
contain exchange sites with a smaller number of neighbouring
sites, as can be deduced from the properties of eqns (5) and
(3).Only one parameter of the appropriate equation, y(i,0),
has its direct physical meaning corresponding to a property
$Y_i P_i$ for the elementary equilibrium.

$$\bar{I}_1(i,0) + I_2 = \bar{I}_2(i,0) + I_1 \tag{6}$$

If no independent data exist, the rest of the parameters found
by fitting , y(i-j,j), are just empirical constants if a model
implies a presence in the ion exchanger a set of exchange sites
with different number of neighbours. It can also be assumed
that all ion exchange sites have equal number of neighbours.
Then parameters y(i-j,j) have meaning as defined in the model
for i=const. Evaluation of the probabilities P_i can be done,
in principle, if the structure of the ion exchanger is known.
We have used the following methods for sulphonic type styrene-
divinylbenzene resin. Consider that its random network can be
represented as a combination of two types of fragments: linear
sulphostyrene chains containing a certain number (in our case
8) monomer units with the fixed at certain distances ends, and
a cross shaped fragment (a four beam star) with a p-DVB molecule in the center, and several (in our case 8) sulphostyrene
fragments joined to it. Analysis of conformation structure was
done using molecular mechanics using modified parameters of
the force field of the COSMIC system(8). The frequencies of
presence of the centers of the sulphonic group (not the end
groups of the fragment) were computed. These frequencies were
assumed to be approximately equal to the probabilities. It was
found that the largest possible distance between neighbouring
sulphonic groups is 13nm. We assume that the chains of the polymer are not completely stretched and have chosen a distance
between sulphogroups equal to 10nm as a "distance of interaction". All exchange sites found in a sphere with radius 10nm
are regarded as "nearest neighbours". This assumption looks
reasonable from the physical point of view and finds its
support in data on properties of soluble polystyrene sulpho-
nates, where i=1 does not describe their properties adequately,
while i=2 was found to be sufficient for a correct description.
The following results were obtained in evaluation of the probabilities P_i:

	P_0	P_1	P_2	$P_{>2}$
Linear chain	0	0,487	0,490	0,023
4-beam star	0	0,022	0,751	0,228

The data presented show that in a linear chain the probabilities of 1 and 2 neighbours are approximately equal and practically only two types of exchange sites occur. Probability of i>2 is small, therefore for ion exchange resins of low cross-linkage dependence log $\tilde{K}$=f($\bar{X}$) can be described by the equation of the second power as shown in (5) and in later works. For ion exchange resins of high cross-linkage the probability of i>2 is much larger because in the "4-beam stars", representing cross-links, probability of i>2 is more than 10 times larger than for the linear chains. In accordance with this the examples given on Fig.2 clearly show that the curves of log$\tilde{K}$=f($\bar{X}$) in this case have inflection points which is a feature of a polynomial of at least the third power. These experimental data fit perfectly with the equation. A more complex example is given in Fig.3 where log$\tilde{K}$=f($\bar{X}$) is presented for exchange of Na^{+} with protonized orthotetracycline (9).

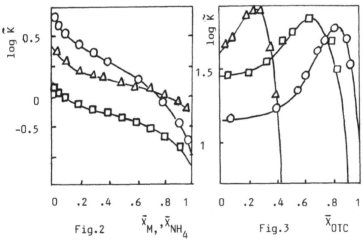

Fig.2 $\bar{X}_M, \bar{X}_{NH_4}$ Fig.3 $\bar{X}_{OTC}$

Figure 2. Logarithm of the apparent constant as a function of equivalent fraction of metal/ammonium ion for ion exchange on sulphostyrene resin with 25% p-DVB, t=25 C, ionic strength 0,100. Points are computed from experimental data, lines are calculated from (i=3) with the parameters given below

Curve	Exchange	lgK(3,0)	lgK(2,1)	lgk(1,2)	lgk(0,3)
1	Na$^+$-H$^+$	0. 540	0. 104	0.320	-0. 36
2	NH$_4^+$-Li$^+$	0. 350	-0. 034	0. 183	-0. 460
3	li$^+$-H$^+$	0. 093	-0. 219	-0. 008	-0. 559

Figure 3. Logarithm of the apparent constant as a function of equivalent fraction of OTC^{+}-ion for exchange OTC^{+}-Na^{+} on sulphostyrene ion exchangers with different DVB contents. Points are computed from experimental data, lines are calculated from eqn (i=4) with the parameters given below.

%DVB	log K(4,0)	log K(3,1)	log K(2,2)	log K(1,3)	logK(0,4)
0.5	0.83	2.22	-0.57	3.38	0.91
2	1.39	1.70	0.68	3.91	-0.61
6	1.58	1.97	0.63	17.5	-105.1

In this case the size of the counter ion is so large that the probability of having more than 3 neighbours in the sphere with radius equal to its largest dimension is very high. It was found that for 14nm $P_{>3}$ is 100% for the linear chain and about 50% for the star.

Acknowledgements. All the work concerning computation of probabilities for the fragments in the cross-linked structures was done by a group of co-workers of the Institute of Physical Organic Chemistry of Belarus Academy of Sciences: A.L.Pushkarchuk, V.M.Zelenkovsky, V.I.Gogolinsky to whom I express my deep gratitude and admiration of their high professionalism.

NOMENCLATURE

a   activity
$\bar{X}$ equivalent fraction of the exchangeable ion I_2 in the resin phase
K   apparent equilibrium constant
Y   molar additive property of the system
y   molar additive property of the microstate

REFERENCES

1. Reichenberg, D., Pepper, K.W., McCaulay, D.J., J. Chem. Soc. 1951, 493.
2. Gregor, H.P., J. Am. Soc., 1951, 73, 3537.
3. Soldatov, V.S., Doklady AN BSSR, 1990, 34, 6, 528-531.
4. Soldatov, V.S., Doklady AN USSR, 1990,314, 3, 664-668.
5. Kuvaeva, Z.I.,Soldatov, V.S., Fredlund, F., Högfeldt, E., J. Inorg. Nucl. Chem., 1978, 40, 1, 103-108.
6. Högfeldt, E., Soldatov, V.S., J. Inorg. Nucl. Chem., 1979, 41, 575-577.
7. Soldatov, V.S., Bichkova, V.A., Reactive Polym., 1983, 1, 4, 251-259.
8. Vinter, J.G., Davis, A., Saunders, M.R., J. of Comp.-Aided Design, 1987, 1, 31-51.
9. Samsonov, G.V., Melenevsky, A.T., "Sorption and Chromatographic Methods of Chemical Biotechnology". - Leningrad: Nauka, 1986, 127.

A GENERAL SOLUTION OF THE NERNST-PLANCK EQUATIONS FOR ION EXCHANGE WITH RATE CONTROL BY LIQUID-PHASE MASS TRANSFER

M. FRANZREB, W. H. HÖLL, H. SONTHEIMER
Kernforschungszentrum Karlsruhe
Institute for Radiochemistry, Water Technology Division
P.O.Box 3640, D-7500 Karlsruhe, Germany

ABSTRACT

A general analytical solution for the Nernst-Planck equations in a Nernst film is presented. The solution reduces to all known analytical solutions in the literature for special cases, e.g., binary or equal-valence ion exchange. In addition, it provides an approximation in the general case of any number of counterions and co-ions with different valences. The approximation is in good agreement with the computationally much more cumbersome numerical solution. The solution presented delivers only the momentary fluxes. To obtain the concentrations and loadings as a function of time, numerical integration is still required.

INTRODUCTION

Although the Nernst-Planck equations have been used for over three decades to describe ion exchange controlled by mass transfer in the liquid phase, no analytical solution for the general multicomponent case has been found. The Nernst-Planck (N-P) equations were first applied to film diffusion controlled kinetics by Schlögl and Helfferich (1). Their solution is restricted to the case of an unselective binary exchange of univalent ions. Later investigations expanded the solution to include selectivity (Copeland et al. (2)) and the case of binary exchange with arbitrary valences (Kataoka (3), Turner and Snowdon (4)). All these authors use the film model first proposed by Nernst. Additionally, the boundary layer model and the penetration model were used to describe the mass transfer in binary systems (5). The main problem with the last two models is that, because of the non-linearity of the flux equations, no analytical solutions can be obtained. There are only a few papers on ternary systems and these are based on the Nernst film model and are restricted to counterions of equal valences (Rahman (6), Kataoka and Yoshida (7)). Kataoka also gives an analytical approximation for the ternary case with counterions of arbitrary valences, but this approximation is complicated and not expandable to the general multicomponent case. Like the approach taken here, all quoted previous solutions require numerical integration for the calculation of conversion as a function of time.

A solution of the N-P equations is presented here for the case of an arbitrary number of counterions with arbitrary valences. This solution is exact for all special cases named above and a good approximation for the general multicomponent case. In addition, some results of shallow bed experiments are given and compared with predictions.

THEORY

Although the use of non-linear flux equations violates the Nernst model's assumption of linearity, the procedure shown here serves to indicate the maximum effects which can result from the electric field.

The film is considered as a planar sheet. The mole fractions at the resin boundary are determined by an arbitrary equilibrium relationship together with the resin loadings. However, the total concentration at this boundary is unknown. The system may contain any number of counterions and co-ions. The fluxes in the film are described by the N-P equations. This section gives only the essential equations of the theory. A more complete version will be the subject of a thesis at the University of Karlsruhe, Germany.

$$J_i = -D_i\left[\frac{\partial c_i}{\partial \zeta} + z_i \frac{c_i F}{RT}\frac{\partial \phi}{\partial \zeta}\right] \qquad i = 1..n \qquad (1)$$

The diffusion through the film is assumed to be a quasi-steady state process, i.e., the rate of change of conditions at the film boundaries is so slow that momentary fluxes and profiles are as they would be at steady state. Accordingly, total differentials instead of partial differentials can be used. Additionally, it is assumed that the fluxes of the co-ions are negligible. This is a good approximation for dilute solutions, likely to give film-diffusion control, because Donnan exclusion then is highly effective. The condition of no net electric current is described by:

$$\sum_{i=1}^{n} z_i J_i = 0 \tag{2}$$

The summations in this and the following equations are only over the n counter-ions. Elimination of the gradient of the electric potential leads to

$$J_i = -D_i \left[\frac{dc_i}{d\zeta} + n_i \frac{c_i}{c_g} \frac{dc_g}{d\zeta} \right] \tag{3}$$

where $n_i \equiv z_i/z_Y$, and c_g is the total equivalent concentration. In the case of several co-ions with different valences, a mean value for the co-ion valence must be defined:

$$z_Y \equiv \left(\sum_{j=1}^{m} z_j^2 c_j \right) \Big/ \left(\sum_{j=1}^{m} z_j c_j \right) \tag{4}$$

Because of the non-linearity of the flux equations the use of this mean value is mathematically not exact. However, the resulting error is small because of the weak influence of the co-ion valence on the fluxes. Most of the known solutions combine eqs. (2) and (3) and integrate the resulting equation to obtain the flux. Here, instead, eq. (3) is differentiated. This leads to a second-order differential equation, which has the disadvantage of requiring more boundary conditions for the integration, but has the advantage of eliminating the constant, but unknown, fluxes J_i.

$$\frac{d^2c_i}{d\zeta^2} + \frac{n_i}{c_g} \frac{dc_i}{d\zeta} \frac{dc_g}{d\zeta} + n_i \frac{c_i}{c_g} \left(\frac{d^2c_g}{d\zeta^2} - \frac{1}{c_g} \left(\frac{dc_g}{d\zeta} \right)^2 \right) = 0 \tag{5}$$

Counterions of equal valence: With a generalization of a proof which Rahman gives for a ternary system (6), it follows that for counterions of equal valences the profile of the total concentration (in equivalents) in the film is linear. Therefore, the first and second derivative of c_i with respect to ζ in eq.(5) can be transformed to derivatives with respect to c_g, and the second derivative of c_g with respect to ζ is zero.

$$\frac{d^2 c_i}{dc_g^2} + \frac{n_i}{c_g}\frac{dc_i}{dc_g} - \frac{n_i c_i}{c_g^2} = 0 \tag{6}$$

Differential equations of this form are known as EULER's differential equations with the general solution:

$$c_i = a_i' c_g + b_i' c_g^{-P} \qquad \text{with } P = n_i \tag{7}$$

If the left-hand side of the solution is expressed in equivalents

$$z_i c_i = a_i c_g + b_i c_g^{-P} \qquad \text{with } P = n_i \tag{8}$$

the values of the parameters a_i and b_i are determined by the boundary conditions as follows:

$$a_i = \frac{1}{c_g^b}\left(z_i c_i^b - b_i \left(c_g^b\right)^{-P}\right) \tag{9}$$

$$b_i = \left(x_i^s - x_i^b\right) / \left[\left(c_g^s\right)^{-P-1} - \left(c_g^b\right)^{-P-1}\right] \tag{10}$$

Counterions of arbitrary valences: Although for this case it is not possible to simplify eq.(5) as before, the exact solution for binary systems with arbitrary valences can be expressed in a form which is identical to eq.(8). The only difference is the expression for the exponent P. Unfortunately, in the general multicomponent case eq.(8) is only a good approximation. Additionally, the exponent becomes a function of the boundary conditions. Substitution of eq.(8) and its derivative in eq.(3) gives:

$$J_i = -\frac{D_i}{z_i}\frac{dc_g}{d\zeta}\left[(a_i - Pb_i c_g^{-P-1}) + n_i(a_i + b_i c_g^{-P-1})\right] \tag{11}$$

The summation of eq.(11) for all counterions together with the condition of no net electric current leads to:

$$\left(\sum_{i=1}^{n} D_i a_i + \sum_{i=1}^{n} n_i D_i a_i \right) + \left(\sum_{i=1}^{n} n_i b_i D_i - P \sum_{i=1}^{n} b_i D_i \right) c_g^{-P-1} = 0 \qquad (12)$$

Eq.(12) is valid at each point in the film and therefore independent of c_g. The only way this condition can be met is if both expressions in parentheses are zero. Equating the second expression to zero yields the desired relationship for the exponent: Substitution of the parameter b_i with eq.(10) and reduction gives an expression for P in which all variables are known.

$$P = \sum_{i=1}^{n} n_i (x_i^s - x_i^b) D_i / \sum_{i=1}^{n} (x_i^s - x_i^b) D_i \qquad (13)$$

After some manipulation the first expression of eq.(12) set to zero and combined with eq.(9) yields an equation for the total equivalent concentration at the resin boundary. The value of this concentration is used in the expression for the parameter b_i.

$$c_g^s = \left[\sum_{i=1}^{n} (1 + n_i) D_i x_i^b / \sum_{i=1}^{n} (1 + n_i) D_i x_i^s \right]^{\frac{1}{P+1}} c_g^b \qquad (14)$$

Now all the parameters in eq.(8) are known. The final step is the integration of eq.(11). After some further manipulation the result is obtained as follows.

$$J_i = -\frac{D_i}{\delta} \left[\left(1 - \frac{n_i}{P} \right) (c_i^s - c_i^b) + n_i \frac{a_i}{z_i} \left(1 + \frac{1}{P} \right) (c_g^s - c_g^b) \right] \qquad (15)$$

The procedure of calculating the fluxes for a given system can be summarized in three steps.

- Calculation of P and c_g^s from eq.(13) and (14).
- Calculation of the parameters a_i for all counterions with eq.(9).
- Calculation of the fluxes with eq.(15)

COMPUTATION AND COMPARISON WITH EXPERIMENTAL DATA

In multicomponent as in binary and ternary systems, the main effects of the electric field turn out to be accumulation or depletion of the co-ions at the resin boundary, and acceleration of the slower ions and retardation of the faster ions. Only one example is shown here: In systems with more than two counterions of very different mobilities, some of the counterions may move against their concentration gradients. This can happen if the second term in eq.(1) has the opposite sign and a larger absolute value than the first term. A conventional mass-transfer equation cannot give this effect.

To verify the prediction of this effect, shallow bed experiments were conducted. The experimental procedure will be described elsewhere. Figure 1 shows the results of experiments with a $H^+ - Ca^{2+} - Mg^{2+}$ system and the weak-acid resin Amberlite IRC 50. The bed initially contained Ca^{2+} and Mg^{2+} ($y_{Ca}^0 = 0.7$ in all experiments) and the feed contained Ca^{2+} and H^+ in different ratios ($x_{Ca}^0 = 0.2$ to 0.8) but at the same total concentration of 0.004 $keq\ m^{-3}$. The solid line is calculated using only physical data of the shallow bed, separately measured equilibrium data, diffusion coefficients from the literature and a film thickness approximation proposed by Kataoka.

$$\delta = \frac{D_r}{k} \quad ; \qquad k = 1.85 \frac{U_L}{e} \left(\frac{e}{1-e} \right)^{1/3} Sc^{-2/3} Re'^{-2/3} \tag{16}$$

In multicomponent systems each counterion has its own diffusivity. Therefore, a representative diffusivity, D_r, has to be defined.

$$D_r = \sum_{i=1}^{n} |J_i \delta| \ / \ \sum_{i=1}^{n} |c_i^s - c_i^b| \tag{17}$$

No fitting was used to match the calculation with the data. The equilibrium line for an infinitely long column and the feed solution line are also plotted in Figure 1. The predictions obtained from the usual driving-force models lie somewhere in the shaded area.

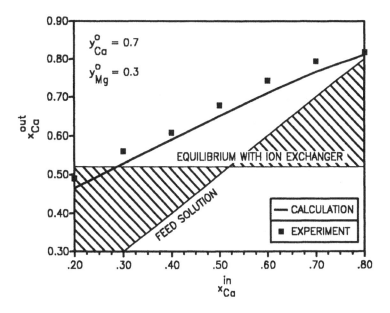

Figure 1. Equivalent fraction of calcium in effluent for various equivalent fractions of calcium in feed.

CONCLUSIONS

The introduced approximation for the solution of the N-P equations offers a new way to account for the liquid-side mass-transfer resistance in ion exchange. The inclusion of the electric field provides a closer approximation than given by the usual driving-force models, but still yields an analytical expression which can be solved directly.

NOTATION

a_i	-	parameter of ion i, defined by eq.(9)
b_i	$(kmol\ m^{-3})^{p+1}$	parameter of ion i, defined by eq.(10)
c_g	$keq\ m^{-3}$	total equivalent concentration
c_i	$kmol\ m^{-3}$	concentration of ion i

D_i	$m^2 s^{-1}$	diffusivity of ion i
D_r	$m^2 s^{-1}$	representative diffusivity
e	-	fractional void volume of bed
F	$C\ mol^{-1}$	Faraday's constant (96487 C/mol)
J_i	$kmol\ m^{-2} s^{-1}$	flux of ion i
n_i	-	$\equiv z_i/z_Y$ relative valence
P	-	exponent
R	$J\ mol^{-1} K^{-1}$	Gas constant (8.314 J/kmol K)
Re'	-	$\equiv U_L\rho d/(1-e)\mu$ modified Reynolds number
Sc	-	$\equiv \mu/(\rho D_r)$ Schmidt number
T	K	temperature
U_L	$m\ s^{-1}$	superficial velocity
x_i	-	equivalent fraction of ion i in solution
y_i	-	equivalent fraction of ion i on resin
z_i	-	electrochemical valence (negative for anions)

δ	m	Nernst film thickness
ϕ	V	electrical potential
ζ	m	space co-ordinate in the film

Subscripts

A,B,C,i	counterions
Y,j	coion

Superscripts

b	bulk solution
s	resin boundary

REFERENCES

(1) Schlögl, R. and Helfferich, F., *J.Chem.Phys.*, (1957), **26**, 5-7
(2) Copeland, J.P., Henderson, C.L. and Marchello, J.M., *AIChE Journal*, (1967), **13**, 449-452
(3) Kataoka, T., Sato, N. and Ueyama, K., *J. Chem. Eng. Japan*, (1968), **1**, 38-42
(4) Turner, J.C.R. and Snowdon, C.B., *J.Chem.Eng.Sc.*, (1968), **23**, 221-230
(5) Van Brocklin, L.P. and David, M.M., *Ind.Eng.Chem.Fundm.*, (1972), **11**, 91-99
(6) Rahman, K., *Chem. Eng. Res. Bull.* (Dacca), (1979), **3**, 27-30
(7) Kataoka, T., Yoshida, H. and Uemura, T., *AIChE Journal*, (1987), **33**, 202-210
(8) Helfferich, F.G., *React. Poly.*, (1990), **13**, 191-194
(9) Kataoka, T., Yoshida, H. and Ueyama, K., *J. Chem. Eng. Japan*, (1972), **5**, 132-136

HEATS OF SORPTION OF GOLD AND PLATINUM COMPLEXES

ON A GRAPHITIC CARBON AND AN ION EXCHANGER

A. J. Groszek and M. J. Templer,
Microscal Ltd., 79, Southern Row, London W10 5AL. U.K.

ABSTRACT

Adsorption and desorption of Gold and Platinum Complexes was studied by Flow Adsorption Microcalorimetery, combined with the analysis of solutions before and after percolation through a graphitic carbon and ion exchanger. The results give information on the reversibility of the sorption phenomena, kinetics of adsorption/desorption and total adsorption or sorption capacity of the adsorbents placed in contact with dilute solutions of metal compounds.

INTRODUCTION

Sorption and heats of sorption of various metal compounds can now be accurately determined by flow adsorption microcalorimetry, in which the dilute solutions of the compounds are percolated through a bed of adsorbent and the heat of sorption, as well as the amount of sorption, continuously monitored (1). Desorption can be similarly monitored leading to simultaneous determination of the heats and amounts of adsorption and desorption. The rates of heat evolution and metal extraction or recovery obtained in flow microcalorimetry can easily be obtained by inspection of the time-heat evolution and time-sorption records. Consequently, the kinetics of the sorption/desorption processes can be accurately evaluated for a given dynamic sorption process.

Flow adsorption microcalorimetery has already been used to study the heats of adsorption of chloroauric acid on graphitic carbons (2). The present paper describes extension of this study to sorption/desorption of the gold and platinum chloride complexes on a graphitized carbon, as well as on a strongly basic anionic exchange resin. The results show differences in the affinity of the sorbents for these complexes and kinetics of sorption and desorption.

EXPERIMENTAL

a) __Adsorbents__: Graphitized Carbon Black, surface area 85 m^2/g (N_2,BET); Ion Exchange Resin Amberlite CG400, sorption capacity-3.8meg/g.

b) __Metal complexes__: Hydrogen tetrachloroaurate hydrate purity > 99.9%; Potassium hexachloroplatinate 98% purity; Hydrogen hexachloroplatinate hydrate, > 99.9% purity.

c) __Apparatus and Procedure__: The Microscal flow microcalorimetric system FMC used in the work is shown schematically in Fig. 1. The Flow Microcalorimeter itself was described previously (1). The procedure used was to fill the calorimetric cell with 0.15-0.17mL of pre-weighed adsorbent and then wet it with a carrier liquid supplied at a constant rate (6ml/h.) from a syringe pump. The resulting heat of wetting is recorded and, subsequently, the liquid percolated at a constant rate through the sample. Once a steady baseline is obtained, the flow of carrier fluid is switched to that of a dilute surfactant solution in the carrier liquid and the resulting heat of adsorption recorded. Solvents and solutions could be interchanged sequentially, permitting the heat of desorption to be measured and the reversibility of various adsorptions evaluated. All the work was carried out in a thermostatted laboratory at $25^{\circ}C \pm 1^{\circ}C$.

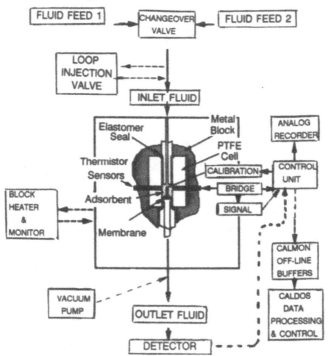

Figure 1. MICROSCAL FLOW MICROCALORIMETER - THE SYSTEM

The amount of adsorption taking place during the heat evolution or adsorption was determined simultaneously with the heats of adsorption by analysing the effluent from the adsorbent placed in the FMC cell using an R.I. detector. The provision of two independent channels of digital processing capability permitted the storage and subsequent processing of the solute concentration data separately or with reference to the heat of sorption from the FMC.

An example of the simultaneous determination of the heats and amounts of adsorption is shown in Fig. 2 for the adsorption of K_2PtCl_6 from its 1 mmolar aqueous solution on resin CG400.

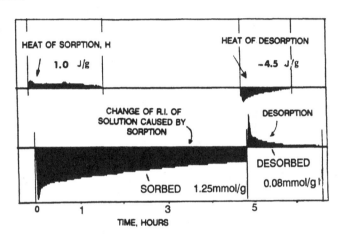

Figure 2. Heat and Amounts of Adsorption of K_2PtCl_6 on Exchange Resin CG400 from 0.001M solution in water.

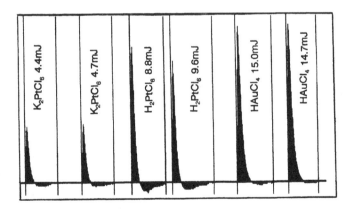

Figure 3. Sequential Heat Effects produced by Injection of 20μL 0.01M K_2PtCl_6 HPtCl$_6$ and HAuCl$_4$ into Water Percolating through Resin CG400.

178

The differential heats of adsorption can also be
determined by injecting small amounts of the compounds to be
sorbed into the stream of carrier liquid percolating through
the sorbent. Sequential injections of 0.2 μmol quantities of
K_2PtCl_6 solution into water percolating through resin CG400
are shown in Figure 3. Heat effects resulting from the
injections of typical eluting agents, such as HCl and thiourea
are shown in Fig. 4.

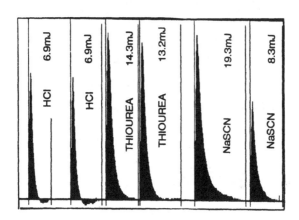

Figure 4. Sequential Heat Effects produced by Injection of 20μL
of Typical Eluting Agents into Water Percolating through Resin
CG400.

RESULTS AND DISCUSSION

Sorption of K_2PtCl_6

The sorption of K_2PtCl_6 from 0.001M aqueous solution was
investigated on a mixture of 5mg of the resin and 90mg of inert
PTFE powder forming a 0.17μL bed in the calorimetric cell. In
this situation the resin occupied only 9% of the bed volume and
the hourly space velocity (HSV) of the liquids that flowed at
the rate of 6mL/h was 392. The sorption occurred initially at
a relatively high rate dropping to a steady decreasing rate, as
shown in Fig. 2.

It is interesting to note that the sorption process is far
from being complete, even after percolation of 1850 bed volumes
of solution. During that time the amount of K_2PtCl_6
supplied to the resin is 0.028 mmol (5.6 mmol/g) and the amount
of sorption 0.0062 mmol (1.25 mmol/g), whereas the capacity of
the 5mg of resin in the calorimeter is only 0.0095 mmol (i.e.
1.9 mmol/g corresponding to the stated capacity of 3.8meg/g).
Clearly the sorption process is being affected by diffusion of
the ions in and out of the resin matrix limiting the amount
that can be exchanged under the fast flow regime.

As expected, the sorption from a more concentrated solution occurred at a much faster rate. This is shown in Fig. 5 recording the sorption from   0.01M K_2PtCl_6. Percolation of the solution for 134 minutes in this case, supplying 22.3 mmol/g of $PtCl_6^=$ ion, led to sorption of only 1.22 mmol/g (or 2.44 meg/g) of the ion, an amount very similar to that determined for the sorption from the 0.001M solution (1.25 mmol/g). Therefore, the sorption capacity of IR CG 400 resin for $PtCl_6^=$ ions is only about 64% of that measured with ions such as Cl^- (3.8 meg/g).

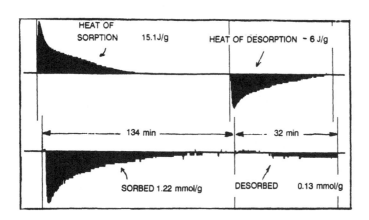

Figure 5. Heat and Amounts of Adsorption of K_2PtCl_6 on Ion Exchange Resin CG400 from 0.01M Solution

Most of the sorbed ions are held very strongly by the resin and only a small amount 0.13 mmol/g is desorbed by percolation of water (168 bed volumes). More effective desorption would, of course, be expected by HCl solutions or solutions of compounds having a high affinity for Pt, such as thiourea and NaSCN.

The integral heat of adsorption measured for the 0.01M solution, 15.1J/g, is much higher than that determined for the 0.001M solution (1.0J/g). The corresponding enthalpies of sorption are 12.6 kJ/mol and 0.9 kJ/mol respectively. This suggests that the mechanism of sorption in the more concentrated solution differs from that in dilute solutions.

The progress of sorption on graphite in relation to that on resin CG400 is shown in Fig. 6.

The rate of extraction of solute from solution is indicated in the figure by continuous measurements of refractive index of the effluent from the FMC, an inert resin (PTFE) being used as a reference non-adsorbing solid.

As can be seen, the resin removed $PtCl_6^=$ ions from
solution somewhat more effectively than the corresponding bed of
graphite (a total of 0.111 mmol/g and 0.033 mmol/g are
extracted by resin CG400 and graphite respectively, out of
0.169 mmol/g percolated through the sorbents in 95 minutes).
Nevertheless, the performance of graphite is surprisingly
effective in view of its total sorption capacity being only 10%
of resin CG400, i.e. 0.19 mmol/g.

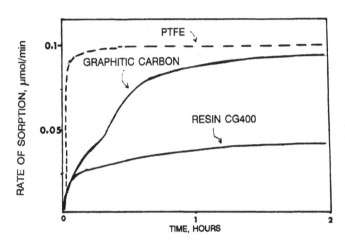

Figure 6. Progress of Sorption on Graphitic Carbon and Resin
CG400 from 0.001M Solution of K_2PtCl_6 in Water

The integral heats of sorption of $PtCl_6^=$ from 0.001M
solution were found to be 1.0J/g for the resin and 1.9 J/g for
graphite. Sorption at much lower saturation levels gave
considerably higher molar heats of sorption for both sorbents,
as shown in Fig 7. The results in this case were obtained by
injecting 20µL of 0.01M solutions of K_2PtCl_6 into water
percolating through the sorbents at 35 HSV. For all the
injections 95% of the injected solute was retained by the
sorbents, saturation of the sorbents increasing from about 0.2%
to 1% for the resin and 1.5% to 9% for graphite after the 6
injections. The heats of adsorption are initially remarkably
high for graphite, indicating that some of the graphite surface
sites have exceptionally high affinity for $PtCl_6^=$ ions.
The heats for the resin were generally much lower for
saturations around 1% and not changing significantly with the
degree of saturation within this range. However, the
relatively high heats of sorption for graphite persisted for
the complete range of surface saturations.
It is probable that the high affinity of the resin, and
especially its basic sites, for water compared with the
relatively low affinity of water for graphite is responsible
for the difference in the heats of sorption (the heat of
wetting of graphite in water is 2.5J/g and for resin CG400 -
14.7J/g).

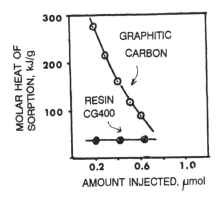

Figure 7. Molar Heats of Sorption of K_2PtCl_6.

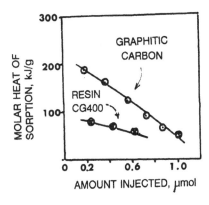

Figure 8. Molar Heats of Sorption H_2PtCl_6

Sorption of H_2PtCl_6

The heats of sorption of H_2PtCl_6 on the resin and graphite at low saturation levels were determined in the same way as for K_2PtCl_6. The results are shown in Fig. 8. Chloroplatinic acid is clearly more strongly adsorbed then its potassium salt on the resin, but for graphite the reverse occurs with the heats of adsorption of H_2PtCl_6 being relatively low.

Saturation of the resin with 0.001M solution of $HPtCl_6$ gave a gradually increasing sorption, similar to that shown in Fig. 6 for K_2PtCl_6, with the rates of sorption being only about 50% of those found for K_2PtCl_6. The heats of sorption were 1.5J/g and 1.0J/g for H_2PtCl_6 and K_2PtCl_6 respectively.

The integral heats of desorption for both platinum compounds are considerably higher than the heats of sorption (-4.5J/g for K_2PtCl_6 and -1.9J/g for H_2PtCl_6), suggesting that they are more related to changes in the structure of water at the resin - solution interface than to sorption of $PtCl_6^=$ ions on the basic sites of the resin.

Sorption of $HAuCl_4$

Chloroauric acid is sorbed by the resin more strongly than H_2PtCl_6 at low saturation levels, as shown in Fig 9. A recent study of $HAuCl_4$ adsorption on graphite (2), indicates that its adsorption capacity from 0.001M solution is 0.25 mmol/g i.e. higher than that for $PtCl_6^=$ ion. The integral heat of adsorption on the graphite surface is 7.3J/g and, therefore, very much higher than that obtained for the resin saturation which is 2.0J/g. The heats of adsorption are significantly higher for both $HAuCl_4$ and H_2PtCl_6 at graphitic carbon and resin saturations below 5%.

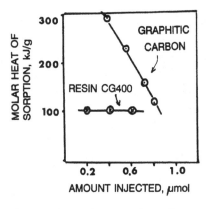

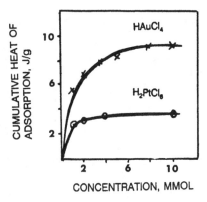

Figure 9. Molar Heats of
Sorption of HAuCl$_4$.

Figure 10. Cumulative Heats of
Adsorption of HPtCl$_6$ and
HAuCl$_4$ on Graphitic Carbon

The high heat of adsorption of HAuCl$_4$ on graphitic carbon is believed to be related to its very high rate of adsorption on this adsorbent (and other carbons such as activated carbons). For example, for the graphitic carbon (Graphon) used in this work, 75% of saturation is achieved in ten minutes of its contact with 0.002M solution. The rate of adsorption is significantly slower for resin CG400, for which only 60% saturation is achieved after 20 minutes of percolation of 0.01M solution. The cumulative heats of sorption of HAuCl$_4$ and H$_2$PtCl$_6$ for solutions of increasing concentration, are shown in Fig. 10, demonstrating the relatively high affinity of graphitic carbon for the gold complex. This affinity is ascribed to the fact that aurochloric acid has a planar structure and can therefore form a strong charge transfer bond with the basal planes of graphite (2).

CONCLUSIONS

Flow adsorption microcalorimetry can rapidly determine capacity and kinetics of sorption on ion-exchange resins and other sorbents under dynamic conditions. The work on sorption of PtCl6$^=$ and AuCl$_4^-$ ions with the use of a strongly basic ion-exchange resin and graphitic carbon, has shown that the sorption of the ions is limited by diffusion of the ions into the resin matrix and the high affinity of the resin for water. Graphitic carbons suffer much less from these effects and in a high surface area form, promise to be more effective adsorbents for the complex ions.

REFERENCES

1.   A.J. Groszek, Extended Abstract 20th Biennial Conf. on
      Carbon, June, 1991.
2.   A.J. Groszek, S. Partyka, and D. Cot, Carbon 29, pp
      821-829, 1991.

FLOW ANALYSIS METHOD FOR TRACE ELEMENTS BY ION-EXCHANGER PHASE ABSORPTIOMETRY

KAZUHISA YOSHIMURA and SHIRO MATSUOKA
Chemistry Laboratory, College of General Education, Kyushu
University, Ropponmatsu, Chuo-ku, Fukuoka, 810 JAPAN

ABSTRACT

The increase in attenuance of a colored species, which has been concentrated on-line onto ion exchanger packed in a flow-through cell, could be measured continuously with a spectrophotometer. The sensitivity of the present method depends both on the sample volume introduced into the flow system and the cross sectional area of the flow-through cell. It was possible to determine sample element concentration in mg/m^3 or lower levels with a few cm^3 of sample solution. The fundamental background and applicability to flow analysis of some trace elements are reviewed.

INTRODUCTION

Ion-exchanger phase absorptiometry is based on the direct measurements of the degree of light-absorption by an ion-exchanger phase which has sorbed a sample component. Since this new technique of solid-phase absorptiometry was proposed in 1976 as "ion-exchanger colorimetry" in the visible region (1) and 1983 in the ultra-violet region (2), a large number of applications have been made for the determination of trace elements in water samples without preconcentration (3-6).

The sensitivity of this method is enhanced by using thicker ion exchanger layers (7). Sensitivity is also enhanced by adopting a system in which the ratio of sample volume to the amount of ion exchanger is high (8): the use of a flow-through cell packed with a very small volume of ion exchanger is very effective. Furthermore, the implementation of on-line detection with ion-exchange retention should further simplify our method and broaden the flexibility for the determination of trace elements in water samples.

EXPERIMENTAL

Spectrophotometer: Light measurements were made with a Nippon Bunko double-beam spectrophotometer, Model UVIDEC-320. The spectrophotometer was mounted vertically so that the top layer of the ion-exchanger beads in the flow-through cell was levelled horizontally. An inside-mirror tube (12 mm in internal diameter and 38 mm in length) was placed between the cell holder and the light-detector window to recover partly the light scattered from the cell. This reduced the background attenuance by about 1. A perforated metal plate of attenuance 2 was placed in a reference beam to balance the light intensities (Fig. 1(a)).
Flow-through cell: A flow-through cell was supplied by Nippon Quartz Glass Co.; it was black-sided, and had a 10 mm path length and 1.5 mm diameter. The cell was blocked with polypropylene (PP) filter tip and filled with the ion exchanger (about 0.01 cm^3) with only 3 - 5 mm in the light-path portion of the cell (Fig. 1(b)).
Flow system: Carrier solution streams were pumped with medium pressure pumps. A sample was introduced into the stream by means of a PTFE six-way rotary valve. A sample loop was made by using a PTFE tube (1 mm i.d.). The increase in attenuance was continuously monitored at a fixed wavelength and recorded on a strip-chart recorder set at 0.5 or 1.0 absorbance full scale. A desorbing agent solution was introduced into the stream by means of another PTFE six-way rotary valve. All tubing was PTFE (1 mm i.d.).

RESULTS AND DISCUSSION

Since this technique was reported in 1987, about ten flow analysis methods using solid-phase absorptiometry have been proposed by other investigators. Here, all systems developed only by the present authors (8-16), listed in Table 1, are described in the individual sections below.

Flow diagram   As shown in Fig. 2(a), the single-line flow system was employed for the determination of simple, colored ions or color-developed analytes (Table 1). For flow injection analysis, to keep the system simple, it was desirable for a coloring agent to be introduced continuously into the flow line (Fig. 2(b)). However, in some cases, this caused a gradual increase in the background attenuance because of accumulation on the ion exchanger of a coloring agent and/or inevitable contamination from reagents used and the flow system, and therefore the coloring agent was introduced into the flow system by using the manifold (c) in Fig. 2, only while the reagent was in contact with the sample analyte to be determined.

Selection of ion-exchanger   It is recommended to use gel-type polystyrene or crosslinked dextran ion exchangers because of their high transparency.

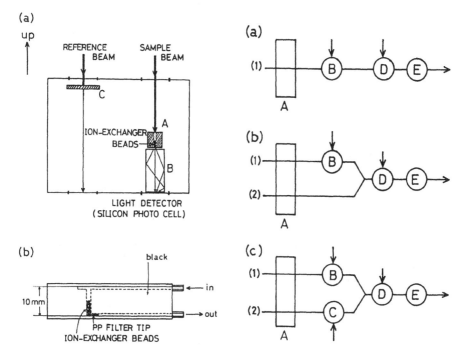

Figure 1. (a) Cell compartment furnished with special accessories. (A) flow-through cell; (B) cylindrical mirror tube (38 mm length, 12 mm i.d.); (C) a commercial perforated metal disc (Hitachi). (b) Schematic diagram of the flow-through cell packed with ion-exchanger.

Figure 2. (a) Schematic diagram of the single-line flow analysis system. (b) and (c) Manifolds for flow injection. Pump (A); six-way rotary valves for introduction for sample (B), for coloring agent (C), and for desorbing agent (D); flow-through cell packed with ion-exchanger beads (E). (1) Sample carrier solution stream; (2) coloring agent carrier solution stream (2). Flow rate at E: 1.2 - 1.5 cm^3/min.

Color development profile.   The color development profile of
a sample component sorbed in a solid phase can be predicted
according to the plate theory (1). For example, if nv cm^3 of
solution—1000v cm^3 of a sample and then a carrier solution—
are introduced into a flow-through cell, an ion-exchanger
column consisting of 11 plates numbered zero to 10, the
fraction Q can be estimated as

$$Q = n/1000 \qquad (n \leq 10) \qquad (1)$$

TABLE 1
Summary of flow methods by ion-exchanger phase absorptiometry

	Solid phase	Chromogenic agent	Carrier solution sample/coloring agent	Desorbing agent
Introduction of color-developed solution				
V	AG 1-X2	PAMB	$HCOONH_4$	NH_4NO_3-acetone
Cr	AG 50W-X2	DPC	H_2SO_4	HNO_3
Cu	AG 50W-X12	Cu^{2+}	HNO_3	HNO_3
Nd	AG 50W-X2	Nd^{3+}	HCl	HNO_3
Flow injection analysis				
Si	Sephadex LH-20	MoSi-MG	H_2O/H_2SO_4	H_2SO_4-acetone
P	Sephadex LH-20	MoP-MG	H_2SO_4/H_2SO_4	H_2SO_4-acetone
Cr	AG 50W-X2	DPC	H_2O/DPC-H_2SO_4	HNO_3
Fe	QAE-Sephadex	BPS	acetate/acetate-NH_2OH	$NaNO_3$
Cu	QAE-Sephadex	BCS	HCl/citrate-NH_2OH	HNO_3
Mo	QAE-Sephadex	Tiron	$EDTA/H_2O$	$NaNO_3$
Bi	QAE-Sephadex	KI	H_2SO_4/KI	EDTA

	Mani-fold[**]	log D	λ (nm)	Sample (cm^3)	SR[***]	SENS (ng)	RSD(sample amount,ng) (%)	DL (ng)	Ref
V[*]	a	2.0	650	-	-	-	-	0.6	13
Cr	a	4.1	550	4.4	160	4.4	4.3 (2.2)	0.22	10
Cu	a	4.8	800	4.4	220	11000	3.4 (3000)	-	8
Nd	a	> 5	795	8.5	450	-	-	-	-
Si	c	-	627	3.7	-	3.7	3.5 (8.5)	0.37	14
P	c	-	627	7.0	35	6.2	5.3 (53)	0.14	12
Cr	b	4.1	550	7.9	310	4.3	4.9 (1.6)	0.07	15
Fe[*]	c	> 4	550	-	-	-	-	0.8	16
Cu	c	> 4	484	8.3	150	3.4	6.9 (4.8)	0.08	-
Mo[*]	c	> 4	410	-	-	-	-	15	11
Bi	b	> 7	473	8.5	480	65	2.5 (36)	2.5	9

[*]Preconcentrated samples were used. [**]See Fig. 2. [***]1 = 1 cm.
PAMB:2-[2-(3,5-dibromopyridyl)azo]-5-dimethylaminobenzoic acid;
DPC: 1,5-diphenylcarbazide; MoSi-MG: ion-associate of molybdo-
silicate and Malachite Green;   MoP-MG: ion-associate for phosphate; BPS: bathophenanthrolinedisulfonate; BCS: bathocuproine-
disulfonate.

and

$$Q = \frac{1}{1000} \sum_{k=n-999}^{n} \sum_{p=0}^{10} \frac{k!}{(k-p)! \, p!} \frac{(1000mD/i)^{k-p}}{(1+1000mD/i)^k} \qquad (n \geq 11) \qquad (2)$$

If n is smaller than 1000, then, k = 11 instead of n-999. Because 1000m/i for a column in which spherical beads are packed is approximately 2, Q vs n curves can be drawn for given D values (Fig. 3). Provided that HETP is 0.05 cm and the inner diameter of the cell is 0.15 cm, v corresponds to about 0.3 mm^3. The D value of the colored species can be a measure of the analyte retention in the flow-through cell. Color development profiles of copper for different D are shown in Fig. 4. After the sample introduction, the attenuance increased because of the copper sorption in the cation-exchanger phase. In high D systems, almost all the copper in the 0.17 cm^3 sample solution was retained after passing at least 20 cm^3 of a carrier solution, whereas the copper in low D systems was eluted from the cell in a fairly short time.

Calibration. Figure 5 shows typical examples for successive color development for chromium(VI) (15). In the system of D = 1.2 x 10^4, it was not necessary to desorb the colored species after each measurement. Calibration curves obtained by using ΔA were straight. The colored species sorbed on the ion-exchanger phase could be desorbed with an appropriate desorbing agent solution and therefore the cell could be repeatedly used for measurements (Table 1).

Sensitivity. The ratio of the two absorbances for sample solutions in the same analyte can be measured to compare the present method sensitivity with the corresponding solution absorptiometry sensitivity. If all the colored species injected are retained in the ion-exchanger phase in the flow-through cell, SR is closely related to:

$$SR = (\bar{\varepsilon} \, V)/(\varepsilon \, S \, l) \qquad (3)$$

This equation implies that the diameter of the flow-through cell and the sample volume are very important factors for

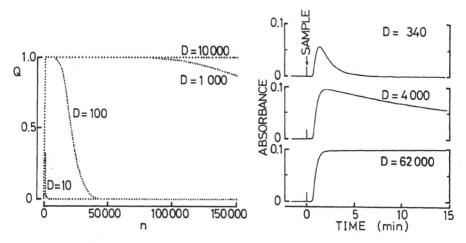

Figure 3.   Retention of components with different D on an   ion-exchanger column in the flow-through cell.

Figure 4.   Color development profiles of hydrated copper for different D.   Sample: 0.17 cm^3, 15 µg of Cu; carrier solution: (A) 0.28, (B) 0.070, (C) 0.014 kmol/m^3 HNO3; flow rate 1.2 cm^3/min.

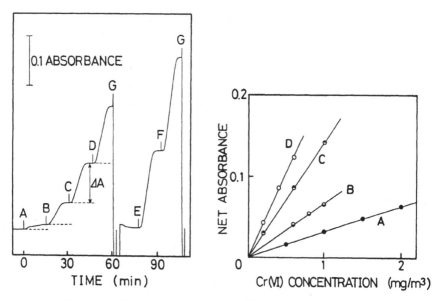

Figure 5. FIA profiles of chromium(VI) obtained using the flow system with ion-exchanger phase absorptiometry.   Sample volume: 7.9 cm^3.   A, blank; B, 0.2; C, 0.4; D, 0.6; E,   0.8;   F,   1.0 mg/m^3; and G, desorbing agent.

Figure 6.   Effect on sensitivity of sample volume introduced (chromium(VI)-DPC system).   Sample volume: A, 1.8; B, 4.0; C, 7.9; D, 11.9 cm^3.

obtaining high sensitivity. As shown in Table 2, the SR values expected from the equation were in reasonably good agreement with the results obtained (8).

For a high D system, the sensitivity of the present method is linearly increased with increase in the volume of the sample solution introduced into the flow system, as shown in Fig. 6 (15). The use of different loop lengths gives a wide concentration range for calibration.

Precision and detection limit.   It might be thought that the precision of the ion-exchanger phase absorptiometry, involving an unorthodox solid-phase optical medium, would be inferior to that obtainable with conventional solution absorptiometry. However, RSD's for given amounts of analytes were within 5 % (Table 1). The detection limits for almost all systems were under mg/m^3 levels, for which the conventional solution methods are inapplicable.

Advantages of the present method.   The concentration and absorptiometric measurements are carried out simultaneously. As a result, the target element can be determined at mg/m^3 or lower levels in 3-6 samples in less than 1 h without preconcentration. The flow technique provides a repeatable measuring system assuring a precision which is often adequate for routine analyses.

The present method can be applied to any systems in which colored sample species can easily be sorbed on an ion exchanger and desorbed from it by an appropriate change in eluting conditions.

TABLE 2
Sensitivity ratio for the hydrated copper system

diameter of flow -through cell (mm)	S (cm^2)	V (cm^3)	SR_{calcd}*	SR_{obsd}**
3.0	0.0707	4.4	62	49
1.5	0.0177	4.4	250	220

*l = 1 cm.   **Carrier solution, 0.014 $kmol/m^3$ HNO_3; linear flow rate, 22 cm/min.

NOMENCLATURE

D   distribution ratio: the quantity of sample component in 1 kg of ion exchanger, divided by the quantity in $0.001 \ m^3$ of solution after equilibration is attained $0.001 \ m^3/kg$

DL   detection limit, defined as the concentration producing an absorbance equal to three times the standard deviation of the background attenuance   kg

ε molar absorptivity in solution $0.1 \ m^2/mol$

$\bar{\varepsilon}$ molar absorptivity in solid phase $0.1 \ m^2/mol$

HETP height equivalent of the theoretical plate   m

i   interstitial solution volume of the ion exchanger m^3

l   light path length for solution absorptiometry   m

λ wavelength at which absorptiometric measurements were done   m

m   ion-exchanger weight in the flow-through cell   kg

Q   fraction of the total amount of the sample species remaining in the flow-through cell

RSD   relative standard deviation   %

S   cross-sectional area (normal to the light beam) of the flow-through cell m^2

SENS sensitivity, defined as sample element giving a final absorbance of 0.1   kg

SR   sensitivity ratio of the present method to the corresponding solution absorptiometry

v   volume of the mobile phase of each plate m^3

V   volume of sample solution introduced m^3

REFERENCES

1. Yoshimura, K., Waki, H. and Ohashi, S., Talanta, 1976, 23, 449-54.
2. Waki, H. and Korkisch, J., Talanta, 1983, 30, 95-100.
3. Capitan, F., Valencia, M. C. and Capitan-Vallvey, L. F., Mikrochim. Acta, 1984, 3, 303-11.
4. Shriadah M.M.A. and Ohzeki, K., Analyst, 1985, 110, 677-79.
5. Yoshimura, K. and Waki, H., Talanta, 1985, 32, 345-52.
6. Nakashima, T., Yoshimura, K. and Waki, H., Talanta, 1990, 37, 735-39.
7. Yoshimura, K. and Waki, H., Talanta, 1987, 34, 239-42.
8. Yoshimura, K., Anal. Chem., 1987, 59, 2922-4.
9. Yoshimura, K., Bunseki Kagaku, 1987, 36, 656-61.
10. Yoshimura, K., Analyst, 1988, 113, 471-74.
11. Yoshimura, K., Matsuoka, S. and Waki, H., Anal. Chim. Acta, 1989, 225, 313-21.
12. Yoshimura, K., Nawata, S. and Kura, G., Analyst, 1990, 115, 843-8.
13. Matsuoka, S., Yoshimura, K. and Waki, H., Mem. Fac. Sci., Kyushu Univ., 1991, Ser. C, 18, 55-62.
14. Yoshimura, K. and Hase, U., Analyst, 1991, 116, 835-840.
15. Yoshimura, K. and Matsuoka, S., New Developments in Ion Exchange, Kodansha, Tokyo, 1991, pp. 377-82.
16. Hase, U. and Yoshimura, K., Analyst, submitted.

^{31}P NMR SPECTROSCOPIC STUDIES ON THE COMPLEXATION EQUILIBRIA IN THE ION-EXCHANGER RESIN PHASE

YOSHINOBU MIYAZAKI and HIROHIKO WAKI
Department of Chemistry, Faculty of Science, Kyushu University,
Hakozaki, Higashiku, Fukuoka, 812 Japan

ABSTRACT

The complexation equilibria of ions sorbed in ion-exchangers were investigated by direct NMR measurements for the solid samples. The stability constant values for the aluminium-phosphinate complex were higher in the cation-exchanger phases than those in the corresponding solutions, this tendency being similar to that for the cobalt(II)-thiocyanate complexes by absorption measurements previously reported(1). On the other hand, the values for cadmium(II)-phosphinate complexes in the cation-exchanger phases were lower than those in the corresponding solution. These results could be explained by considering properties of the ion-exchangers, such as low dielectric constant, high internal pressure, low ionic mobility, low water activity and spatial restriction.

INTRODUCTION

An ion-exchanger can be considered as a concentrated polyelectrolyte solution of a special kind. Ion-exchange studies on complex formation have so far mainly been restricted to the solution phase and the study concerning the ion-exchanger solid phase itself has been little noticed, though it is important for elucidating ion-exchange behavior. It is known only that the very high complexation of metal ions in the anion-exchanger phase is due to the high ligand anion concentration in the phase and that the very low complexation in the cation-exchanger phase is due to the exclusion of ligand anions. Undoubtedly there should be more other intrinsic factors. One of the reasons for the lack of ion-exchanger phase study may be ascribed to a technical difficulty in the direct observation of such solid particle samples. We have been studying this problem

by direct spectroscopic measurements on the solid samples, such
as visible or ultraviolet absorptiometry and NMR spectroscopy.
In this paper the stability constants of aluminium-phosphinate
and cadmium(II)-phosphinate complexes in the cation-exchanger
phases by ^{31}P NMR are reported and compared to those in ordinary
solutions.

EXPERIMENTAL.

Chemicals

All reagents used were of commercially available reagent
grade. Strong acid type cation-exchangers of sodium form,
Muromac AG (purified "Dowex") 50W-X2, -X4 and -X8(100-200 mesh)
were used. All ion-exchangers were purified by the usual
regeneration procedure before use.

NMR spectral measurement

All NMR spectra were recorded on a JEOL JNM-GX 400 at a
probe temperature of 22 ± 1 $^{\circ}$C. ^{31}P NMR spectra were recorded at
161.858 MHz. The NMR parameters were chosen so that the best
quantitative spectra can be obtained, which are flip angle ~90
$^{\circ}$(20 μS), pulse repetition time 3 S, spectral width 30000 or
6500 Hz, and data points 16384 or 32768. The chemical shift
was recorded in ppm with respect to 85 % H_3PO_4 as an external
reference, taking negative sign for downfield shift. The
field/frequency lock was achieved with D_2O contained in a 2 mm
tube. The ion-exchanger beads retaining sample species were
packed into a 10 mm NMR sample tube with a small volume of
equilibrium solution and the NMR spectrum was measured in the
same way as that for conventional solution NMR. The sample tube
containing only solution was rotated but that containing ion-
exchanger beads was not rotated during measurement. The
integral spectral intensity was measured by weighing a paper
cut-out corresponding to the spectral peak area.

RESULTS AND DISCUSSION

Stability constants of aluminium-phosphinate complex $AlPH_2O_2{}^{2+}$

Since the ligand exchange rate of aluminium ion is sufficiently slow for NMR time scale, the NMR signals can be observed as separate peaks for each species (Fig. 1).

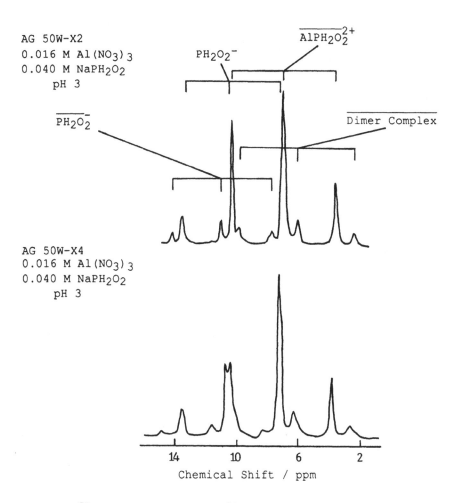

Fig. 1 ^{31}P NMR spectra of $Al^{3+}-PH_2O_2{}^-$ complex system for cation-exchanger bed.

Using the peak areas from $AlPH_2O_2^{2+}$ complex and free $PH_2O_2^-$ anion, their concentrations in the cation-exchanger phase were determined with a $PH_2O_2^-$ calibration curve. The free Al^{3+} concentration in the cation-exchanger phase was calculated by subtracting the complex concentration from total aluminium concentration determined by a batch distribution experiment. From these three values the stability constant $[\overline{AlPH_2O_2^{2+}}]/[\overline{Al^{3+}}]\,[\overline{PH_2O_2^-}]$ was evaluated. The stability constants in the corresponding solutions at the same ionic strength were evaluated in the same way (Table 1).

It was found that the stability constant values for cation-exchanger phases are higher than those for each corresponding solution in all cases. This behavior was in agreement with that for cobalt(II)-thiocyanate complex system previously reported(1).

Stability constants of cadmium(II)-phosphinate complex $CdPH_2O_2^+$

The ligand exchange rate of cadmium(II) ion is so fast for NMR time scale that only one averaged ^{31}P NMR signal is observed in all cases (Fig. 2). Here the chemical shift can be expressed by the following equation.

$$\frac{\overline{\delta_{obs}}}{[\overline{Cd^{2+}}]} = -\overline{K}_1\overline{\delta_{obs}} + \overline{K}_1\overline{\delta}_{CdPH_2O_2^+}$$

The free cadmium(II) concentration in the cation-exchanger phase $[\overline{Cd^{2+}}]$ was calculated by subtracting the concentration of $CdPH_2O_2^+$ from that of total cadmuim(II) in the ion-exchange phase by a successive approximation treatment. The plot of $\overline{\delta_{obs}}/[\overline{Cd^{2+}}]$ vs. $\overline{\delta_{obs}}$ gives $\overline{K}_1$ from the slope (Fig. 3). The same treatment was applied to the solution complexation at the same ionic strength. An example of the solution analysis is also given in Fig. 3 and the stability constant values are listed in Table 1. As can be seen from the table, the stability constants for the cation-exchanger phases were always lower than those in the corresponding solutions, this behavior being opposite to that of the Al^{3+}-$PH_2O_2^-$ system or other complex systems so far investigated.

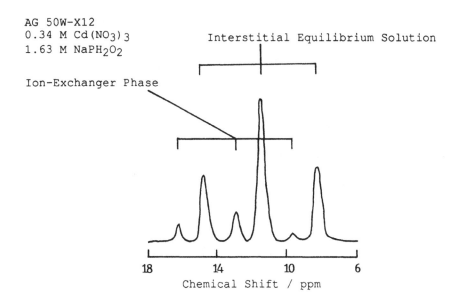

AG 50W-X12
0.34 M Cd(NO$_3$)$_3$
1.63 M NaPH$_2$O$_2$

Interstitial Equilibrium Solution

Ion-Exchanger Phase

18   14   10   6
Chemical Shift / ppm

Fig. 2 ^{31}P NMR spectrum of Cd^{2+}-PH$_2$O$_2^-$ complex system for cation-exchanger bed.

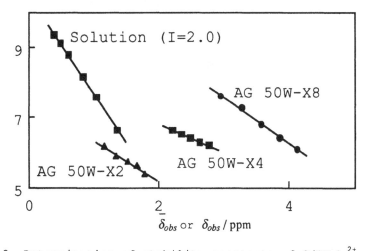

Solution (I=2.0)

9

7

5

AG 50W-X8

AG 50W-X4

AG 50W-X2

0   2   4

δ_{obs} or δ_{obs} / ppm

Fig. 3   Determination of stability constants of CdPH$_2$O$_2$$^{2+}$ complex in cation-exchanger phase and the corresponding aqueous solution by ^{31}P NMR.

Several important factors may be related to the ionic interaction in the solid phase. The lower dielectric constant, high internal pressure and lower water activity may enhance the metal-anion attraction, that is, an ion-exchanger phase has more factors favorable to complexation than ordinary aqueous solution. The only unfavourable factor for the solid phase may be lower mobility of ions, since this may lead to a lower probability of collision between metal and ligand ions. In the case of the aluminium complex, the liberation of water molecules from the hydration sphere of aluminium ion accompanied by complexation with phosphinate anion may be very large and this effect could overcome the opposite factor. In the case of the cadmium(II) complex, on the other hand, the hydration of divalent cdmium(II) ion in the ion-exchanger phase is much less than that of aluminium, thus the stabilization effect during the liberated water molecules transfer from solid phase to solution is very small and the low mobility effect may remain as the most important factor.

Table 1. Stability constants of one-to-one complexes in cation-exchanger phases and the corresponding solutions

Complex System	Ion-Exchanger	$\overline{K}_1$	Solution	K_1
CoNCS$^+$	AG 50W-X2	8.9	$I = 1.7$	5.3
	-X4	11	$= 2.0$	5.2
	-X8	28	$= 3.4$	6.0
AlPH$_2$O$_2^{2+}$	AG 50W-X2	130	$I = 1.6$	96
	-X4	150	$= 2.3$	110
	-X8	190	$= 3.4$	160
CdPH$_2$O$_2^+$ *	AG 50W-X2	1.1	$I = 1.2$	2.8
	-X4	2.0	$= 2.0$	3.0
	-X8	1.9	$= 2.9$	2.9

* Standard deviation ≤ 0.02

NOMENCLATURE

$[\overline{Al^{3+}}]$	free aluminium ion concentration	
	in ion-exchanger phase	mol/dm^3
$[\overline{AlPH_2O_2^{2+}}]$	$AlPH_2O_2^{2+}$ complex concentration	
	in ion-exchanger phase	mol/dm^3
$[\overline{Cd^{2+}}]$	free cadmium(II) ion concentration	
	in ion-exchanger phase	mol/dm^3
$[\overline{PH_2O_2^-}]$	free phosphinate ion concentration	
	in ion-exchanger phase	mol/dm^3
$\overline{\delta}_{obs}$	observed ^{31}P chemical shift for a	
	complex system in ion-exchanger	
	phase referring to the shift for	
	$PH_2O_2^-$ ion as standard	ppm
$\overline{\delta}_{CdPH_2O_2^+}$	^{31}P chemical shift for $CdPH_2O_2^+$	
	ion in ion-exchanger phase	
	refering to the shift for $PH_2O_2^-$	
	ion as standard	ppm
I	ionic strength in solution	
K_1	stability constant of the 1:1 complex	
	in solution	
$\overline{K}_1$	stability constant of the 1:1 complex	
	in ion-exchanger phase	

REFERENCES

(1) Waki H. and Miyazaki Y., Polyhedron, 1989, **8**, 859.

SYNTHESIS AND CHARACTERISATION OF RESIN SORBENTS FOR CEPHALOSPORIN C RECOVERY

BRIAN ROWATT AND DAVID C. SHERRINGTON
Department of Pure and Applied Chemistry,
University of Strathclyde,
Thomas Graham Building, 295, Cathedral Street,
Glasgow G1 1XL United Kingdom

ABSTRACT

A number of conventional styrene-divinylbenzene bead type polymeric sorbents have been synthesised. The degree of crosslinking and porogen to monomer ratio has been varied systematically and in some cases more polar functionality introduced. A matrix of novel highly porous polymers have also been synthesised again with varying degrees of crosslinking, porogen to monomer ratio and percentage void volume. The porosity characteristics of the two groups of resins have been determined using mercury porosimetry, N_2 sorption and solvent imbibition. The sorption of cephalosporin C by the two groups of resins has been investigated and an attempt made to correlate these results with their porosity characteristics.

INTRODUCTION

An ever increasing number of organic molecules are being produced by bio-technological processes and inevitably these appear as dilute aqueous solutions. Recovery of the organic products from such solutions is not always straightforward, and particularly with sensitive molecules, significant losses can occur when conventional isolation technologies are employed. The ability to sorb organic molecules onto hydrophobic sorbents (polymeric or inorganic species) under very mild conditions, is therefore a highly attractive alternative recovery procedure (1-4). The overall objective of this work was to synthesise a range of polymeric sorbents with known chemical structures and well defined porosity characteristics, and to employ these in the isolation of cephalosporin C from dilute aqueous solutions. In principle such a study should produce a structure performance correlation from which some basic design principles and understanding should emerge.

Cephalosporin C

EXPERIMENTAL

Synthesis of Sorbents

Conventional non-functional styrene-divinylbenzene based sorbent resins were prepared by suspension polymerisation using procedures developed previously in the authors' laboratory (5). The first group employed commercial divinylbenzene as the crosslinking agent. This is known to be a mixture of monomers comprising approximately 55 vol% divinylbenzene m- and p- isomers, and 45 vol% ethylstyrene m- and p- isomers. A second group of sorbents were prepared from a special grade of divinylbenzene, the composition of which is 80 vol% divinylbenzene m- and p- isomers, and 20 vol% ethylstyrene m- and p- isomers. Finally, in order to achieve resins with the highest possible content of divinyl-benzene, pure p-divinylbenzene was synthesised using conventional Wittig chemistry, starting with terephthaldicarboxaldehyde. Table 1 lists the sorbents produced and the chemical composition of the polymerisation mixtures from which they were prepared. To introduce more polar binding sites into the resin structure, two resins containing 4-vinyl-pyridine residues and two containing methacrylic acid residues were prepared, the preparation conditions of which are summarised in Table 2. For all these resins, toluene was used as the porogen. Resin yields were always very high, and the co-monomer feed ratio can be taken as a reasonable measure of the co-monomer segment composition of the resins produced.

A novel type of porous polymeric sorbent was also prepared from a high internal phase emulsion consisting of water dispersed in an organic monomer phase. In these systems the internal phase, water constitutes > 75 vol% and after polymerisation of the organic continuous phase, the water can be extracted from the rigid highly porous crosslinked polymer which results (6). The crosslink ratio and the phase volume of water have been adjusted and a porogen has been introduced to produce a range of sorbents with different porosity characteristics, the relevant polymerisation mixtures are shown in Table 3.

Sorption Experiments

A typical batch sorption experiment was performed by gently agitating a sample of resin sorbent (0.1 g) in an aqueous solution (10 mL) of cephalosporin C at pH 7.0 and 25°C for 24 h. The reduction in concentration was determined from UV absorbance measurements of the

TABLE 1
Polymerisation mixtures used to produce conventional styrene/
divinylbenzene sorbents

Resin Designation	Divinylbenzene type (vol%)		Ethylstyrene (vol%)	Porogen	
				Type	Ratio Porogen/ Monomers (vol/vol)
PS20X1EtH[a)]	m-, p-	20	20	2-ethylhexanol	1/1
PS50X2T	m-, p-	50	50	toluene	2/1
PS55X3A	m-, p-	55	45	cyclohexanol/ dodecanol 9/1	3/1
PS55X4C	m-, p-	55	45	cyclohexanol	4/1
PS55X0.5T	m-, p-	55	45	toluene	0.5/1
PS55X1T	m-, p-	55	45	toluene	1/1
PS55X2T	m-, p-	55	45	toluene	2/1
PS55X3T	m-, p-	55	45	toluene	3/1
PS80X0.5T	m-, p-	80	20	toluene	0.5/1
PS80X1T	m-, p-	80	20	toluene	1/1
PS80X2T	m-, p-	80	20	toluene	2/1
PS80X3T	p-	80	20	toluene	3/1
PS100X0.5T(1)	p-	100	-	toluene	0.5/1
PS100X0.5T(2)	p-	100	-	toluene	0.5/1
PS100X1T	p-	100	-	toluene	1/1
PS100X2T	p-	100	-	toluene	2/1
PS100X3T(1)	p-	100	-	toluene	3/1
PS100X3T(2)	p-	100	-	toluene	3/1

a) Balance of composition = 60 vol% styrene.
b) The first four resins were available from previous projects.
c) PS100X0.5T(1) and (2) and PS100X3T(1) and (2) were different preparations of nominally the same two resins.
d) Vol% refers to the % volume of total comonomers.

TABLE 2
Polymerisation mixtures involving polar comonomers

Resin Designation	divinylbenzene (vol%)	ethylstyrene (vol%)	styrene (vol%)	polar monomer (vol%)
PS50X5MA2T	50	40	5	5 MA
PS50X30MA2T	50	12.5	7.5	30 MA
PS50X5VP2T	50	40	5	5 VP
PS50X30VP2T	50	17.5	7.5	30 VP

Notes: MA = methacrylic acid; VP = 4-vinylpyridine
Porogen = toluene (Porogen/monomer = 2/1 v/v)

TABLE 3
Polymerisation mixtures used to prepare highly porous polymeric sorbents

Porous sorbent designation	divinylbenzene (vol%)	ethylstyrene (vol%)	styrene (vol%)	functional monomer (vol%)	water (vol%)	monomer/toluene porogen vol. ratio
X20PV90	20	16.5	63.5	-	90	-
X20PV95(1)	20	16.5	63.5	-	95	-
X20PV95(2)	20	16.5	63.5	-	95	-
X20PV90 50 CMS	20	16.5	13.5	50 CMS	90	-
X20PV95 50 MS	20	16.5	13.5	50 MS	95	-
X20PV95 50 EHM	20	16.5	13.5	50 EHM	95	-
X20PV95 50 LMA	20	16.5	13.5	50 LMA	95	-
X55PV90 0.5T	55	45	-	-	90	0.5
X55PV90 1.0T	55	45	-	-	90	1.0

Notes: a) Vol% of monomers refers to total volume of monomers = 100%.
b) Vol% H_2O refers to total volume of all monomers plus H_2O = 100%.
c) X20PV95(1) and (2) refer to two separate preparations of nominally
the same porous polymers.
d) CMS = chromomethylstyrene; EHM = 2-ethylhexylmethacrylate;
MS = methyl styrene; LMA = laurylmethacrylate.

supernatant (λ_{max} for cephalosporin C ~ 267 nm). Initial cephalosporin C concentrations in
the range 0.001-0.025M were employed and a sorption isotherm constructed for each sorbent.
The saturation capacities for each sorbent were determined from the isotherms, the values for
which are shown in Tables 4 and 5.

Recovery of Cephalosporin C
The ability to recover sorbed cephalosporin C from various resins was examined simply by
filtering each saturated resin from its batch sorption solution, washing with deionised water
and then agitating the resin overnight in an appropriate "stripping solvent" (10 mL). Table 6
shows some typical results obtained.

Characterisation of Surface Area and Porosity of the Two Groups of Sorbents
Surface areas were measured by generating nitrogen adsorption isotherms using a Micromeritics Accusorb 2100E instrument and subjecting the data to a Brunauer, Emmett and
Teller (BET) treatment. Pore volumes, average pore diameters and pore size distributions
were measured by mercury intrusion using a Micromeritics Autopore 9220 instrument. Some
idea of the pore volume of the sorbents in the wet state was obtained by measuring the

TABLE 4

Surface area, porosity and Cephalosporin C sorption data
for conventional polymeric sorbents

Resin Designation	Surface[a] area $(m^2 g^{-1})$	Average Pore[b] Diameter (nm)	Pore Volume[c] (mL Hg g^{-1})	Pore Volume[c] (mL CH_3OH g^{-1})	Cephalosporin C sorption capacity (g g^{-1} sorbent)	Surface area[d] accessible to Cephalosporin C(m^2g^{-1})
XAD-2	315	11.7	0.53	0.91	0.021	40
XAD-4	493	11.6	0.61	1.22	0.094	179
XAD-16	986	14.9	0.99	2.29	0.056	106
XAD-1180	401	14.0	0.89	2.20	0.051	97
PS20X1EtH	68	13.8	0.33	1.24	0.018	107
PS50X2T	685	8.4	0.55	2.37	0.057	34
PS55X3A	247	93.7	2.88	3.60	0.088	167
PS55X4C	272	67.9	2.64	3.32	0.074	141
PS55X0.5T	103	8.7	0.18	0.34	≈0	0
PS55X1T	609	9.8	0.32	0.79	0.020	38
PS55X2T	655	10.9	0.41	1.62	0.024	46
PS55X3T	759	11.5	0.55	2.45	0.039	74
PS80X0.5T	561	8.8	0.25	0.37	≈0	0
PS80X1T	687	9.3	0.22	0.65	0.032	61
PS80X2T	738	11.7	0.49	1.59	0.045	85
PS80X3T	870	13.2	0.79	2.24	0.042	80
PS100X0.5T(1)	583	38.7	0.32	1.40	≈0	0
PS100X0.5T(2)	477	28.4	0.26	0.83	0.006	11
PS100X1T	370	18.5	0.08	0.78	0.020	38
PS100X2T	452	10.9	0.37	2.04	0.034	62
PS100X3T(1)	487	9.2	0.32	1.66	0.011	21
PS100X3T(2)	495	11.0	0.22	2.20	0.009	18
PS50X5MA2T	645	11.1	0.45	1.56	0.022	42
PS50X30MA2T	604	9.4	0.26	2.29	0.007	13
PS50X5VP2T	604	10.4	0.42	1.26	0.039	74
PS50X30VP2T	60	7.2	0.13	0.69	0.023	44

Notes: a) from N_2 sorption - B.E.T.; b) from Hg intrusion; c) from methanol imbibition; d) calculated assuming one molecule occupies 130 $Å^2$; e) XAD-2, 4, 16 and 1180 are commercially available sorbents from Rohm and Haas - no definitive information is available about their chemical make-up, other than they are styrene-divinylbenzene type sorbents.

imbibition of methanol using a technique previously reported (7). A summary of the surface area and porosity data for both sets of sorbents is given in Tables 4 and 5. Full pore size distribution curves were also generated but are not reproduced here.

DISCUSSION

It now seems very likely that the recovery of cephalosporin C from fermentation broths by sorption onto a polymeric sorbent is an industrial process (8,9,10). Although a number of commercially available resins have clearly been screened for this activity, there remains a dearth of information relating sorption ability to resin structure (9).

TABLE 5
Surface area, porosity and Cephalosporin C sorption data for
highly porous polymeric sorbents

Resin Designation	Surface[a] area (m^2g^{-1})	Average Pore[b] Diameter (nm)	Pore Volume[b] $(mLHg\ g^{-1})$	Pore Volume[c] $(mL\ CH_3OH\ g^{-1})$	Cephalosporin C sorption capacity $(g\ g^{-1}\ sorbent)$	Surface area[d] accessible to Cephalosporin C $(m^2\ g^{-1})$
X20PV90	21	1259.0	5.66	4.95	0.064	121
X20PV95(1)	23	803.5	10.43	6.65	0.083	158
X20PV95(2)	33	2291.5	16.22	6.43	0.080	152
X20PV90 (50CMS)	18	320.4	3.92	3.29	0.070	133
X20PV95 (50EHM)	19	2892.6	5.46	5.99	0.093	177
X20PV95 (50MS)	18	393.1	7.54	2.93	0.011	21
X20PV90 (502MA)	18	1074.7	15.0	8.17	0.028	53
X55PV90 (0.5T)	137	824.6	9.55	3.50	0.028	53
X55PV90 (1.0T)	264	405.6	13.70	5.40	0.031	59

a) - d) see Notes Table 4

TABLE 6
Recovery of Cephalosporin C from some selected sorbents

Sorbent	Saturation Capacity $(g\ g^{-1}\ sorbent)$	Recovery[a] (%)	Eluting Solvent
XAD-4	0.094	80.6	isopropanol/water 1:1
PS50X2T	0.057	90.0	isopropanol/water 1:1
PS50X2T	0.057	48.5	isopropanol/water 3:1
PS50X2T	0.057	74.7	methanol/water 1:3
PS80X2T	0.045	71.0	isopropanol/water 1:1
X20PV95(1)	0.083	~100	methanol
X20PV95(50EHM)	0.093	~100	methanol

Note: a) ± 5%

One obvious model for the sorption of any water-soluble molecule from aqueous solution is that a simple surface adsorption process occurs, with hydrophobic forces being responsible for the binding process. Undoubtedly such a simple mechanism does indeed make a contribution but the picture is clearly much more complex. This mechanism, for example, would predict a relatively simple relationship between the surface area and the sorption capacity (g^{-1} of sorbent). For surface areas evaluated by nitrogen adsorption (BET treatment) such a correlation exists neither for conventional resins nor for the novel highly porous sorbents.

Conventional Sorbent Resins

In the case of conventional resins a superficial correlation did emerge with the pore volume of the resin measured by methanol imbibition for the first few resins examined. As more resins were examined however the correlation became somewhat distorted with some resins deviating markedly from the average behaviour. One of these is very important (that of XAD-4) because it exhibits a significantly high saturation sorption capacity.

Examination of the data for average pore diameters and pore size distributions does not produce any other clear correlations, and the only conclusion that can be made is that surface area, pore volume, pore size and pore size distribution must all play an important role in controlling sorption capacity.

If the cross-sectional area of a cephalosporin C molecule is taken as 1.3 nm^2 (determined from molecular graphics) then saturation sorption capacities enables figures for "accessible" surface area to be deduced. The results of this manipulation are shown in Table 4. Clearly these figures suggest that only a fraction of the nitrogen BET surface area is capable of being exploited in the case of conventional resin sorbents. The results add further weight to the argument that pore volume and pore size i.e. "accessibility", and "connectivity" play a vital role in the sorption process. Limitations in these two factors almost certainly arise at the molecular level, since crushed samples of resins show no rise in sorption capacity whatsoever.

Novel Highly Porous Sorbents

It is not clear how the behaviour of the novel highly porous sorbents relate to that of conventional resins. The overall structures of these species is very different and is characterised by very low overall densities and associated very high pore volumes, large pore sizes (often referred to as cells because of their large size - typically 10 μm) and comparatively low nitrogen BET surface areas. As with conventional resins there seems to be no particular correlation with cephalosporin C sorption capacity and any of the porosity characteristics measured. An intriguing enigma arises when the "accessible" surface area for cephalosporin C (calculated as before) is compared with the nitrogen BET areas (Table 5). This shows that the area occupied is apparently significantly larger than the area available, except in the case of X55PV90 (0.5T), X55PV90 (1.0T) and X20PV95 (50MS). It is possible that the sorbate succeeds in finding binding sites and therefore surface area not available to nitrogen. While this is extremely unlikely, the two measurements are indeed made under different conditions - one is the dry state and one with saturation by water - and the presence of water might induce major structural changes. A more likely possibility is that multi-layer sorption occurs and that the large cell sizes in these materials actually encourage this in some way. It might even be that micro-crystallisation of the sorbate occurs. Indeed, evidence has already been published regarding the apparent association of cephalosporin C into aggregates inside conventional resins. (11). Overall, the sorption capacities of these materials is remarkably high and compares favourably with the best conventional resins. Furthermore, coupled with the very high pore volume and large cell sizes, the materials might well show considerably improved resistance to fouling with broth mixtures. Also, the results re-emphasise the observation with conventional sorbents that surface area considerations alone can be very misleading in trying to understand sorbent behaviour and in designing sorbents for particular applications.

Recovery of Cephalosporin C

The data shown in Table 6 shows that elution of sorbed cephalosporin C from a selection of conventional sorbents using the stripping solvents shown is relatively efficient. The data also suggest that the elution process depends only on the nature of the stripping solvent. No specific dependence on sorbent structure has been identified, nor does there seem any reason why this should be the case.

REFERENCES

1.   Abrams, I.M., Ind. Eng. Chem. Prod. Res. Dev., 1975, **14**, 108.

2.   Miller, J.R., J. Polym. Sci. Polym. Symp., 1980, **68**, 167.

3.   Carrington, R., "A Review of Antibiotic Isolation Techniques" in "Bioactive Microbial Products 3 - Downstream Processing", Eds. J.D. Stowell, P.J. Bailey, and D.J. Winstanley, Academic Press, 1986, Chapter 4, 45-58.

4.   Boothroyd, B., "Recovery of Antibiotics Using Column Extraction Methods" in (see 3 above), Chapter 5, 59-76.

5.   See Appendix in "Polymer-supported Reactions in Organic Synthesis", Eds. P. Hodge and D.C. Sherrington, J. Wiley and Sons, Chichester, U.K. 1980.

6.   Barby, D. and Haq, Z., Europ. Pat. 0,060,138; 1982 (to Unilever).

7.   Greig , J.A. and Sherrington, D.C., Polymer, 1978, **19**, 163.

8.   Voser, W. U.S. Patent 3,725,400; 1973 (to Ciba Geigy).

9.   Pirotta, M. Angew. Makromol. Chem., 1982, **109/110**, 197.

10.   Bernasconi, E., Murador, E., Glacoma Rosa, O. and Varesio, C. "Recovery of Antibiotics Using Polymeric Adsorbent Resins", Paper presented at S.C.I. International Symposium on "Development and Use of Ion Exchange for Industry", Churchill College, University of Cambridge, U.K. July, 1988.

11.   Huxham, I.M., Rowatt, B. and Sherrington, D.C., Makromol. Chem., **192**, 1695 (1991).

ION EXCHANGE EQUILIBRIUM AND TRANSPORT OF BIOLOGICALS IN RESINS FOR PHARMACEUTICAL PROCESSING

GIORGIO CARTA and IDA L. JONES
Department of Chemical Engineering
University of Virginia
Charlottesville, Virginia 22903-2442, USA

ABSTRACT

We consider equilibrium and transport rates of biologicals in ion exchange resins, using amino acids as model compounds. The uptake of these molecules by sulfonated polystyrene-DVB resins occurs via the stoichiometric exchange of amino acid cations. The overall sorption behavior is thus dependent upon ion exchange and solution equilibria, which determine the extent of ionization of the amino acids. Intraparticle transport rates are controlled by the slow diffusion of amino acid cations in the resin, and are strongly dependent upon the hydration of the resin and the size of the diffusing molecule.

INTRODUCTION

Ion exchange resins are often used in pharmaceutical processes for recovery, purification, and bulk separations. The performance of such operations is intimately related to (i) equilibrium and (ii) diffusional mass transfer processes. Selectivity differences can be exploited to obtain bulk separations, and, thus, it is important to determine how molecular properties and resin characteristics affect the ion exchange selectivity. Intraparticle transport rates, on the other hand, determine the sharpness of concentration boundaries in fixed bed operations, and determine the efficiency of resin utilization. Experimental results and theoretical considerations on these effects are provided in this paper.

EQUILIBRIA

Since amino acids are amphoteric, differently charged species, formed according to the following reactions

$$NH_3{}^+CHRCOOH \rightleftharpoons NH_3{}^+CHRCOO^- + H^+ \qquad (1)$$

$$NH_3{}^+CHRCOO^- \rightleftharpoons NH_2CHRCOO^- + H^+ \qquad (2)$$

are present. With a cation exchange resin in H^+-form, the following exchange reaction takes place

$$R'H + NH_3{}^+CHRCOOH \rightleftharpoons R'NH_3CHRCOOH + H^+ \qquad (3)$$

With an anion exchange resin in the hydroxyl form, amino acid anions are exchanged for hydroxyl ions. In either case, however, the zwitterionic species interacts only very weakly with the resin, while the species with the same charge of the resin is excluded by the Donnan potential effect (1-4).

The equilibrium for reaction 3 may be described in terms of the mass action law as

$$y = \frac{q_A}{Q} = \frac{S_{A,H}\, x}{1 + (S_{A,H} - 1)\, x} \qquad (4)$$

where q_A is the concentration of amino acid in the resin, Q the ion exchange capacity, $S_{A,H}$ the selectivity coefficient, and

$$x = \frac{C_{A^+}}{C_{A^+} + C_{H^+}} \qquad (5)$$

the ionic fraction of amino acid cations in solution. For an anion exchange resin eq 4 also applies, if x is defined as the ionic fraction of amino acid anions. In practice, one is interested in the amount of amino acid taken up by a resin for a given total amino acid concentration in solution given by

$$C_A = C_{A^-} + C_{A^\pm} + C_{A^-}$$ (6)

In this case x is obtained from the solution composition using the dissociation constants of the amino acid and the electroneutrality condition (1-4).

The uptake of phenylalanine from dilute hydrochloric acid solutions was determined experimentally for the sulfonated polystyrene-DVB resins in Table 1. All of these resins are

TABLE 1
Resin Properties

Resin	%DVB	Capacity, Q (meq/g dry)	Hydr. water, γ (g/g dry)	Radius (cm)
Dowex 50WX4	4	4.8	2.10	0.042
XUS-40232[+]	6	5.6	1.70	0.029
Dowex 50WX8	8	5.6	1.20	-
HCR-W2	8	5.3	1.10	0.041
XUS-40260[+]	8	5.6	1.20	0.032
HGR-W2	10	5.2	0.90	0.040
XUS-40197[+]	10	5.1	0.86	0.032
Amberlite 252[^]	12	5.0	0.75[*]	0.040

[+] "EP" resin, Dow Chemical Company
[^] Rohm & Haas, macroreticular
[*] gel-phase only

gel-type, with the exception of Amberlite 252, which is macroreticular (3). The hydration water, γ, was determined gravimetrically while the ion-exchange capacity and phenylalanine uptake were obtained from a mass balance method (4). The volume average particle radius was obtained from microphotographs of swollen resin in H^+-form. Typical experimental results are given in Figure 1a and b, in the form suggested by eq 4. It is evident that uptake data obtained for different pH and amino acid concentrations are correlated by a single line in terms of ionic fractions. Thus, non-ionic adsorption is negligible in these resins.

The limiting values of the selectivity coefficient $S_{A,H}$ at infinite dilution are shown in Figure 2 for the different resins. $S_{A,H}$ increases for phenylalanine with the hydration water of the resins, indicating a trend opposite to that of

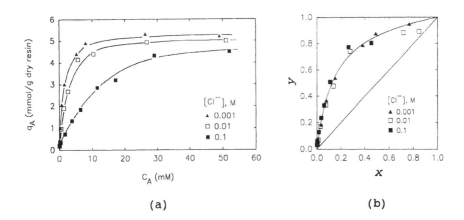

(a)   (b)

Figure 1. Uptake of phenylalanine by the H^+-form of XUS-40260
  resin, T=25 °C. (a) iso-coion uptake curves; (b) x-y plot.

small inorganic cations. As previously shown, the selectivity
for the exchange of various amino acids is correlated with
their hydrophobicity (4), indicating that hydrophobic
interactions play an important role in the selectivity of the
exchange process.

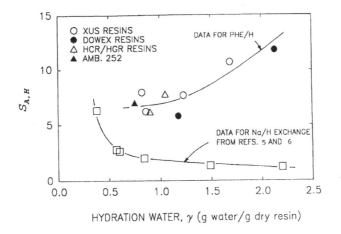

Figure 2. Selectivity for exchange of phenylalanine cations and H^+
  at infinite dilution vs hydration water, T=25 °C.

TRANSPORT RATES

We consider the case of a spherical resin particle initially loaded with amino acid, A, and then brought in contact with a NaOH solution. If the NaOH is sufficiently concentrated, the ionic fraction of amino acid cations becomes zero at the particle surface, and outward transport of the amino acid occurs via the counter-diffusion of H^+, Na^+, and amino acid cations in the resin bead. Neglecting the penetration of negatively charged species into the resin, the conservation equations for this ternary diffusion problem are

$$\frac{\partial q_i}{\partial t} = \frac{1}{r^2} \frac{\partial}{\partial r} \left[r^2 \sum_{j=A,Na} \left(D_{i,j} \frac{\partial q_j}{\partial r} \right) \right], \quad i=A, Na \qquad (7)$$

where

$$D_{i,j} = -\frac{D_i (D_j - D_H) \, q_i}{\sum\limits_{k=A,Na,H} (D_k q_k)} \qquad (8)$$

$$D_{i,i} = -\frac{D_i (D_i - D_H) \, q_i}{\sum\limits_{k=A,Na,H} (D_k q_k)} + D_i \qquad (9)$$

with initial and boundary conditions

$$t=0, \quad q_A = q_A{}^i, \; q_{Na} = 0, \; q_H = q_H{}^i \qquad (10)$$

$$r=0, \quad \partial q_A/\partial r = \partial q_{Na}/\partial r = \partial q_H/\partial r = 0 \qquad (11)$$

$$r=r_o, \quad q_A = 0, \; q_{Na} = Q, \; q_H = 0 \qquad (12)$$

Here r_o is the bead radius and D_i the ionic diffusivity of species i. Superscript i indicates the initial conditions.

Intraparticle concentration profiles, obtained from the numerical solution of these equations by orthogonal collocation on finite elements, are shown in Figure 3. H^+, initially present in the resin, is rapidly exchanged for Na^+, and, within a few minutes, the H^+ concentration is reduced to zero. Since little desorption of the amino acid occurs during this initial phase, the desorption process may be approximated by considering only the slow counter-diffusion of A^+ and Na^+.

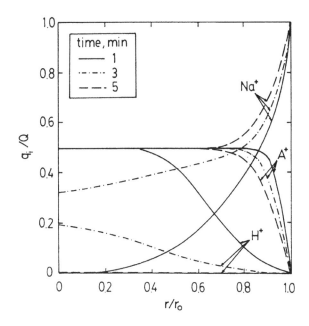

Figure 3. Calculated intraparticle concentration profiles for desorption of phenylalanine with $D_A=1.4 \ 10^{-8}$, $D_{Na}=8.0 \ 10^{-7}$, $D_H=2.3 \ 10^{-6}$ cm^2/s, $r_o=0.04$ cm, $q_A{}^i/Q=0.5$, $q_{Na}{}^i=0$.

Helfferich and Plesset (7) have provided a useful analytic approximation for this case, and their equations may be conveniently used to determine D_A from a fit of the experimental data, using values of D_{Na} obtained from isotopic exchange experiments. It should be noted that contacting the resin with concentrated NaOH removes external mass transfer limitations and allows complete desorption of the amino acid.

Batch desorption curves were obtained in a recirculated differential bed apparatus similar to the one described by Saunders et al. (3). The resin samples were pre-equilibrated with acidic phenylalanine solutions and then contacted with 50-200 mM aqueous NaOH. Experimentally it was found that the desorption profiles for the gel-type resins were independent of the NaOH concentration in this range, and the resulting curves could be accurately described by the Helfferich and Plesset model. For the macroreticular resin, however, the desorption rate was affected by the penetration of hydroxide in the resin pores, and the desorption profiles were different for different

NaOH concentrations (8). In this case the gel-phase diffusivity was determined as shown by Saunders et al. (3).

The diffusivity values obtained with this technique for different resins are shown in Figure 4. The data show a nearly exponential decrease of the phenylalanine diffusivity with degree of crosslinking. Since the hydration water of the resins is inversely proportional to the degree of crosslinking (see Table 1), the trend can be explained in terms of the reduced presence of free water in the resin gel as the %DVB is increased. Interestingly, the diffusivity of phenylalanine cations is ~ 100 times smaller than the diffusivity of Na^+.

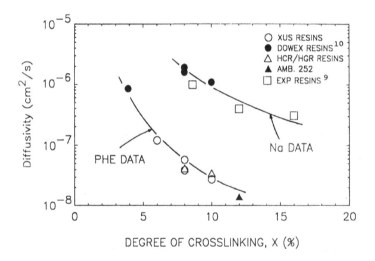

Figure 4. Diffusivities of phenylalanine cations and sodium ions in sulfonated polystyrene-DVB resins, T=25 °C.

CONCLUSIONS

The ion exchange equilibria and intraparticle diffusion of amino acids in ion-exchange resins have been studied, using phenylalanine as a model compound for biologicals. The uptake of these amphoteric molecules can be described taking into account ion exchange and solution equilibria. The selectivity of the exchange of phenylalanine for hydrogen ions depends upon the hydration of the resin. Contrary to what is observed for the exchange of inorganic cations, however, the selectivity increases as the degree of crosslinking is reduced. The intraparticle diffusivity of phenylalanine cations is also strongly affected by the hydration water of the resin, and decreases in a nearly exponential manner with the degree of crosslinking of the resin.

Acknowledgement

We are grateful for financial support by the Dow Chemical Company.

REFERENCES

1. Carta, G., Saunders, M.S., DeCarli, J.P.,II and Vierow, J.B., AIChE Symp. Ser., 1988, **84**, 54.

2. DeCarli, J.P, Carta, G. and Byers, C.H., AIChE J., 1990, **36**, 1220.

3. Saunders, M.S., Vierow, J.B. and Carta, G., AIChE J., 1989, **35**, 53.

4. Dye, S.R., DeCarli, J.P and Carta, G., Ind. Eng. Chem. Research, 1990, **29**, 849.

5. Reichenberg, D., Pepper, K.W. and McCauley, D.J., J. Chem. Soc.-London, 1951, 493.

6. Reichenberg, D. and McCauley, D.J., J. Chem. Soc.-London, 1955, 2741.

7. Helfferich, F. and Plesset, M.S., J. Chem. Phys., 1958, **28**, 418.

8. Carta, G., Saunders, M.S. and Mawengkang, F., in Fundamentals of Adsorption, Mersmann, A.B. and Scholl, S.E. (eds.), Engineering Foundation, New York, 1991.

9. Boyd, G.E. and Soldano, B.A., J. Am. Chem. Soc., 1954, **75**, 6091.

10. Yoshida, H. and Kataoka, T., Ind. Eng. Chem. Research, 1987, **26**, 1179.

THE APPLICATION OF MODELLING TO THE PREDICTION OF ADSORPTION IN BATCH-STIRRED TANKS, PACKED-BED AND FLUIDISED-BED COLUMNS IN BIOTECHNOLOGICAL SEPARATIONS

J.B. NOBLE, G.H. COWAN, W.P. SWEETENHAM
BIOSEP
B353, Harwell Laboratory, Oxfordshire OX11 0RA, UK
and
H.A. CHASE
Department of Chemical Engineering
University of Cambridge
Pembroke Street, Cambridge CB2 3RA, UK

ABSTRACT

The basis of a set of computer programs for the performance prediction of adsorption and chromatography units, or for the derivation of appropriate capacity and rate parameters is outlined. Applications to the optimisation of yield in the scale-up of an ion exchange process to recover an intracellular enzyme from a clarified cell homogenate, and as an aid in isocratic elution chromatography to select conditions to achieve the desired product purity are described.

INTRODUCTION

Ion exchange processes are used extensively in the purification of pharmaceutical products particularly those of a proteinaceous nature. However, full process optimisation is often restricted by the value of the products involved, the cost of the specialist adsorbents, and time constraints imposed by the commercial environment. Nevertheless, previous work (1) has shown that computer simulation can be used to target key experiments in process design, and can hence reduce the laboratory time and costs involved in process optimisation and scale up, thus leading to the simultaneous optimisation of product purity and yield. Furthermore, the authors believe that the use of computer simulations can have widespread applications in staff training and mal-performance diagnosis.

Although the theoretical worth of computer aided design is unquestioned the approach will not receive widespread use until it has been fully validated by application to real multicomponent biological systems. To this end a proprietary design package incorporating computer programs has been developed and validated by application to real biological ion exchange systems. Full details of the procedure are given in a Design Report available to industrial members of the Biotechnological Separations Club, BIOSEP, and outlined in reference (9).

Information on the theory and solution techniques used in the programs and initial applications have been described previously ((1) to (10)). The codes are of two types, those which can be used for process simulation and those which can be used for deriving physical parameters by fitting to experimental data to determine capacity and rate characteristics. The programs may be applied to single and multicomponent adsorption (up to six components, with or without competition for sites on the absorbent) in batch-stirred tanks, packed-bed and fluidised-bed columns. They can be used for simulation of both frontal analysis and elution chromatography modes of operation in the prediction of packed-bed column performance. Two types of model of the mass transfer process are used within the codes. The programs based on the simple kinetic model describe the resistance to mass transfer by analogy with a reversible second order chemical reaction implying a Langmuir isotherm ((2), (5), (7), (8), (9)). For simulation, values are required for the backward and forward rate constants appropriate to the model. In the liquid film plus pore diffusion model the resistance to mass transfer is taken as a combination of liquid film resistance at the surface of the adsorbent particles together with diffusion resistance within spherical porous particles ((1), (9), (11)). Input data required include the mean adsorbent particle diameter, the intraparticle porosity of the adsorbent, the liquid film mass transfer coefficient and effective pore diffusivities. Axial dispersion effects can be computed using column codes based on both models. With the latter model the appropriate column simulation program may be used with isotherms other than Langmuir, including the user's own equilibrium equations.

Details are given in the present paper of the use of the programs as an aid to the development of a packed-bed separation for the recovery of the intracellular enzyme leucine dehydrogenase (LDH) from a clarified cell homogenate using ion exchange media. Examples are also presented to demonstrate the effectiveness of the use of the programs in the assessment of resolution in the separation of components in isocratic elution chromatography.

OVERVIEW OF DESIGN PACKAGE

The sophistication of modern automated computer controlled chromatography systems makes them ideal for optimising adsorption conditions. However, the time and volumes of media required for optimisation of pilot and process scale loading/elution flowrate, load volume and elution step length restricts their utility. Hence it is in these areas that numerical simulation can offer the greatest benefits.

In Figure 1 an overview of the modelling approach is detailed. The first step is to generate estimates of the parameters necessary for process simulation. Having done this assessment can be made of the major limitations to mass transfer and the appropriate level of model complexity selected. Simulations can then be performed to select the most suitable contacting system (packed-bed, fluidised-bed or stirred tank), and first estimates of the operating variables (flowrate, bed height, particle size etc.). Once the above task has been completed experiments must be performed to validate the model and where necessary diagnostic (fitting) routines used to refine the simulation parameters or model complexity. Then having produced a reliable system model small scale optimisation can be undertaken. Scale-up parameters such as liquid mixing or changes in adsorbent size can thus be investigated and their effects quantified, thus allowing process scale optimisation.

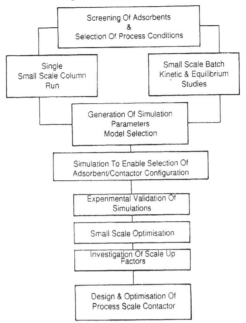

Figure 1. Overview of Approach

EXPERIMENTAL VALIDATION OF THE APPROACH

The above approach has been validated by application at both laboratory and industrial scale to real system purifications. The technique is well demonstrated by consideration of the extraction of the intracellular enzyme leucine dehydrogenase from a clarified cell homogenate at a pH of 8 and a conductivity of 16 mS. The target molecule represented 0.6% of the total protein at a concentration of .25 Enzyme Activity Units/ml. The adsorbent used was a silica based Diethylaminoethyl (DEAE) anion exchange material.

The simulation parameters and model complexity were defined as in Figure 1. For this high molecular weight enzyme pore diffusion was found to limit the rate of uptake. Using the chosen model it proved possible to run simulations which enabled the selection of optimum particle size and flow rate

based upon throughput and dynamic capacity considerations. In
Figure 2 a plot is shown of the effect flow rate has on packed
column breakthrough, with higher flow rates producing lower
dynamic capacities and increased asymmetry in the breakthrough
curve.

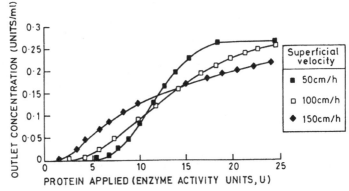

Figure 2. Effect of Flow rate on Packed Column Performance
LDH Onto 1 cm diam x 4 cm high DEAE Spherodex M Grade Column

The next stage in the process is to select the most
promising simulations and attempt experimental validation.
Figure 3 shows the suitability of the chosen model and
simulation parameters for predicting column performance for
superficial velocities of 100 cm/hr and 150 cm/hr respectively
(no comparable results are available for a superficial
velocity of 50 cm/hr). From here detailed design can take
place allowing optimisation of the loading stage to produce
the required product yield, and further simulation to define
the optimum elution regime/load volume combination.

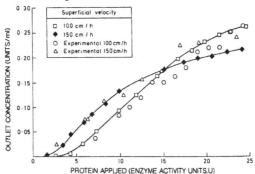

Figure 3. Effect Of Flow Rate On Packed Column Performance
Comparison Of Experimental With Predicted

In a similar manner the codes can be applied to the optimisation of eluate step length and flow rate to enable the desired product purity to be obtained.

ASSESSMENT OF ELUTION CHROMATOGRAPHIC PROCEDURES

In isocratic elution chromatography an appropriate sample is applied to the packed bed and the bed is irrigated with a liquid phase of constant composition until the peaks of the applied components emerge from the bed. The key feature to operating and optimising any form of elution chromatography is to ensure that the required resolution of components is achieved during chromatography. Hence peak shape and position are particularly important. The codes can be used to investigate the influence of some of the operational parameters on the resolution that can be obtained in such separations.

For example, if the amount of material in a sample is increased whilst the volume of the sample is held constant (i.e. the concentration is increased) computer simulations indicate that the position of the maximum of the leading eluted peak occurs earlier in the chromatogram as can be seen in Figure 4 illustrating the resolution of a binary mixture. The shape of the peak changes with concentration, with the peak becoming more asymmetric, with a sharp front and a trailing tail as the concentration is increased. Such spreading has a detrimental effect on the resolution of the two species, and the resolution can be observed to be less as the concentration of material in the sample is increased.

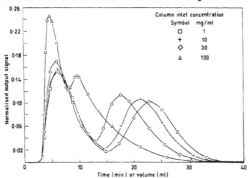

Figure 4. Effect of Sample Concentration on Resolution

As the liquid flow through the bed is increased the resolution of two components is predicted to decrease as depicted in Figure 5.

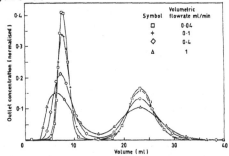

Figure 5.   Effect of Volumetric Flow Rate on Resolution

It can be seen that the peaks spread out as the flow rate increases thus diminishing the separation achieved. Qualitatively this is a consequence of the increased flow rate resulting in less time for mass transfer to occur between the absorbing species and the adsorbent in the bed.   Thus the simulations support the generally accepted technique of reducing liquid flow rate to improve separation.

A common solution often proposed to increased resolution between two components is to use a longer bed.   Such a suggestion is supported by the simulations shown in Figure 6, which indicate that as the length of the bed is increased the resolution of the two components is also increased.   Although the resolution between the components is indeed improved, it should be noted that the peaks are broader when the bed is longer and hence the material appears in a greater volume at a lesser concentration.

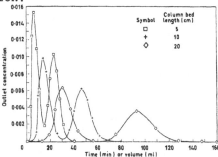

Figure 6.   Effect of Column Bed Length on Resolution

CONCLUSION

The dovetailing of small scale experiments (to determine capacity and rate parameters) with computer simulation codes provides powerful tools in application to the design and optimisation of preparative and process-scale adsorption and chromatography for biotechnological separations. The approach gives a better understanding of the effects of changing process variables to lead to more economic processes and minimisation of investment risk.

REFERENCES

(1)   Horstmann, B.J. and Chase, H.A. Chem. Eng. Res. Des., 1989, 67, 243-254.
(2)   Thomas, A.C. J. Am. Chem. Soc., 1944, 66, 1664-1666.
(3)   Liapis, A.I. and Rippin, D.W.T. Chem. Eng. Sci., 1977, 32, 619-627.
(4)   Liapis, A.I. and Rippin, D.W.T. Chem. Eng. Sci., 1978, 33, 593-600.
(5)   Chase, H.A. J. Chromatography, 1984, 297, 179-202.
(6)   Arve, B.H. and Liapis, A.I. J. AIChE, 1987, 33(2), 179-193.
(7)   Cowan, G.H., Gosling, I.S., Laws, J. and Sweetenham, W.P. J. Chromatography, 1986, 363, 37-56.
(8)   Cowan, G.H., Gosling, I.S. and Sweetenham, W.P. in Separations for Biotechnology, Verrall, M.S. and Hudson, M.J. (eds), Ellis Horwood Ltd., 1987, Chap.10, 152-175.
(9)   Cowan, G.H. in Adsorption: Science and Technology, NATO ASI Series E: Applied Sciences, Rodrigues, A.E., LeVan, M.D. and Tondeur, D. (eds), Kluwer, 1989, 158, 517-537.
(10) Cowan, G.H. Interphex 88 Conference, 1988, 2B, 33-45.
(11) Cowan, G.H., Gosling, I.S. and Sweetenham, W.P. J. Chromatography, 1989, 484, 187-210.

ACKNOWLEDGEMENTS

Acknowledgement is made to BIOSEP, Harwell Laboratory, Oxfordshire, OX11 0RA, UK for permission to publish the work presented in this paper. The copyright in the paper remains the property of the UKAEA, AEA Environment and Energy.

DETERMINATION OF ADSORPTION/DESORPTION KINETICS OF PROTEINS ON ION EXCHANGE MEDIA

L.X. TANG, R.W. LOVITT, J.R. CONDER and M.G. JONES
Biochemical Engineering Research Group
Department of Chemical Engineering
University College of Swansea
University of Wales
Swansea SA2 8PP

ABSTRACT

An automated computer controlled analyser has been developed for the determination of adsorption and desorption characteristics of ion exchangers used for protein separations. The analyser is capable of performing a range of tasks including continuous monitoring and control of solution parameters such as pH, conductance and temperature while measuring concentration of substrates with an integrated UV detector. Measurements with protein adsorption and desorption kinetics are illustrated.

INTRODUCTION

Separation by selective adsorption methods such as chromatography plays a significant role in the downstream processing of biotechnology products. A range of proteins from human or animal blood and from milk whey have been separated using chromatographic stages (1-2). The importance of separating biological macromolecules by such selective adsorption methods has created an increasing need for understanding the mechanism and kinetics of the adsorption/desorption process to permit prediction and optimisation of the performance of the adsorption separation.

Several studies of the performance of the process have been conducted (3-9). Studies of the kinetic aspects of performance can be generally divided into two types according to the mode of operation: (a) stirred tank operation and (b) packed bed column operation. In the first type, the protein solution is stirred using a magnetic stirrer or a shaking bath and its concentration during adsorption is measured by spectrophotometry using periodical sampling or continuous recording. In packed column operation the adsorption is determined by feeding the bed with different concentrations of protein and monitoring the UV absorbance at the outlet. In both types of operation, the pH and ionic strength of all solutions and slurries used is pre-adjusted using buffer (typically tris, phosphate or acetate) and salt solutions.

There are many problems associated with current methods of studying adsorption kinetics. These include slow response, change in solution concentration due to addition of solutes or discrete sampling, and variation in solution pH, temperature and conductivity during operation. Traditionally, buffer has been used for controlling solution pH. This, however, may also have side effects on the adsorption, as the buffer itself is a charged species and it, or its associated metal ion will interact with the ion exchange material. Such problems may strongly affect the experimental data obtained and therefore the final interpretation of the adsorption kinetics. It is advantageous to develop automated equipment which possesses the ability to carry out continuous, "real-time" monitoring, not only of the decrease/increase in the concentration of the biological macromolecular adsorbate such as protein but also of the change in the actual values of solution parameters such as pH and temperature during adsorption/desorption operations. More importantly, the equipment should have the capacity to control or adjust the solution parameters within the desired range of interest.

In this paper, we describe a system which possesses these features, and illustrate its capabilities with some preliminary results of a protein adsorption/desorption study using the equipment in batch mode.

EXPERIMENTAL

A diagram of the experimental equipment is shown in Figure 1. It consists basically of three parts: separation cell, measuring and controlling unit and computer unit. The separation cell for batch adsorption/desorption is a 100mL glass vessel with a water jacket for maintaining the desired temperature. The measuring and controlling unit contains components itemised in Table 1. The total liquid volume circulating through the UV detector is approximately 3mL during operation. The computer unit carries out parameter control (through the measuring and controlling unit) and data acquisition and processing. The associated computer program allows rapid reading of experimental data; a set of five parameter data can be read, scaled and displayed every 30 seconds. It also allows change of the frequency of data acquisition to suit different requirements of study.

Table 1. Components of the measuring and controlling units and performance data

Component	Function	Performance data
Measuring:	Measure:	
UV detector	Absorbance of protein	$0 - 1 \pm 0.001$
Temperature probe	Solution temperature	$0 - 100 \pm 0.1$ °C
pH electrode	Solution pH	$0 - 14 \pm 0.05$
Conductivity cell	Specific conductance	$0 - 100 \pm 0.01$ mS/cm
Balance	Mass of solute added	$0 - 410 \pm 0.001$ g
Controlling:	Control:	
Heater/circulator	Temperature	$20 - 100 \pm 0.1$ °C
Stirrer	Agitation	$0 - 2500$ min^{-1}
UV pump	Flow into UV detector	$0 - 42.5$ mL/min
Acid pump	Flow of HCl solution	$0 - 6.0$ mL/min
Base pump	Flow of NaOH solution	$0 - 6.0$ mL/min
Salt pump	Flow of NaCl solution	$0 - 6.0$ mL/min
Filter (8μm)	Passage of resin	only liquid through

The results illustrated were obtained with cation exchange resin which was a sulphopropyl-derivatised regenerated cellulose material sieved to a 250-300μm fraction. The feed was bovine serum albumin (BSA) solution at a concentration of 10g/L. NaOH (0.05M) and HCl(0.05M) solutions

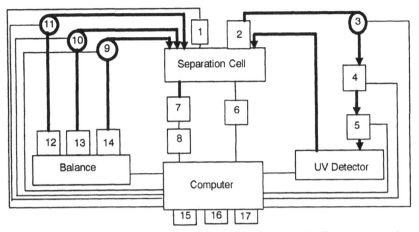

1.Stirrer; 2.Filtration; 3.UV pump; 4.pH electrode; 5.Conductivity cell; 6.Temperature probe; 7. waterbath/circulator; 8.heater; 9-11 additon pumps for salt, acid & base reservoirs, 12-14; 15. VDU; 16. Data storage; 17 Printer. **Bold** lines show liquid flow lines. Normal lines indicate signal and control connections

Figure 1. Schematic diagram of the automated system for the study of protein adsorption/desorption processes

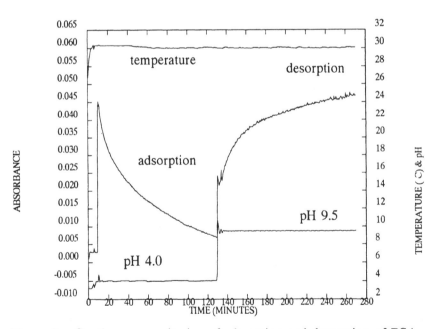

Figure 2. Continuous monitoring of adsorption and desorption of BSA under controlled temperature (30 °C) and pH (4.0 and 9.5)

were employed to control the reaction pH.

A 8μm membrane filter (Millipore, type SC) was placed in the filtration compartment. 0.5g wet ion exchange resin was weighed into the separation cell and 40-60mL of freshly distilled water was added. The protein adsorption program was started and the values of pH and temperature to be maitained were chosen (30°C, pH 4). When the parameters stabilised at the chosen values, 0.5-20mL 10g/L BSA solution was added to the cell to begin the adsorption process. Throughout the process the computer monitors and, if necessary, controls the five solution parameters (UV absorbance, temperature, pH, conductance and mass of the solutes added). When adsorption reaches the required degree, the solution pH to be controlled is re-selected (pH 9.5) on the computer to start the desorption process.

RESULTS AND DISCUSSION

Table 1 shows basic performance data for the automated equipment. Calibration of the UV detector at a wavelength of 254nm gave a linear, but pH-dependent, plot against BSA concentration (0-10g/L), fitted by the equation

$$A = 10^{-4} C (523.464 + 171.04 \text{ pH} - 37.65 (\text{pH})^2 + 2.515 (\text{pH})^3)$$

where A is the UV absorbance and C the BSA concentration (g/L).

Figure 2 shows a continuous BSA concentration trace under controlled conditions of temperature and pH. At pH 4, with addition of BSA solution, the UV absorbance undergoes a sharp step change, then decreases progressively as adsorption proceeds. When the solution pH is automatically adjusted to 9.5 by computer controlled pumps, the desorption process starts immediately.

The protein concentration profiles in the liquid phase and solid phase are calculated using the above equation and the results for the adsorption process are shown in Figure 3. The kinetic profiles can also be used to derive equilibrium isotherms by extrapolating the profiles to infinite time. The quantity and the quality of the data obtained means that data

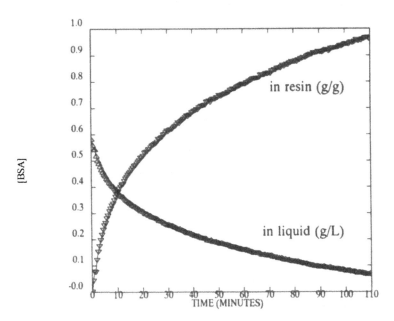

Figure 3.   Change of BSA concentration in both liquid and resin
phases during adsorption.

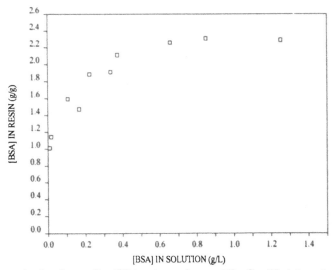

Figure 4.   Isotherm for BSA adsorption at 30 °C, pH 4.0 and specific
conductance 10 mS/cm.

can be analysed with a high degree of confidence and precision. Excellent fits are achieved with a sum of two exponential terms and the reasons for this are now under investigation. The kinetic method of determining equilibrium isotherms may be less precise than direct analysis of equilibrated batch samples but gives satisfactory results, as seen in Figure 4. A significant advantage of the kinetic method of obtaining equilibrium data is that there is no reliance on equilibrium having been achieved. This reduces the risk of protein denaturation during long equilibration times required when using soft gel ion exchange media.

CONCLUSIONS

This preliminary study has demonstrated the capabilities of newly developed computer-automated equipment using multiple-parameter control, fast data acquisition and programmed data-processing for analysis of protein adsorption and desorption. Studies are underway to test other ion exchange media and proteins as well as column operation.

ACKNOWLEDGEMENT

The authors would like to thank SERC for financial support of this work

REFERENCES

1. Janson, J.C. and Hedman, P., *Adv. Biochem. Eng.*, **25** (1982) 43-99.
2. Leaver, G., Conder, J.R. and Howell, J.A., *Sep. Sci. Technol* **22** (1987) 2037-2059.
3. Graham, E.E. and Fook, C.F., *AIChE J.*, **28** (1982) 245-250
4. Chase, H.A., *J. Chromatogr.* **297** (1984) 179-202
5. Arnold, F.H., Chalmers, J.J., Saunders, M.S., Croughan, M.S., Blanch, H.W. and Wilke, C.R., *ACS Symp. Ser.* **271** (1985) 113-124.
6. Cowan G.H, In: *Adsorption: Science and Technology*, Rodrigues A.E. *et al*(eds.)1989, 517-537.
7. Skidmore G.L., Horstmann B.J. and Chase H.A., *J. Chromatogr.* **498** (1990) 113-128
8. Skidmore G.L. and Chase H.A., *J. Chromatogr.*, **505** (1990) 329-347
9. Leaver, G., Howell J.A. and Conder J.R., paper presented at 8th International Symposium on Preparative Chromatography, Washington D.C., 1991; *J. Chromatogr.*, 1992, in press.

A MONOLITHIC ION-EXCHANGE MATERIAL SUITABLE FOR DOWNSTREAM PROCESSING OF BIOPRODUCTS

ROB NOEL, ALAN SANDERSON, LES SPARK
bps separations ltd
Green Lane Industrial Estate,
Spennymoor,
Co.Durham DL16 6YL, UK

ABSTRACT

A monolithic ion-exchange material which addresses the practical problems of large scale biochromatography is described. Its particular combination of protein capacity, fast kinetics of adsorption\desorption and high flow rates at low pressures (7.3 m/h, at 2 bar with a 11.5 cm bed depth) permits productivities with biological feedstocks in the order of 85 $kg/m^3/h$. Pre-treatment of feedstocks with fine filters was found to be unnecessary, creating the possibility of primary protein purifications directly from 'dirty feedstocks' such as cell homogenates and cheese whey. Repeated cleaning with 60°C 0.5M NaOH is shown to have no effect in repeated process cycling.

LIMITATIONS OF CURRENT PROCESS MEDIA

Chromatographic media have been extensively used in protein chemistry and have enabled the isolation of pure protein products by nature of their binding capacity and resolution. However their transposition to downstream processing in the biotechnology industry has high-lighted some inherent problems, such as slow flow rates with high back pressures, packing difficulties, blockage of frits and pre-filters and difficult cleaning-in-place.

These problems lead to low throughputs, caused by slow flow rates, in the order of 0.3 - 1 m/h (1,2). The effect of this is to lengthen process times, reduce product yield through protein degradation and therefore reduce overall productivity. Some of the best suited beaded matrices are susceptible to shrinkage and swelling of the bed volume with increases/decreases in the ionic strength or pH of the liquid phase. This phenomenon leads to channelling in process columns, necessitating re-packing and slowing of production. These problems are further compounded by the need to contain materials of 10-100μm in columns by tight frits which become blocked and regularly need replacement, therefore the very nature of the column matrix limits the feedstocks that can be processed. Pre-column filtration in the 0.2 to 10μm range becomes essential thereby increasing capital and consumable costs but is necessary to prevent column fouling (3).

Another major source of dissatisfaction with beaded matrices is the generation of fines, an unacceptable contaminant to food and clinical products.

MONOLITHIC MATERIALS - AN INTRODUCTION

Chromatographic media are required that address these diverse problems and make large scale adsorption chromatography of proteins a simple and more practical proposition. Such materials should, however, also retain the high protein capacity and recovery associated with the currently available media.

We have sought solutions to these large-scale problems through the application of monolithic structures produced at

bps separations ltd. The materials are of regenerated cellulose and have an interconnected open-pore framework with the texture of sponge and exhibit excellent fluid flow properties (Fig.1). The material can be cut and packed into suitable columns, in which there is no channelling, bed volume or bed resistance change with a variety of high or low pH solutions.

Protein capacities are in the range of 80 mg/mL for lysozyme and 50 mg/mL for conalbumin on the CM and SE-derivatives. While the DE and QM-derivatives bind 50 mg/mL of bovine serum albumin and 70 mg/mL for soybean trypsin inhibitor.

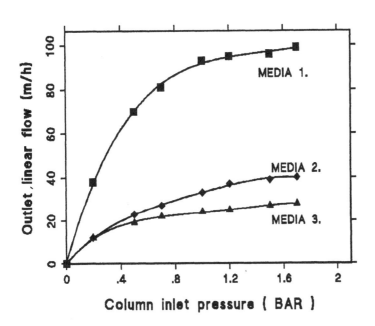

Fig.1. Flow/pressure graph comparing monolithic media (MEDIA 1) with amorphous (MEDIA 2) and crystalline (MEDIA 3) cellulose media. Column dimensions 4 mm x 43 mm i.d.

HIGH FLOW RATE SEPARATIONS

The linear flow velocities attainable with monolithic matrices enable kinetic studies at high flow rates to be performed (4,5). Similarily using an industrial feedstock, egg white, conalbumin and lysozyme can be separated from diluted solution in a two step elution procedure using a carboxymethyl-derivative (Fig.2). Varying the linear flow from 5-18 m/h (pressure 0.8 bar at 18.4 m/h) this separation gives a linear increase in productivity from 20 to 73 $kg/m^3/h$ for conalbumin and 6 to $12kg/m^3/h$ for lysozyme.

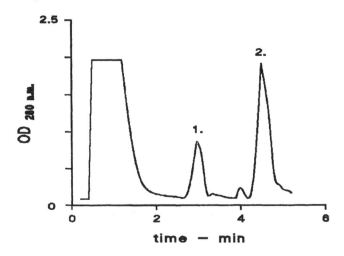

Fig.2. The separation of egg white proteins using PRODUCTIV™ CM monolithic matrix. Flow rate 12.2 m/h, 2 cm bed depth. Loading buffer 50mM acetate pH 4.5. Elution buffers; 50mM phosphate pH 6.8 - conalbumin (peak 1) and 50mM carbonate pH 10.6 - lysozyme (peak 2) specific activity 32 000 Units/mg. Column loading 50 mL feedstock onto a 10 mL column.

APPLYING PROCESS CRITERIA TO MONOLITHIC MATERIAL

Reproducibility of separations during continuous processing is vital and so 30 cycles of operation were performed using egg white feedstock, without cleaning. The deviation in the eluate peak heights was only 4% from the mean. Furthermore, a 0.5M NaOH solution was recirculated through matrix for 24h at 22°C and subsequently for 1 h at 90°C. After each treatment an egg white separation was performed, no deterioration in total productivity occurred, on the contrary, a small increase was observed from 82 kg/m³/h to 85 kg/m³/h.

In the preparation of feedstocks for large scale chromatographic processing considerable costs can be incurred by pre-column filtration steps. Egg white feedstock was prefiltered through a 100-200μm coarse filter instead of the 20-25μm filters previously used. There was no change in the productivity of lysozyme or conalbumin between the two feedstocks.

This result prompted investigations into processing 'dirty feedstocks' with minimal pre-treatment.

PRIMARY PURIFICATION OF DIRTY FEEDSTOCKS

Table 1 lists the feedstocks applied to monolithic media and details the treatment required to perform a primary separation.

A good example of a dirty feedstock was prepared using a yeast extract. Yeast cells (*Saccharomyces cerevisiae*) were homogenised as previously described (6) and centrifuged at 1000g for 10 min to pellet undisrupted cells. The cloudy supernatant was spiked with human serum albumin (HSA) to 2 mg/mL and the pH adjusted to 7.5. 5 mL of the solution was applied onto a 10 mL column of DE-derivatised monolithic matrix. A 12.5% SDS-PAGE analysis of the eluate shows a primary purification of the HSA occured (Fig.3). The whole protocol took 20 min in total. Processing such feedstocks usually requires a minimum of two or more of the following techniques - precipitation, flocculation, centrifugation, diafiltration, ultrafiltration and batch debris removal (7).

TABLE 1.
The feedstocks and minimal pre-treatments required for their processing on monlithic material

Feedstock	Pre-treatment
Micrococcus sp. lysate	none
Cheese whey	none
E.Coli lysate	sonication
Blood plasma	diluted 5-10 fold
Yeast homogenate	centrifuge 1000g, 10 min
Egg White	ovomucoids precipitated
	100-200μm filtered

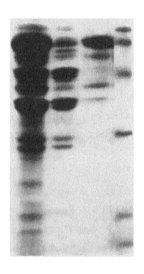

Fig.3. 12.5% SDS-PAGE analysis of a separation of HSA from a yeast homogenate on PRODUCTIV™ DE. From left to right: Lane 1. yeast + HSA, lane 2. breakthrough, lane 3. eluate (HSA enriched), lane 4. low molecular weight markers (78,66,43,30,17 and 12Kd from top to bottom)

Another industrial feedstock is cheese whey - a protein rich source but with impurities such as fat, lactose and salt. The processing of whey may involve either the total uptake of protein from the whey or a fractionation of useful proteins such as immunoglobulins (IgG's), lactoferrin and lactoperoxidase from beta-lactoglobulin (beta-LG), bovine serum albumin (BSA) and alpha-lactalbumin (alpha-LA). A column of dimensions 11.5cm X 2.5 cm i.d. with no supporting frits was packed with DE-monolithic matrix and run at a flow rate of 7.3 m/h at a pressure of 2 bar maximum. Whey obtained from skimmed cow's milk at pH 5.6 and at $30^{\circ}C$ using calf rennet, was adjusted to pH 7.6 with NaOH. At a pH of 6.7 a fine white precipitate of calcium phosphate forms, however this was not found to effect column performance. 150 mL of feedstock was processed in two column runs over a total of 12.5 min. Lactoferrin and IgG's were shown to have been separated from the alpha-LA, beta-LG and BSA by SDS-PAGE analysis. From Kjeldahl protein analysis the product yield of eluates was a minimum of 80%, the recovery of protein from the column was a minimum of 94%. The productivity of beta-LG, alpha-LA and BSA was determined to be 34 $kg/m^3/h$.

CONCLUSIONS

This paper has illustrated the use of a monolithic chromatographic adsorption media to separate protein from feedstocks with process times much reduced when compared with conventional media. This was achieved by:

1) high flow rates combined with fast uptake of protein from solution, yielding column run times in the order of minutes rather than hours.

2) little or no sample pre-treatment. Columns did not block with particle sizes equal to or below 100-200μum diameter. Column frits were not required.

The robustness of the matrix to caustic cleaning, even at high temperatures and flow rates provides fast and efficient cleaning-in-place giving the media an extended useful life.

There is also the exciting possibility of performing primary purifications with dirty feedstocks by direct processing of bacterial and fungal homogenates, milk whey, fermenter broths and tissue culture supernatants.

REFERENCES

1) Janson,J.C. and Hedman,P (1982) "Large Scale Chromatography of Proteins" Advances in Biochem. Eng. ,**25**, pg 44-97 Ed. A.Fiechter.

2) Pharmacia Fine Chemicals "Scale-Up to Process Chromatography"
IBF biotechnics (Villeneuve La Garenne, France), 1987.
Fractogel TSK - E.Merck (Darmstadt,Germany) .

3) Amicon (Danvers,MA) "A Practical Guide Industrial-Scale Protein Chromatography" (1982)

4) Ming,F. and Howell,J (1991) "Kinetic behavior of a novel matrix ion exchanger, carboxymethyl-HVFM operated at high flow-rate" Journal of Chromatography,**539**, 225-266.

5) Johansson,G. and Joelsson,M. (1986) "Specifically increased solubility of enzymes in polyethyleneglygol solutions using polymer-bound triazine dyes." Analytical Biochem. ,**158**,104-110.

6) Ming,F. PhD Thesis, University of Bath 1989.

7) "Protein Purification Application - a Practical Approach" Ed. E.L.V.Harris and S.Angal 1990 Oxford University Press.

DIRECT RECOVERY OF PROTEIN PRODUCTS FROM WHOLE FERMENTATION BROTHS:
A role for ion exchange adsorption in fluidised beds.

PHILIP H MORTON AND ANDREW LYDDIATT
Biochemical Recovery Group, School of Chemical Engineering,
University of Birmingham, Edgbaston, Birmingham B15 2TT UK

ABSTRACT

Five commercial ion exchange adsorbents were evaluated in settling tests in respect of suitability for fluidised bed adsorption. Comparison was made of DEAE Spherodex LS performance in fixed bed and fluidised bed recoveries of an acidic protease from clarified and unclarified broths generated in pilot-scale fermentations of Yarrowia lipolytica. Overall protease recoveries were similar in both contactors, but fluidised bed treatments yielded enzyme with the highest specific activity and purity. The findings encourage the generic development of continuous adsorptive recovery of proteins in fluidised beds integrated with fermentations.

INTRODUCTION

Ion exchange adsorption can be implemented as one of a sequence of unit operations applied to the recovery and purification of extracellular proteins produced in microbial fermentations. Conventionally, the process is operated in fixed bed or batch suspension contactors and typically requires a clarified feedstock free of microbial cells and associated debris. Liquid fluidised beds can accommodate the flow of microbial suspensions as a result of the increased voidage generated by bed expansion in upward liquid flows exceeding critical velocities. Such a strategy obviates the requirement for feedstock clarification, thereby foreshortening purification schemes and potentially increasing process efficiencies. Liquid fluidised beds have been reported in antibiotic recovery using polystyrene adsorbents (1), but application with macromolecular products is compromised by the hydrophobic character and microporous structure of such solid phases. Recently, Wells et al (2) have studied the performance of macroporous agarose composites (Macrosorb DEAE K6AX, whilst Draeger and Chase (3) have compared the performance of homogenous agaroses (Q-Sepharose FF) in fixed and fluidised beds.

A major constraint upon widespread adoption of the technique has been the limited number of solid phases with suitable geometries, densities and hydrodynamic behaviour coupled with an appropriate range of surface chemistries designed for ion exchange, hydrophobic or affinity adsorption of proteins. Contemporary commercial development in adsorption has concentrated upon the refinement of fixed bed chromatography (small diameter, monodisperse, macroporous particles) to the exclusion of particles suited to other contactors and process situations (4). In addition, few successful applications of liquid fluidised bed adsorption are cited in the literature. We report here preliminary findings concerning the recovery and partial purification of an acidic protease from whole broths generated in fermentations of the yeast Yarrowia lipolytica. Comparison is made of the performance of fixed and fluidised beds of DEAE Spherodex LS.

MATERIALS AND METHODS

The comparative settling velocities of potential ion exchange solid phases (10 mL) were estimated in a column (3 x 20 cm) of potassium phosphate buffer (pH 6.5). The settling velocity between two points was corrected for particle hindrance by the relationship $U_h = U_t (1-h)^{4.65}$, where: U_h = hindrance velocity, U_t = settling velocity, and h = solid phase volume fraction (5).

Yarrowia lipolytica (formerly Candida olea (6); the gift of Dr Tom Young, School of Biochemistry) was maintained at pH 4.5 and 30°C on 0.3% malt extract, 0.3% yeast extract, 2% glucose and 0.5% bacteriological peptone. Glucose solutions were sterilised at 121°C for 15 min, whilst protein supplements were microfiltered and subsequently sterile filtered (Sartobran) at 10-fold strength. Batch fermentations (15 L) were conducted in 1% glucose, 1% whey protein (Biopro, Bio-Isolates) and 0.17% yeast nitrogen base (Difco; amino acid and ammonia free) at pH 4.2, 30°C and 15 L/min sterile air. PPG 2025 was added as antifoam, and dissolved oxygen maintained at 40% saturation. Growth was stopped at maximum protease accumulation (about 2000 Units/mL at 45 h), and broth was used directly in fluidised beds or clarified by microfiltration (6).

Clarified broth (1 L; adjusted to pH 6.5) was pumped at 1 mL/min through
a fixed bed (7.6 x 1 cm; Figure 1) containing DEAE Spherodex LS
(100-300 µm; IBF) equilibrated in 20 mM potassium phosphate at pH 6.5.
When protease concentrations in the bed breakthrough approached 10% of
load values, the bed was washed with 30 mL equilibration buffer and
eluted using a two-step gradient of 0.18 M and 0.5 M NaCl in
equilibration buffer. Fractions (2 mL) were collected for assay of
protease and total protein content. Whole fermentation broth (2 L;
protease concentration as above, adjusted to pH 6.5) was introduced to a
recirculating fluidised bed (10 x 2.1 cm; Figure 1) of DEAE Spherodex LS
(12 ml) at 30 ml/min yielding a bed expansion of 100%. Flow distribution
was achieved through a conical fitting at the base of the bed and a 100 µm
mesh. At adsorption equilibrium, the bed was washed free of cells with
equilibration buffer at 60 mL min^{-1} (200% bed expansion). Protein elution
from 6 mL batches of adsorbent was achieved by flow reversal and the step
method outlined above for fixed bed contactors.

Biomass was estimated from spectrophotometric absorbance at 650 nm, and
by dry weight determinations. Acidic protease activity was quantified by
a modified (6) method of Kassel and Meitner (7) using acid-denatured
hemoglobin as substrate and pepstatin as standard, and expressed as
pepstatin equivalents (Units/mL). Total protein content was estimated by
reaction with Coomassie blue (8) using Bipro as standard (0-30 µg ml^{-1}).
Protease purity was assessed by SDS-PAGE in a Pharmacia Phast System
using the protocols recommended by the manufacturer.

RESULTS AND DISCUSSION

Settling velocities were estimated for ion exchange adsorbents in order
to evaluate their suitability for fluidisation. Low values for DEAE-52
Cellulose and Q Sepharose FF (U_h = 0.30 and 0.54 x mm/s respectively)
were confirmedby poor performances in a recirculating fluidised bed,
although the latter has been used in single pass contactors (3).
Amberlite CG-50 and DEAE Macrosorb K6AX were better (U_h = 3.4 and
8.2 mm/s), but were incompatible both with the protease product and a
process requiring sterile operation. DEAE Spherodex LS, a dextran-coated

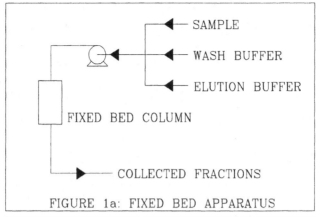

FIGURE 1a: FIXED BED APPARATUS

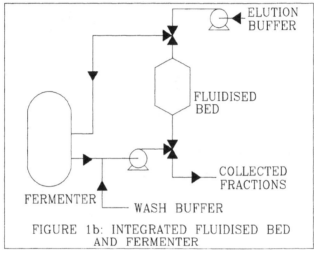

FIGURE 1b: INTEGRATED FLUIDISED BED
AND FERMENTER

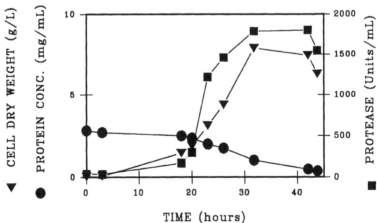

FIGURE 2: GROWTH CURVES FROM A 201 FERMENTATION

silica, possessed a moderate settling velocity (U_h = 4.9 x mm/s) and good fluidisation characteristics. This material was chosen for further study.

Figure 2 illustrates a typical growth curve for a 20 L batch fermentation of Yarrowia lipolytica. Maximal biomass was reached at 30 h, and concomitant accumulation of protease reached a maximum at about 45 h. Whey protein, utilised as the yeast nitrogen source, was significantly depleted by protease action. The PPG 2025 added to combat foaming of the protein-rich broth at the start of the fermentation severely reduced permeate fluxes during microfiltration used for broth clarfication prior to fixed bed recoveries (data not shown; discussed in 6).

The comparative performance of DEAE Spherodex LS in the recovery and purification of acidic protease was studied in fixed and fluidised beds contacted with equivalent concentrations and volumes of protease in clarified and unclarified fermentation broths. The ionic strength of the broth was low and did not compromise ion exchange adsorption. Protein adsorption capacities were similar for both contactors (about 0.14 g/mL), but specific adsorption of protease activity was greater in the fixed bed (3.4×10^5 Units/mL) than in the fluidised bed (2.9×10^5 Units/mL).

Figures 3 and 4 illustrate the fixed bed elution profiles of protease and total protein recovered from DEAE Spherodex (6 mL batches) contacted with clarified or unclarified broth in fixed and fluidised beds respectively. Identical volumes were chosen for the desorption studies to facilitate a direct comparison of elution profiles. Treatments with 0.18 M NaCl in fixed beds served to displace loosely bound impurities (largely whey protein), whilst protease activity predominately eluted in 0.5 M NaCl. The overall recovery of protease in the 0.5 M NaCl was similar for fixed and fluidised systems (50 and 48% of the respective loads), but the product specific activity recovered from the former was less than for the fluidised contactor (9,000 versus 13,000 Units/mg protein). The higher percentage of bound protease eluted from the fluidised bed adsorbent (74% as against 64% for the fixed bed) may reflect product autolysis during the extended time-scales required for adsorption to the fixed bed (2.5

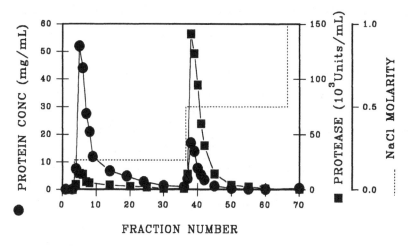

FIGURE 3: ELUTION PROFILE FROM A FIXED BED

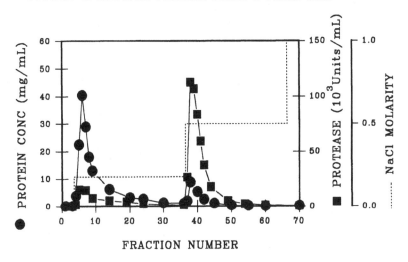

FIGURE 4: ELUTION PROFILE FROM A FLUIDISED BED

FIGURE 5:

SDS-PAGE ANALYSES OF THE
PROTEASE PURIFICATION

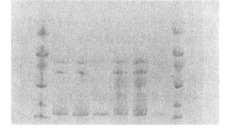

- 94k
- 67k
- 43k
- 30k
- 21k
- 14k

TRACKS: 1 2 3 4 5 6 7 8

times greater). Autolysis of purified and partially purified protease solutions in refrigerated storage in solution is common (data not shown).

Figure 5 is a print of SDS-PAGE analysis of crude and partially purified protease recovered from fixed and fluidised bed contactors. Tracks 1 and 8 contain protein standards of known molecular mass (14-94k Daltons). Tracks 2 and 3 contain two samples taken from the 0.5M NaCl elution following fluidised bed adsorption, whilst track 4 contains original fermentation feedstock clarified for analysis. Tracks 5 and 6 contain material similarly recovered from the fixed bed contactor, and Track 7 contains the original feedstock. Comparison of Tracks 4 and 7 with 2,3,5 and 6 confirm that both processes significantly concentrate the protease. However, all purified samples are contaminated with residual whey β-lactoglobulin. This results from the volume of the step gradient (minimised to maximise product recovery) and the close proximity of isoelectric points for β-lactoglobulin (pI = pH 5.2) and the protease (pI = pH 4.2). The observed difference in specific activity between fixed and fluidised bed products is confirmed by Coomassie blue visualisation of additional impurities in the molecular weight range 14 to 35k Daltons in Tracks 6 and 7. Purified protease has a molecular mass of 39k Daltons. It is not presently known if the co-purifying molecular species of 53k Daltons visible in Tracks 2,3,6 and 7 has proteolytic activity.

CONCLUDING DISCUSSION

The performance of DEAE-Spherodex has been compared in fixed and fluidised beds for the recovery of an acidic protease from Yarrowia lipolytica grown in pilot scale fermentations. Enzyme recoveries were similar for both contactors, but purity and specific activity was highest for enzyme recovered from the fluidised bed. Such product quality might be sufficient for some applications (e.g. diagnostics, food processing etc.), but further purification by conventional means (eg affinity chromatography) is possible (data not shown). High recirculation rates enabled adsorption to be completed in the fluidised bed at least 2.5 times more rapidly than in the fixed bed. The accommodation of whole broth in the fluidised bed eliminated the broth clarification required of fixed bed contactors. When required, this is best achieved in a microfiltration process, but throughput is severely restricted by the presence of PPG antifoam in the feedstock (6).

Direct product adsorption in recirculating fluidised beds has proved very useful for the primary separation of bulk protein fractions on agarose-, agar-, or cellulose-ceramic composites contacted with suspensions of disrupted yeast (6,9). However, such composites are not easily sterilised and may be unsuitable for certain products. This preliminary study of DEAE-Spherodex LS has demonstrated a suitability for fluidised bed adsorption of protein products. The material is robust, and may be readily sanitised and/or sterilised (data not shown). Integration of a recirculating fluidised bed with a batch fermenter (Figure 1b) will uniquely facilitate the continuous recovery of acidic protease throughout the productive fermentation (Figure 2). This will serve to enhance system productivity by eliminating concentration-dependent feedback control of protease synthesis. In addition, reduced opportunities for autolytic degradation of the product, and elimination of solid-liquid separation as a mandatory first step of product recovery, will benefit process productivity. The acidic protease from Yarrowia lipolytica is an excellent experimental model to demonstrate the virtues of the direct recovery of labile products from whole broths by fluidised bed adsorption.

Acknowledgement: We gratefully acknowledge the support of SERC and ICI Pharmaceuticals in the form of a CASE Research Studentship for PHM.

REFERENCES

1. Bartels, C. R., Kleiman, G., Korzun, N. and Irish, D. B. (1958). Chem. Eng. Prog. **54**, 49-52.
2. Wells, C. M. Patel, K. and Lyddiatt, A. (1987). In: Separations for Biotechnology (eds Verrall, M. S. and Hudson, M. J.), Ellis-Horwood pps 217-224.
3. Draeger, N. M. and Chase, H. A. (1990). I. Chem. E. Symp. Ser. **118**, 129-140.
4. Gibson, N. B. and Lyddiatt, A. (1990). In: Separations for Biotechnology 2 (ed Pyle, D. L.), Elsevier Applied Science pps 152-161.
5. Maude, B. and Whitmore, S. (1958). Br. J. Appl. Phys. **9**, 477-482.
6. Sansome-Smith, A. W., Huddleston, J. G., Young, T. W. and Lyddiatt, A. (1989). I. Chem. E. Symp. Ser. **113**, 209-226.
7. Kassel, B. and Meitner, P. A. (1970). Methods in Enzymol. **19**, 337-341.
8. Bradford, M. M. (1976). Anal. Biochem. **72**, 248-252.
9. Gibson, N. B. (1991). Ph.D. Thesis, University of Birmingham.

MICROBIOLOGICAL TREATMENT OF RADIOACTIVE WASTE ION EXCHANGE RESINS

RISTO JÄRNSTRÖM
Imatran Voima Oy
Loviisa NPP
07900 Loviisa, Finland

ABSTRACT

Loviisa Nuclear Power Plant has been operating for 25 reactor-years and has produced 180 m^3 of radio-active waste ion exchange resins, which is considered to be intermediate radioactive waste. For final deposition all this mass, produced during reactor life-time, has to be solidified in concrete. To get a required strength for the final product the ratio between resin and concrete is 1 : 10, which means that the volume of the waste is increased more than tenfold. The activity of the resin can be rather high, but the mass in milligrams of the active components is low. In Loviisa NPP we have for several years studied the possibility of treating ion exchange resins microbiologically and converting the matrix to gas and water and thus releasing the activity into the liquid phase, wherefrom it can easily be removed in a concentrated form by ion exchange.

INTRODUCTION

In nuclear power plants the produced radioactivity originates from two main sources: from the fuel and from impurities dissolved in the primary coolant. From the fuel originates fission products as Cs-134, Cs-137, iodines, noble gases and so on with varying half-lives from milliseconds to several years. Transuranic elements such as Pu-isotopes and higher ones come from the same source as does tritium. Impurities in the coolant come from the base materials and chemical additives, which are dosed to maintain the optimum chemical conditions to minimize the corrosion rate. Some of these impurities are neutron activated in the core as Co-59 to Co-60 and therefore give a high contribution to the dose-rates in the plant. To keep the dose-rates low in the primary system it is very important to continuously remove the coolant from the circuit for purification with ion exchanger systems and thus avoid a high build-up of activity on the surfaces of primary components.

In pressurized water reactors (PWR) and boiling water reactors (BWR) there are in use several ion exchanger systems for the clean-up purpose and these use both powdered resins and resins in bead form. The cycles of the resins in different systems varys from some days to about one year. This means that the nuclear power industry is a big consumer of resins and it uses ion exchanger resins from some cubic-meters to several tens counted on the year basis. It is obvious that all these resins classified as intermediate active waste must be deposited in a proper and safe way until the activity decays to the natural background level. The present idea is to reduce the final waste volume to a minimum and thus save depository space and reduce costs.

BACKGROUND

The idea of treating organic material with microbes arose in the early eighties when there was a fear that low level waste could produce in the depository explosive conditions by microbial production of methane. The whole question was put up-side-down;

"Why not feed the waste to microbes before deposition so there will be nothing to eat anymore and thus avoid production of explosive gases".

Successful tests with ordinary burnable waste were performed in pilot-plant tests and a volume reduction factor of about 15 was achieved.

During these tests a new idea arose. Why not try the microbiological method on waste resins? It was commonly known that plastics and especially resins were slightly affected in nature by microbiological processes, but the decomposition speed was too slow. Anyway we took the advice and recipe from **"The big cookbook"**:

Within time all organic material will disintegrate !

"Ashes to ashes, dust to dust"

We had to find synergism and conditions to speed up the process of natural decay of resins.

LABORATORY AND BENCH-SCALE TESTS

In 1986 laboratory tests with resins were started, which showed that the BOD-demand was suprisingly high. Furthermore it was noticed that "biological regeneration" of resins was easy to perform. This means that the functional groups were biologically decomposed and thus releasing the activity to the effluent, wherefrom it could be easily separated by ion exchange. But our goal was to get a complete disintegration of the organic matrix and to produce as endproduct only gas and water. Ion exchange resins are ideal, because their consistency is homogenous and almost no ash is left after decomposition - everything is decomposed to gas and water and activity is released in the water phase.

This activity released in the water phase - mainly the longlived isotope Cs-137 - can be specifically separated by a cesium removal process developed in Loviisa with a decon- tamination factor of over 1000. This process is now in full scale operation in removing Cs-137 and Cs-134 from evapo- rator bottoms and thus decreasing the activity level so low that the effluents can be released from activity control.

PILOT PLANT TESTS

In summer 1988 we started the first tests at pilot-plant scale with two 1,5 m^3 reactors. At first we chose the anaerobic method for disintegration of resins, which had shown hopeful prospects in smaller-scale operation. Much

work was done to develop the equipment to a technically
satisfactory state; many different pre-treatment modifications were made to optimize the process conditions. The
microbes chosen were a naturally appearing spectrum of
different species which were cultivated for this purpose in
these reactors. To survive they have to obtain energy and
"building bricks" and to achieve this one can force them to
eat almost anything. **"Eat or die"** was the main concept in
this procedure.

After a couple of years of testing, harassing the
microbes and combining different options we were getting
very promising results. After realizing some important facts
about microbe behavior we were able to achieve stochiometric
gas production in correlation with supplied food. One
important fact was the surface active properties of powdered
resins in making flocs; due to surface forces repelling the
microbes the microbiological activity was initially low.

Later during the investigation several new process
parts were added and combined as pre-treatment, intermediate treatment and an aerobic process to optimize the
decomposition and to increase the yield.

We have had these two reactors in operation for three
and a half years, one for styrene based resins and the other
for acrylic resins.

During this time there has been a feed of more than $1m^3$
of resin in each reactor into a liquid volume of 1.5 m^3. The
resin residue is for the moment less than 10 % of the feed
amount and the calculated half-life for the resins in this
process is 7 - 10 days. For the reactor capacity we
estimate 1 - 2 kg of resins per cubicmeter per day.

BENEFITS OF MCROBIOLOGICAL SYSTEM

When the developement of the system was started, several
comparative analyses were made. When comparing this
microbiological system with an incinerator process or "wet-
ox" process for disintegration of waste resins we found
several disadvantages in these processes compared with the
microbiological approach. The explosion risk with the "wet-

ox" process and the enormous off-gas handling in both processes increase costs. In the microbiological process everything happens in modest chemical conditions; no strong acids or oxidizing agents are needed and an expensive off-gas treatment is not needed. The only end products produced are water and non-active methane - carbon dioxide gas mixture which is non-active and can be led out through the ventilation stack or to a burner. Combining with the Cs-removal process the final deposition volume can be reduced by a factor of 1000. This means saving money and environmental problems are reduced. This method can of course be applied for normal inactive waste resins which are in many countries regarded as problem waste.

REFERENCES

E.H.Tusa, Microbial Treatment of Radioactive Waste at the Loviisa NPP., Proceedings of the Joint International Waste Management Conference, Kyoto, Japan, October 22-28,1989.

RESOLUTION OF POTATO TUBER ESTERASE GLYCOFORMS ON FAST-FLOW CELLULOSE ION EXCHANGERS

QING-WU XIE AND L. JERVIS
Department of Biological Sciences, Polytechnic South West
Drake Circus, Plymouth, Devon, PL4 8AA, U.K.

ABSTRACT

Potato (Solanum tuberosum) tubers contain a family of
soluble glycoproteins with esterase activity. These can be
resolved by FPLC on a Mono Q column. Partial resolution of
the glycoforms can be achieved using low cost fast-flow
cellulosic ion exchangers (Whatman DE 92). The ability of
ion exchangers, designed for process scale chromatography,
to separate glycoforms illustrates the excellent kinetic
characteristics of these adsorbents.

INTRODUCTION

Potato tubers contain up to 45% of their soluble protein as
a heterogeneous group of dimeric glycoproteins with esterase
activity that have been given the trivial name patatin (1-
3). Patatin has been purified by a number of different
procedures (4-6) and the purified protein has been subjected
to detailed structural analysis (1,4,7). Different isoforms of patatin have been resolved from one another and
have been shown to be very similar with considerable
polypeptide sequence homology and strong immunological
cross-reactivity (1,2). Patatin glycoforms provide a useful
model system for the assessment of ion exchange adsorbents.
They are available in large quantities and can be detected
specifically after resolution by polyacrylamide gel
electrophoresis (8). The separation of protein isoforms can

be achieved using modern high resolution chromatographic media designed for HPLC or FPLC (9,10). Such media may consist of porous microbeads, pellicular materials, micropellicular aggregates (perfusion media), or membranes (9-11). Regardless of type, high resolution media are characterised by short diffusion pathways, rapid mass transfer rates and fast adsorption-desorption kinetics. In general such media are expensive. Media based on porous, medium or large-sized particles usually have high capacity and may be designed for fast flow but they have relatively long diffusion pathways, slow mass transfer rates and may have slow adsorption-desorption kinetics (12). These media are usually relatively inexpensive and their characteristics are appropriate to medium or process-scale protein purification.

The 90-Series fibrous cellulose ion exchangers produced by Whatman were designed for process-scale protein purification. The method of manufacture produces media with excellent flow properties and fast adsorption-desorption kinetics (16). Whilst using these media for large scale purification of potato tuber proteins, we detected partial fractionation of patatin glycoforms. The degree of resolution achievable was subsequently examined using small scale conditions and the results were compared with those obtained by FPLC on Mono Q.

METHODS

Potato tubers (Solanum tuberosum) var Maris Piper) (5 g) were washed, peeled diced and macerated in cold 10 mM pH 5.0 (for FPLC) or pH 5.5 (for DE 92) piperazine/chloride buffer (10 mL). The resulting slurry was filtered through muslin and centrifuged at 12000 g for 10 minutes. The clear supernatant was dialysed against two changes of the appropriate buffer and recentrifuged before application of 5.0 mL samples to either a Mono Q (5.0 cm x 0.5 cm) or DE 92 (8.0 x 0.5 cm) column. Columns were eluted as described in

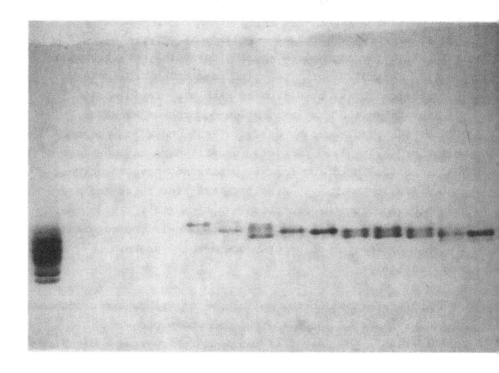

Figure 1. Resolution of patatin glycoforms using Mono Q FPLC

The plate shows a polyacrylamide gel slab stained for esterase activity. The lane on the extreme left is a sample of potato tuber extract. Other lanes are consecutive fractions eluted from a Mono Q column with a linear salt gradient of 0 to 200 mM NaCl in 10 mM pH 5.0 piperazine – chloride buffer. Flow rate was 0.4 mL min^{-1}.

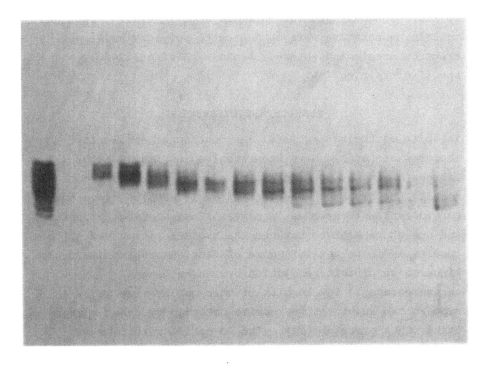

Figure 2.   Resolution of patatin glycoforms using DE-92 ion
exchange chromatography

The plate shows a polyacrylamide gel slab stained for
esterase activity. The lane on the extreme left is a sample
of potato tuber extract. Other lanes are consecutive
fractions eluted from a DE-92 column with a stepwise salt
gradient of 20 to 200 mM NaCl in 10 mM pH 5.5 piperazine –
chloride buffer. Gradient steps were 10 mM up to 120 mM
NaCl and 20 mM thereafter. Flow rate was 1.0 mL min^{-1}.

the legends to Figures 1 and 2. Fractions (0.5 mL) were collected and assayed for esterase activity (3). Fractions containing esterase were subjected to polyacrylamide gel electrophoresis and esterase bands were visualised by specific staining (3).

RESULTS AND DISCUSSION

In spite of their structural and functional similarity, the isoforms of patatin have been resolved partially by DE 92 fibrous cellulose ion exchanger and more fully, but not totally, resolved by Mono Q (Figures 1 and 2). Given the considerable differences in physical and chemical properties and design parameters between the two separation media, the results indicate the influence of fast adsorption-desorption kinetics on protein resolution by ion exchange chromatography. The results of this qualitative study support the quantitative results obtained by other workers using other proteins (12). The 90-Series fast flow ion exchange celluloses clearly represent an advance in media design and should prove valuable aids to high resolution protein purification at medium and process scale.

REFERENCES

1. Paiva, E., Lister, R.M., & Park, W.D., Plant Physiol., (1983), **71**, 161-168.

2. Park, W.D., Blackwood, C., Mignery,G.A., Hermodson, M.A., & Lister, R.M., Plant Physiology, (1983), **71**, 156-160.

3. Racusen, D., Can. J. Bot., (1983), **61**, 370-373.

4. Racusen, D., & Foote, M., J. Food Biochem., (1980), **4**

5. Kosier, T., & Desborough, S.L., Plant Physiol., (1981), **67**, S-92.

6. Jervis, L., Shepherd, A.L., & De, Maine, M.J., Biochem. Soc. Trans., (1986), **14**, 1096-1097.

7. Racusen, D., & Weller, D.L., J. Food Biochem., (1984), **8**, 103-107.

8. Racusen, D., (1986) Can. J. Bot., (1986), **64**, 2104-2106.

9. Regnier, F.E., & Gooding, K.M., Anal. Biochem., (1980). **103**, 1-25.

10. Richey, J., Amer. Lab., (1982), **10**, 104-129.

11. Sofer, G.K. & Nyström, L.-E. (1989), Process Chromatography - A Practical Guide, Academic Press, London.

12. Levison, P.R., and Clark, F.M. in "Pittsburgh Conference Abstract Book", Pittsburg Conference, Pittsburgh, (1989) p. 752.

ISEP CONTINUOUS CONTACTOR FOR CONTINUOUS ION EXCHANGE

STEVE WEISS
Advanced Separation Technologies,
5315 Great Oak Drive, Lakeland, FL 33801, U.S.A.

ABSTRACT

This paper explains how truly continuous ion exchange can be achieved using the ISEP contactor and its advantages over the traditional fixed bed system are highlighted. Correct design of the resin chamber configuration leads to greatly reduced resin inventories with all the resin continuously in use. Two case studies illustrate the application of continuous ion exchange to the extraction of lysine from fermentation broth and the large scale production of potassium nitrate from potassium chloride.

INTRODUCTION

A truly continuous contactor can be used for ion exchange and other adsorption applications. It eliminates the need for valves and controllers which are inherent to fixed bed and simulated moving bed systems. It allows the best possible ionic separation to be achieved with the minimum amount of ion exchange resin and regenerating chemicals. For complex separation applications, this contactor can commercially duplicate any technology developed in a laboratory on a fixed bed column. This includes countercurrent and split flow adsorption and regeneration. One example is the ISEP® Contactor developed by Advanced Separation Technologies Inc.

Specifically, ISEP continuous operation brings the following advantages:

(1) continuous steady-state, non-interrupted process flows enter and exit the ISEP,
(2) decreased sorbent inventory requirements, as all sorbent is active in a zone,
(3) higher desorption product fluid (eluate) concentrations from countercurrent contacting,
(4) higher eluent efficiency, decreased eluent usage from countercurrent contacting,
(5) high tolerance to suspended solids in feed liquids, upflow contacting possible
(6) minimised wash-solution requirements resulting from countercurrent operation
(7) zone lengths are individually optimised connecting beds in series,
(8) zone contact areas are variable, connecting beds in parallel.

The ISEP® Contactor consists of 30 fixed bed columns which rotate (constant rotational speed) on a carousel arrangement. The columns are connected to an upper and lower portion of the distributor which also rotate with the carousel arrangement. Each of the two distributors has a stationary segment consisting of 20 openings called ports. Liquid flows from a port into the column and exits through the other port. As a column leaves a port area, another column enters the same port area. This geometry produces **uninterrupted** flow of fluid through the port and into the columns. (See Fig. 1).

Column allocation depends on the adsoprtion and regeneration properties of the product and the required performance target. Columns can be arranged to have parallel or series operation and countercurrent regeneration can be performed to optimise regenerant usage and reduce osmotic shock on the resin.

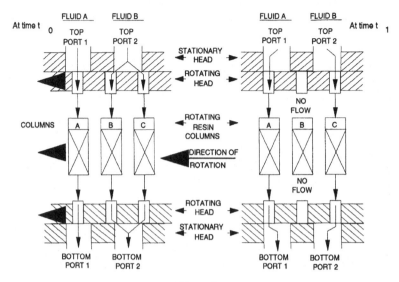

Fig. 1

COMPARISON WITH FIXED BED OPERATION

Consider a fixed bed of solid sorbent as shown in Fig. 2.

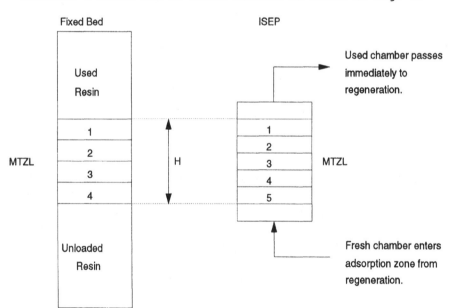

Fig. 2

From Fig. 2 we see that the Mass Transfer Zone Length (MTZL) is divided into slices with each slice representing an adsorption chamber on the ISEP machine. Each chamber has a height H typically 0.3 - 1.2 m. in length. The total theoretical MTZL is h = 4h. In Fig. 3 the chambers move from right to left and become increasingly loaded as they move through the adsorption zone. The resin flow rate, R, determines the speed of rotation of the carousel. As a chamber moves beneath the first adsorption point it receives fluid flow and adsorption commences. The ideal ending profile should represent a saturated sorbent and will be a vertical line at the right hand edge of the Pass 1 chamber. To maximise the sorbent capacity, the leading (leftmost) chamber must exit port #1 saturated and for this reason we require 5 h bed height in this five port case, to always contain the mass transfer zone length within the adsorption zone resin bed. It is, therefore, advantageous to divide the MTZL into as many small segments as is reasonable.

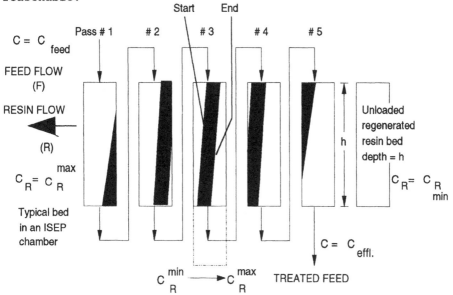

Fig. 3 - Mass transfer profile: ISEP

(1) Mass balance $F \times (C_{feed} - C_{effl}) = R \times (C_R^{max} - C_R^{min})$
(2) Number of passes = $n = H/h) + 1$
(3) Shaded area represents the movement of the adsorption front, left to right, in a chamber while it is in a particular pass.

In a fixed bed system the total cycle includes steps for adsorption, regeneration, washing and rinsing. There must be sufficient inventory in the bed to ensure that the mass transfer zone length (MTZL) does not exit the bed during the regeneration time of the 'offline' bed. In the case of the continuous ISEP system the time for regeneration and sorbent inventory are independent and can be optimised separately.

Elution efficiency in countercurrent systems comes from operating the elution at a single point on the elution profile curve shown in Fig. 4. Imagine a column of exhausted resin that is put into regeneration and the eluent peak beginning to emerge from the column effluent; now imagine moving the resin in the opposite direction at such a rate as to maintain the peak effluent level. This is how the ISEP elution zone is designed to work.

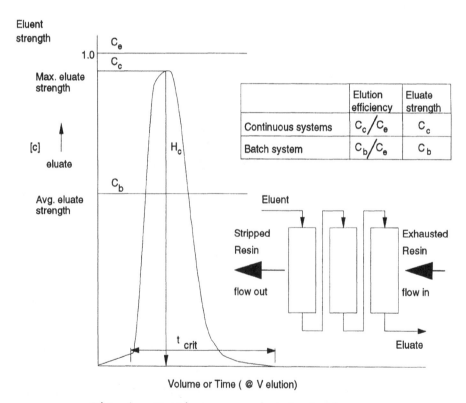

	Elution efficiency	Eluate strength
Continuous systems	C_c/C_e	C_c
Batch system	C_b/C_e	C_b

Fig. 4 – Continuous vs. batch elution

Large reductions in sorbent inventory result from proper design of continuous ion exchange systems like the ISEP as the inventory is usefully employed at all times.

The following two cases illustrate the benefits of the ISEP continuous system over the fixed bed type.

Case No. 1 - Potassium Nitrate Production from KCl and HNO3

Previously potassium nitrate was commercially made by solvent extraction techniques. Ion exchange techniques were known but found to be uneconomical because 150,000 ppm of ion transfer was necessary. Now the ISEP® Continuous Contactor is commercially producing high purity potassium nitrate because of substantially reduced capital costs vs. solvent extraction. The ISEP® uses 1/10 the amount of ion exchange resin used in fixed bed columns. The ISEP scheme illustrated in Fig. 5 produces potassium nitrate at a strength of 15-20%.

Potassium Nitrate Production ISEP Flowsheet

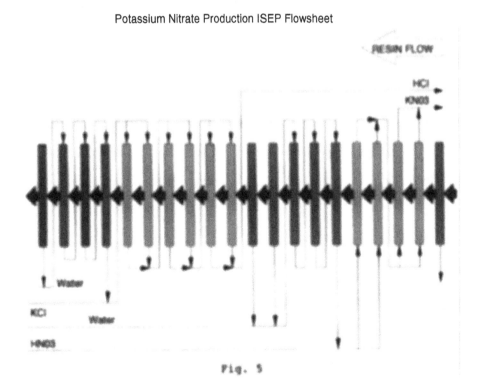

Fig. 5

Table 1	Fixed Bed	ISEP
Cycle time (h)	6	1
Bed depth (m)	5	0.6
Resin volume (m³)	147	4.6
Capital cost ($)	810,000	650,000
Resin cost ($)	441,000	14,000

Case No. 2 - Purifying lysine from Fermentation Broth

This is a traditional ion exchange application. The ISEP® uses 1/22 the amount of ion exchange resin as fixed bed column systems and produces a product solution that is 20% higher in concentration. Total plant capital investment is 2/3 less than fixed bed columns. A 45,000 TPY lysine production plant with ISEP® Contactors started operation in 1990. Fig. 6 illustrates the ISEP® process scheme for this application. Note pH adjustment is carried out in the adsorption zone to maximise lysine recovery.

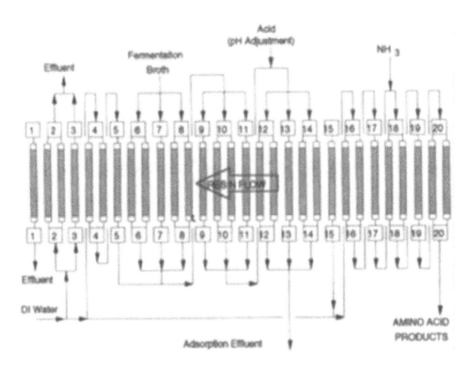

Fig. 6

Table 2	Fixed Bed	ISEP
Cycle time (h)	4	1.4
Resin volume (m^3)	549	32
Capital cost ($)	1,600,000	1,500,000
Resin cost ($)	4,555,000	128,000
Production concentration (g/L)	100-130	180-200
Ammonia eluent efficiency (%)	30-35	65-70
Product recovery (%)	85-90	95 min.

NOMENCLATURE

Cfeed	Concentration of adsorbate in feed	eq/L
Ceffl	Concentration of adsorbate in effluent	eq/L
CRmax	Concentration of adsorbate on resin (final)	eq/L
CRmin	Concentration of adsorbate on resin (initial)	eq/L
Ce	Maximum theoretical eluate strength	eq/L
Cc	Maximum eluate strength (continuous ISEP)	eq/L
Cb	Maximum eluate strength (batch)	eq/L
F	Liquid feed flow rate	L/s
H	Total height of theoretical mass transfer zone length	m
h	Height of each chamber resin bed	m
n	Number of chamber passes	
MTZL	Mass transfer zone length	m
R	Resin flow rate	L/s

ADSORPTION OF TRACE METALS ON MODIFIED ACTIVATED CARBONS

M. STREAT
Department of Chemical Engineering
Loughborough University of Technology
Loughborough, Leicestershire, LE11 3TU

J.K. NAIR
The Ralph M. Parsons Company Limited,
Kew Bridge Road,
Brentford, Middlesex, TW8 0EH

ABSTRACT

Chemical and high temperature oxidation of the surface functional groups of activated carbon significantly enhances the weakly acidic ion exchange capacity. This treatment increases the sorption of trace metals from aqueous solutions at near-neutral and alkaline pH values. Distribution coefficients of Sr, Cs and Co are greater than 10^4 cm^3/g at trace concentration levels of 0-1 mg/L in solution. The selectivity series for multivalent ions is $Eu^{3+}>Am^{3+}>Sr^{2+}>UO_2^{2+}>Cs^+$ in the pH range 2-8.

INTRODUCTION

Activated carbon is widely used for the treatment and decontamination of aqueous solutions, especially for the removal of colour, odour and trace contamination from potable water, domestic water and industrial waste waters. Particular applications include the removal of organic compounds and dyestuffs from effluent solutions. Separately, the removal of toxic metals has been investigated with some success. For example, Dobrowolski, Jaroniec and Kosmulski (1986) have recently studied the adsorption isotherm and the kinetics of the sorption of Cd from aqueous solutions using surface modified activated carbons in the pH range 2-8. However, the application of activated carbon to trace metal decontamination is still limited. By far the most widely developed large-scale application of activated carbon is the recovery of gold from dilute cyanide leach solutions. This process has been extensively reviewed by Bailey (1987).

The mechanism of adsorption of soluble molecular species is by physisorption, though chemisorption is more likely for ionic species such as trace metals in dilute salt solutions. Undissolved or finely divided particulates are often sorbed by colloidal precipitation or by electrostatic interactions at the surface or within the pores of the adsorbent material.

Activated carbon is a generic term applied to a family of porous carbonaceous materials, none of which are characterised by structural formula or by chemical analysis. They are usually distinguished by their source material and by the method of activation. Most commercial active carbons are derived from either coal, wood or coconut shell, though there is an enormous range of source materials that could be used. Activated carbons used in liquid phase applications are generally microporous, with a large surface area of about 1000 m^2/g and a pore volume of about 0.5 cm^3/g. The surface chemical structure of activated carbon is dependent on the method of activation, especially the activation temperature. Temperatures well below 700°C will produce weakly acidic functionality and this is attributed to carbon/oxygen groups on the surface and within the pores. The precise surface chemical structure is complex and has recently been reviewed by Bansal, Donnet and Stoekli (1988).

There is considerable evidence in the literature to suggest that activated carbons behave as typical weakly acidic ion exchange materials in aqueous solution. Garten and Weiss (1957) have described the ion exchange and electron exchange properties of activated carbon and they reviewed the wide ranging potential applications that had emerged in the extensive literature on the subject in the first half of the twentieth century. This important field of ion exchange seems to have remained dormant thereafter because of the advent of the now familiar synthetic ion exchange resins based on crosslinked polystyrene/divinylbenzene and poly(methylmethacrylate) co-polymers. It is the need to develop cheap disposable ion exchange materials for the decontamination of effluent solutions and the requirement to recover precious metals from tailings that has prompted a renewed and lively interest in activated carbon at the present time.

Sigworth and Smith (1972) and House and Shergold (1984) have shown that polyvalent metals can be adsorbed from aqueous solution under carefully controlled pH conditions. The effective pH range is the near - neutral to alkaline region and this is typical of conventional ion exchange materials containing carboxylic and phenolic functional groups. There is strong evidence to suggest that these functional groups are also to be found on the surface of activated carbon, together with other weakly acidic groups such as quinones and lactones. The weak acid exchange capacity of commercial activated carbons is about

1-2 m equiv/g in the pH range 2-10. This compares unfavourably with a typical commercial polymeric weak acid ion exchange material based on a poly(methylmethacrylate) structure (e.g. Amberlite IRC 50 manufactured by the Rohm and Haas Company) which possesses a carboxylic cation exchange capacity of about 10 m equiv/g in alkaline conditions.

Some early work has shown that it is possible to modify the functional groups on activated carbon by surface oxidation. Puri (1962) has shown that elevated temperature treatment of activated carbons in the presence of air will enhance the acidic surface functionality by the addition of further carbon/oxygen groups and he pointed out that this could lead to improved cation exchange capacity. Puri, Singh and Mahajan (1965) also showed that surface oxidation was achieved by chemical treatment of carbons in nitric acid solution and that this produced an increase in cation exchange capacity. It was this early work which led us to the present study of surface modification of activated carbon in order to'tailor-make materials for use in the treatment of effluents and waste waters for the removal of toxic and harmful metals.

EXPERIMENTAL

Two separate samples of coconut shell activated carbon, designated HW (Hopkin and Williams Ltd) and Pica (Le Carbone S.A.), have been studied in the 'as-received' condition and after oxidation by heat treatment in air and digestion in concentrated nitric acid. The surface area, pore volume and average pore diameter of carbon samples was measured using a Micromeritics ASAP 2000 automatic analyser. Carbon samples were heated in air in a furnace at temperatures up to 450°C for variable periods of time. This resulted in weight loss and the residual carbon was recovered, washed and stored in a controlled atmosphere prior to use. Other carbon samples were digested in 8 M HNO_3 at 100°C for variable time. This also caused some weight loss and the residual carbon was recovered, washed and stored prior to use. We also tested a sample of Amberlite IRC 50 weak acid resin for comparison.

Aqueous solutions were buffered to particular pH values and spiked with radio-active tracers of Sr, Cs and Co in order to facilitate rapid analysis by gamma spectrometry. Trace metal concentrations were usually in the range 0 - 10 mmol/L. The selectivity of modified activated carbon was checked with a simulated nitrate effluent solution containing trace levels of Sr, Cs, Co, Eu, U and Am. This solution was also analysed by gamma spectrometry.

Adsorption isotherms were obtained by performing batch experiments contacting mg amounts of activated carbon in the size range 250-350 μm with 30 cm^3 of spiked aqueous solutions at 21 ± 0.5°C. Batch samples were left in contact for 24 h to reach equilibrium.

RESULTS AND DISCUSSION

The physical properties of 'as-received' and modified activated carbons are given in Table 1. Surface oxidation of the carbon at elevated temperature does not reduce the overall surface area significantly but at 400°C after 16 h the micropore area is reduced and there is an increase in pore volume and average pore diameter. Surface oxidation by digestion in 8M HNO$_3$ does seem to reduce total and micropore surface area, and the pore volume but the average pore diameter is hardly affected. These changes in physical properties may in part explain some of the marked improvements in metal adsorption, though it is more likely that surface oxidation is the reason for enhanced ion exchange capacity.

Figs. 1-3 show the adsorption isotherms for Sr, Cs and Co on 'as-received' samples of HW and Pica carbons as a function of liquid phase pH value. Metal ion uptake is unfavourable at a pH value of 3 since this is below the dissociation pK value of the surface acidic functional groups e.g. carboxylic and phenolic. At near-neutral and alkaline pH values, i.e. close to and above the pK of carboxylic and phenolic groups, the sorption of univalent and divalent metals is appreciable. The distribution coefficients for Sr, Cs and Co are greater than 10^4 cm^3/g at trace concentration levels of 0-1 mg/L in solution. The distribution coefficient is expressed as mg of metal/g activated carbon divided by mg of metal/cm^3 of solution, i.e units of cm^3/g.

The Sr ion exchange isotherms of air oxidised samples of activated carbons are compared with untreated carbons in Figs 4 and 5. There is a marked increase in sorption capacity at pH 9.7 and this is most striking with Pica samples heated at 400°C in air for several hours. Surface oxidation with HNO$_3$ produces an even greater number of weakly acidic functional groups on the carbon and we have detected an order of magnitude increase in the sorption of Sr from a buffered aqueous solution at pH 9.7 (see Fig 6). The effect of pH value on the sorption of Sr is pronounced as shown in Fig 7. The amount of Sr sorbed as a function of pH is given for an HNO$_3$ oxidised carbon in comparison to Amberlite IRC 50. The polymeric ion exchange resin contains carboxylic functional groups with an apparent pK value in the range 4-6. Notice how the Sr uptake falls at all pH values below about 7. The oxidised carbon HW (8M/8), however, sorbs

268

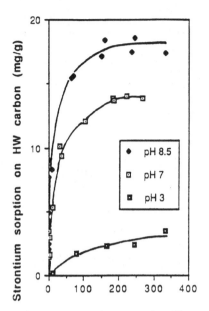

Figure 1 Effect of pH on Sr sorption on HW (as received).

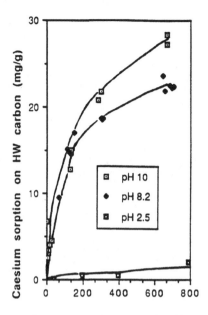

Figure 2 Effect of pH on Cs sorption on HW (as received).

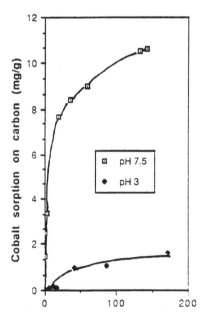

Figure 3 Effect of pH on Co sorption on HW (as received).

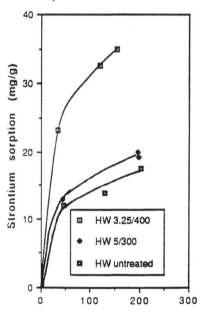

Figure 4 Adsorption isotherms for Sr sorption from buffered solution (pH = 9.7) on heat treated HW carbon.

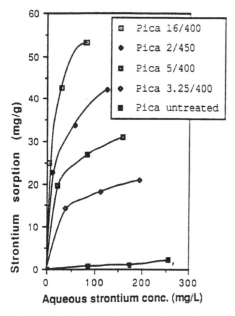

Figure 5 Adsorption isotherms for Sr sorption from buffered solution (pH=9.7) on heat treated Pica carbon.

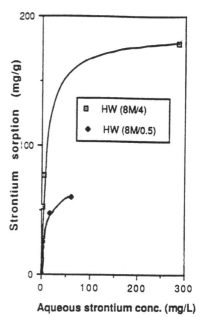

Figure 6 Adsorption Isotherms for Sr sorption from buffered solution (pH=9. on HNO3 treated carbon.

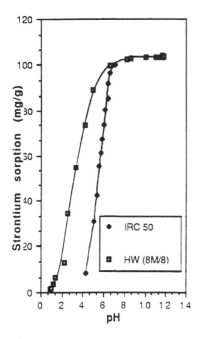

Figure 7 Effect of pH on adsorption of Sr from Sr(NO3)2 0.004 M solution on IRC 50 and HW (8M/8)

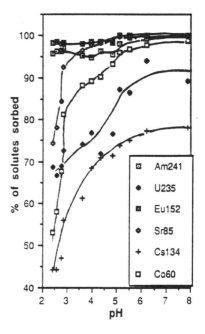

Figure 8 Effect of pH on uptake of solutes from simulant solution on HW(8M/8)

Sr to some extent in the pH range 2-12 indicating the existence of other weakly acid groups capable of exchanging cations at low pH values. Phenolic groups (pK about 9-10) contribute to the ion exchange capacity in alkaline media (i.e pH range 8-12). The remarkably good cation exchange performance of nitric acid oxidised carbon is illustrated in Fig 8, which shows the selective recovery of traces of radionuclides from a simulated aqueous nitric acid effluent solution as a function of pH value. Am^{3+} and Eu^{3+} are strongly sorbed in the pH range 2-8 and Sr^{2+}, UO_2^{++}, Co^{2+} and Cs^+ show a fall-off at pH values below about 6. The selectivity series for multivalent ions appears to be $Eu^{3+}>Am^{3+}>Sr^{2+}>UO_2^{2+}>Cs^+$ in the pH range 2-8.

CONCLUSIONS

The sorption of metal ions at trace level concentration in aqueous solution is greatly enhanced by the use of surface oxidised activated carbon. The concentration of the contaminant metals onto carbon will facilitate pollution control in the nuclear, water, waste water and effluent treatment industries. Methods of recycling the contaminated carbon are presently being developed. One possibility is to incinerate spent carbon in order to produce a highly concentrated metal oxide ash for immobilisation in cement prior to ultimate waste disposal.

REFERENCES

Bailey, P.R., (1987), Application of Activated Carbon to Gold Recovery, in The Extractive Metallurgy of Gold in South Africa, Ed. G.G. Stanley, Volume 1, South African Institute of Mining and Metallurgy, Johannesburg, 379-614.

Bansal, R.S., Donnet, J.B., and Stoekli, F (1988), Active Carbon, Marcel Dekker, New York.

Dobrowolski, R., Jaroniec, M., and Kosmulski, M., (1986), Carbon, 42, (7), 15-20.

Garten, V.A., and Weiss, Đ.E., (1957), Reviews of Pure and Applied Chemistry, 7, 69-122.

House, C.I., and Shergold, H.I., (1984), Trans. Instn. Min. Metall., 93, C19.

Puri, B.R., (1962), Surface Oxidation of Charcoal at Ordinary Temperatures, 5th Carbon Conference, Pennsylvania State University, Pergamon Press.

Puri, B.R., Singh, S., and Mahahjan, O.P., (1965) J. Ind. Chem. Soc. 42, (7), 427-434

Sigworth, E.A., and Smith, E.B., (1972), Journal AWWA, June, 386.

271

ACKNOWLEDGEMENTS

The authors wish to acknowledge financial assistance from British Nuclear Fuels plc for this work.

Table 1
Properties of Activated Carbons used in this Work

	Surface Area (m²/g)	Micropore Area (m²/g)	Pore Vol. (cm³/g)	Av Pore Dia (nm)
Pica (as received)	1244	1052	0.463	1.49
Pica (3/400)*	1222	1273	0.591	1.46
Pica (2/450)*	1135	1240	0.614	1.37
Pica (16/400)*	1270	361	0.704	2.22
HW (as received)	1450	1213	0.528	1.46
HW (8M/1)**	1017	1230	0.560	1.39
HW (8M/2)**	649	772	0.360	1.39
HW (8M/9)**	215	243	0.120	1.38

* heat treated sample, e.g. Pica (3/400) was heat treated in air for 3 h at 400°C

* all samples were treated in 8M HNO_3 at 100°C, e.g. HW (8M/1) was reacted with 8M HNO_3 at 100°C for 1 h. HW (8M/2) and HW (8M/9) were reacted for 2 and 9 h respectively.

ELECTROCHEMICAL ION EXCHANGE

PAULINE M. ALLEN, NEVILL J. BRIDGER, CHRISTOPHER P. JONES,
FABIENNE M.-T. MENAPACE, MARK D. NEVILLE AND ANDREW D. TURNER
Applied Electrochemistry Department, AEA Industrial Technology
Building 429, Harwell Laboratory, Didcot,
Oxfordshire, OX11 0RA, England.

ABSTRACT

Electrochemical ion exchange (EIX) is a novel separation
process which has been developed at Harwell over a number of
years. In this technique, ion exchange material is
incorporated into an EIX electrode by means of a binder.
Absorption of ions is controlled by application of a potential
between the EIX electrode and a counter electrode. Elution is
achieved by polarity reversal - eluant chemicals are not
required. This enables multiple use of ion exchange capacity.
EIX has been demonstrated to absorb a wide variety of cations
and anions. Initially, work focused on treatment of nuclear
waste streams. This versatile technique is currently being
developed for a range of non-nuclear applications, such as
removal of heavy metals from industrial effluent, recycling of
precious metals, and water de-ionisation. This paper gives a
summary of the scientific principles involved, as well as
examples of operating experience.

INTRODUCTION

Electrochemical ion exchange (EIX) was first investigated as a

process for desalination of brackish water. Subsequently this

novel separation process has been developed at Harwell over a

number of years. In this technique, ion exchange material is

incorporated into an EIX electrode by means of a binder.

Absorption of ions is controlled by application of a potential

between the EIX electrode and a counter electrode. Elution is

achieved by polarity reversal - eluant chemicals are not

usually required. This enables multiple use of ion exchange capacity, through repeated absorption/elution cycling.

Initially, work focused on the treatment of nuclear waste streams. Here it is important not only to reduce the concentration of radionuclides to below legally-defined limits, but also to minimise the overall volume of waste produced. This versatile technique is currently being developed for a range of non-nuclear applications. These include removal of heavy metals from industrial effluent, recycling of precious metals, water de-ionisation, corrosive anion removal, and removal of nitrate from potable water. EIX is unique, in that exchange processes are controlled electrochemically. The use of regenerating chemicals is therefore minimised and it is possible to achieve large volume reduction factors (VRF's) since elution may be carried out in a single bed volume (BV).

APPROACH

Principles of EIX

In cation EIX, a weak acidic exchanger is bonded to a mesh electrode by means of an elastomeric binder. During the absorption cycle, the EIX electrode is made cathodic. Electrolysis of water generates a local alkaline environment within the electrode structure:

$$2H_2O + 2e^- \rightarrow H_2 + 2OH^-$$

The hydroxyl ions deprotonate the exchanger, generating active sites:

$$RCOOH + OH^- \rightarrow RCOO^- + H_2O$$

which then undergo exchange:

$$RCOO^- + M^+ \rightarrow RCOOM$$

The applied potential also induces migration into the ion exchanger, enhancing the kinetics, and enabling high utilisation of capacity.

Elution of cations from the EIX electrode is achieved by
reversing the polarity. This produces a local acidic
environment within the electrode:

$$2H_2O \rightarrow O_2 + 4H^+ + 4e^-$$

Electrogenerated protons displace metal ions from the weak
cation exchanger:

$$RCOOM + H^+ \rightarrow RCOOH + M^+$$

The presence of the electric field gradient enables virtually
complete elution.

An analogous set of reactions is invoked when an anion is
absorbed and desorbed within a weak base anion exchanger.

It should be noted that whenever the EIX electrode is
cathodic, the counter electrode is anodic (and vice versa).
In the absence of the ion exchanger, recombination of H^+ and
OH^- would simply occur.

Electrode Manufacture

A simple, reproducible method has been developed for the
fabrication of EIX electrodes. Dried ion exchanger is ground
in a centrifugal mill to give a powder (typical particle
size = 0.1 mm). The powder is then mixed with a solution
(18.75 wt.%) of a synthetic rubber (Kraton) in 1,1,1-
trichloroethane. The resulting slurry is then poured into a
mould containing a platinised titanium mesh electrode. The
electrode is left overnight, in order to allow the solvent to
evaporate.

Uncoated mesh is used for the counter electrode. Plastic
clips clamp the electrodes together, forming the EIX module.
Two experimental configurations have been used. In the first,
there is a counter electrode on either side of the EIX
electrode, whereas in the second, there is a counter electrode
on the front side of the EIX electrode, and a sheet of inert
material (e.g. polythene) on the back. The resulting module

can be immersed in stirred electrolyte (batch mode) or contained in a flow cell through which electrolyte can pass.

RESULTS

NUCLEAR APPLICATIONS

Caesium/Sodium Separation

A radionuclide commonly found in nuclear waste streams is ^{137}Cs. This element is accompanied by a second alkali metal, sodium, which is present at higher concentration. Routine ion exchange is unsatisfactory, because the sodium causes premature breakthrough. This difficulty may be overcome, however, by use of a selective ion exchange material.

Exceptional caesium decontamination results have been obtained with EIX electrodes constructed from amorphous zirconium phosphate. A feed solution containing 100 g/m^3 sodium and 10 g/m^3 caesium was passed through a cation EIX flow cell with the cell current maintained at 30 A/m^2. The concentrations of both species were dramatically reduced. The decontamination factors (DF's) for caesium and sodium were 5000 and 133 respectively. Even after Na$^+$ breakthrough, the concentration of Cs$^+$ in the effluent was below the level of detection (mg/m^3). Elution on polarity reversal resulted in a volume reduction factor (VRF) greater than 100.

Cobalt/Lithium Separation

Treatment of high concentrations of cobalt (100 g/m^3) by conventional EIX resulted in precipitation of Co(OH)$_2$ on the electrode, which caused poor kinetics. By using the radio-isotope ^{60}Co, it was demonstrated that low levels of cobalt could be removed effectively using EIX. Subsequently, modules incorporating amorphous zirconium phosphate have been applied to Co(II) solutions at concentrations below the solubility limit of the hydroxide. Cobalt has been successfully removed from a solution containing 100 mg/m^3 Co. The concentration of Co(II) in the effluent stream was below the level of detection (µg/m^3). DF's of $\sim$ 1000 were measured in some experiments.

Elution of the EIX module was found to be pH dependent.
Co(II) did not elute into distilled water (pH 7). However,
when the pH was adjusted to $\sim$ 2 by addition of nitric acid,
quantitative recovery of Co(II) was achieved.

In PWR streams, Li^+ is present as well as Co(II). Both
are removed by EIX with amorphous zirconium phosphate.
However, selective desorption is possible. Lithium hydroxide
is freely soluble in water, whereas the solubility of cobalt
hydroxide is only 0.0032 g/L. By maintaining a high pH
(11-12) during desorption, only Li^+ was eluted. The pH was
then reduced (1.5-2.5) to allow regeneration of cobalt. A VRF
in excess of 200 was achieved for a 100 mg/m^3 Co initial feed.

Zirconium Phosphate Hydrolysis

Amorphous zirconium phosphate possesses excellent cation
exchange properties for both caesium and cobalt.
Unfortunately, ion exchange is accompanied by phosphate
elution, which indicates hydrolysis of the exchanger:

$$Zr(HPO_4)_2 + 6OH^- \rightarrow Zr(OH)_4 + 2PO_4^{3-} + 2H_2O$$

The product, hydrous zirconia, also functions as a cation
exchanger, but of inferior performance. Consequently,
hydrolysis of zirconium phosphate leads to slower absorption
kinetics and reduced capacity, limiting electrode lifetime to
$\sim$ 6 years.

A number of possibilities have been examined, in order to
minimise this problem. It has now been demonstrated that
hydrous zirconia can be converted to zirconia phosphate within
an EIX electrode. This is achieved by absorption of
phosphoric acid, whilst the electrode is poised at anodic
potential. A clear end-point is observed when uptake of H_3PO_4
is complete. In one experiment, an EIX electrode containing
30 g of hydrous zirconia absorbed 3 g of phosphoric acid. The
resulting electrode removed Cs^+ ions with enhanced
efficiency.

This method has now been applied to zirconium phosphate
electrodes, depleted by hydrolysis. Ion exchange behaviour
was restored.

NON-NUCLEAR APPLICATIONS

As described above, EIX was originally developed as an
effluent treatment process for the nuclear industry.
Attention is now being focused on non-nuclear applications.
Use of water in industrial processes inevitably gives rise to
waste streams requiring treatment before discharge. Energy
efficient processes are needed to concentrate the toxic
contaminant in such a way as to minimise discharges to the
environment and to reduce the volume of any waste for landfill
disposal.

To this end, a screening exercise has been undertaken, to
establish whether EIX is a suitable technique for the removal
of heavy metals from solution.

Cadmium Removal

Excellent cadmium decontamination results have been obtained
by cation EIX, using electrodes constructed from IRC50. A
feed solution containing cadmium sulphate (10 g/m^3 Cd) was
passed through the flow cell with the current maintained at
30 A/m^2. Dependence on flow rate was examined. Even at
moderately high throughputs (7 BV/h) the cell was able to
remove > 99.99% of the cadmium present.

Cadmium elution was found to be pH dependent (c.f.
cobalt). At pH 7, the cadmium remained in the EIX electrode.
Lowering the pH by addition of sulphuric acid led to partial
release of cadmium ions into solution. However, the situation
was complicated by plating of cadmium metal onto the bare
counter electrode. In these preliminary experiments < 50% of
the cadmium was released from the EIX electrode into
solution.

Chromium Removal

The solution chemistry of chromium is complex: a number of
species both cationic and anionic may be present
simultaneously. For the purposes of this study, solutions
containing chromium as the chromate anion (CrO_4^{2-}) were
treated using anion EIX (IRN78L exchanger).

A feed solution containing potassium chromate (10 g/m³ Cr) was passed through the flow cell at varying flow rates (fixed current density) and at varying current density (fixed flow rates). This procedure allowed us to establish the optimum conditions for chromium removal. The absorption data for this feed were very encouraging. A DF in excess of 150 was observed at 0.52 L/h ($\sim$ 3.5 BV/h) over a range of current densities (30-60 A/m³). It is clear that a high current density and a low flow rate leads to a high degree of solution decontamination.

CONCLUSIONS

EIX has been demonstrated to absorb a wide variety of cations and anions from aqueous streams. Although originally applied to nuclear waste streams, it is clear that this technique has much wider applicability in industrial effluent processing and metal recovery operations.

Important features of EIX are high utilisation of the IX capacity, the electrical elution of ions, and multiple use of the exchange capacity through repeated absorption/elution cycling.

ACKNOWLEDGEMENTS

This research was funded by AEA Technology (Corporate Research), CEC (Research Programme on Radioactive Waste Management and Disposal), Nuclear Electric, Her Majesty's Inspectorate of Pollution, and members of the Effluent Processing Club (based at Harwell Laboratory).

HYDROXYAZO-DYES IN METAL IONS PRECONCENTRATION BY ION EXCHANGE

CORRADO SARZANINI*, ORNELLA ABOLLINO, EDOARDO MENTASTI

Department of Analytical Chemistry, University of Torino

Via Giuria 5, 10125 Torino, Italy

ABSTRACT

Eleven different sulphonated azo-dyes were investigated in their ability to chelate metal ions and were evaluated for an application to an ion exchange preconcentration technique. Cu (II) was chosen as a reference metal ion and its uptake efficiencies as a function of experimental parameters were evaluated. The different behaviours were explained in terms of the structure of each ligand. Some of the ligands were subsequently tested in the preconcentration of trace elements and natural water analysis. The suitability of the procedure studied for evaluating stability constants of metal chelates was also demonstrated.

INTRODUCTION

Several azo-dyes have been extensively used as metal complexing agents for preconcentration procedures such as the immobilization of metal chelates on solid sorbents, under adsorption or ion-exchange mechanisms, for trace metal ions determination (1-4). In this work eleven sulphonated azo-dyes, namely Acid Orange 8 (AO8), Orange II (O2), Orange G (OG), Acid Red 4 (AR4), Acid Red 8 (AR8), Chromotrope 2R(C2R), Acid Alizarin Violet N (AVN), Calmagite (CA), Palatine Chrome Black (PCB), Plasmocorinth (PL), and Nitrazine Yellow (NY) were evaluated for metal ion preconcentration techniques based on ion-exchange. All these dyes are characterized by a naphthyl group bound through the azo moiety to a phenyl or to another naphthyl ring. The aromatic rings carry substituents such as hydroxide, methyl, methoxy or halogen groups. Moreover the presence of negatively charged sulphonato moieties, located apart from the co-ordinating sites, allows to solubilize the ligands and their complexes in water and also to immobilize them on an anion exchange resin. Choosing Cu(II) as a reference metal ion, the uptake efficiency of the complexes on the resin, and therefore their subsequent recovery, were investigated as a function of the ligand structure, and of experimental parameters such as the ligand-to-metal ratio, the pH and the presence of interfering species. The results and differences were explained in terms of the structure of each ligand. The proposed method was tested on natural samples. The suitability of the procedure to evaluate the stability constants for the metal complexes was also demonstrated.

EXPERIMENTAL

Apparatus

A Plasma 300 Allied Analytical Systems (Waltham, MA, USA) inductively coupled plasma atomic emission spectrometer (ICP-AES) was used throughout for the determinations of the metal ions in all solutions. An Orion EA 920 pH meter equipped with a combined glass-calomel electrode was employed for pH measurements. All solutions were prepared with high purity water HPW (Millipore Milli-Q). High purity acids and ammonia were obtained with a quartz sub-boiling still (K. Kurner, Rosenheim, West Germany). The solutions were prepared and stored in high-density polyethylene vessels (Nalgene), previously cleaned with warm HNO_3 and rinsed with HPW. The resin was packed into Bio-Rad polypropylene Econo-Columns (5.0 mL). All sample preparations and manipulations were performed under a class-100 laminar flow hood.

Reagents

AO8, AR4, AR8, C2R, CA, AVN, PCB, PL and NY were purchased from Aldrich. O2 and OG were obtained from Kodak. The solid sorbent was a macroporous anion exchange resin (Bio-Rad AG MP-1, 100-200 mesh, chloride form). Stock metal ion solutions, 1000 mg/L (Merck Titrisol) were diluted for obtaining reference and working solutions. The ammonium acetate buffer was prepared by mixing the appropriate amounts of acetic acid and ammonia. Mixtures of mono- or trichloroacetic acid and ammonia were employed for buffering at pH lower than 4.

PROCEDURES

Uptake experiments procedure

0.5 g aliquots of resin were slurry-loaded in the columns, purified with a few milliliters of methanol and 2M HNO_3, which also produced conversion to the nitrate form, rinsed with water and conditioned at the pH of the next experiment. Solutions of the desired metal ion (100 mL, 0.1 mg/L, unless otherwise stated) were added with the proper amount of the ligand and buffered to the desired pH. In some experiments, an interfering agent was added (see below). The samples were driven through the resin, which was consequently washed with a few milliliters of water at the same pH. The flow rate was approximately 1.5 mL/min. The metal immobilized on the resin was then recovered with the aid of 10 mL of nitric or perchloric acid (see below). Metal ion concentrations in the aqueous solutions after elution through the resin and in the acid eluate were determined by ICP-AES in order to evaluate recovery yields.

Evaluation of Stability Constants

At low concentrations the dye is essentially monomeric and the reaction taken into account between Plasmocorinth B, (HL^{4-}), and a metal ion, (M^{n+}), is :

$$M^{n+} + HL^{4-} \underset{}{\overset{K}{\rightleftharpoons}} MHL^{(n-4)+} \tag{1}$$

from which, taking into account the dissociation equilibria, the conditional stability constant is defined :

$$K_{cond} = \frac{f_{MHL} \cdot [MHL^{(n-4)+}]}{f_M \cdot [M^{n+}] \cdot f_{LH} \cdot [H_3L^{2-}]} \cdot \alpha_M \cdot \alpha_L \tag{2}$$

The activity coefficients were calculated according to Debye-Huckel or Davies equations, depending on ionic strength and considering pH values at which no hydrolysis reactions involving the metal M^{n+} would have a significant effect; the equation becomes:

$$logK_{cond} =$$

$$pK_{1a} + pK_{2a} + log\frac{MHL^{(n-4)+}]}{[M^{n+}] \cdot [H_3L^{2-}]} + \frac{0.51\,I^{1/2}\,(Z_M^2 + Z_{HL}^2 - Z_{MHL}^2)}{(1+1.5\,I^{1/2})} - 2\,pH \tag{3}$$

Starting from the above assumptions and knowing pH, ionic strength, ionization constants of hydroxo-species and of the dye [5] the quantity to be evaluated was the ratio :

$$[MHL^{(n-4)+}] / [M^{n+}] \cdot [H_3L^{2-}] \tag{4}$$

In order to evaluate this ratio the proposed ion-exchange method considers a complexation condition at a pH value (pH_{eq}) at which approximately half the metal is in the complexed form MHL. The method is based on the assumption that when a solution containing metal complex $MHL^{(n-4)+}$, free ligand H_3L^{2-}, and free metal ion M^{n+} is put into contact with the anion exchange resin only the free metal ion is unretained. The separation and determination of the species involved may be performed following the procedure : solutions (100 mL) containing metal ion $(1.57 \times 10^{-5}\,M)$ and ligand $(1.57 \times 10^{-5}\,M)$, were prepared, brought to the proper pH and were passed $(1.5\,mL\,min^{-1})$ through a column packed with 0.5 g of AG MP-

1 resin, previously equilibrated at the desired pH and ionic strength. The free metal ion, in collected solutions, and the metal complex, recovered with acidic elution, were then determined by ICP-AES.

The mass balance was verified by evaluation of metal released with acid elution :

$$M_{tot} = M_{free} + MHL^{(n-4)+} \qquad (5)$$

M in the form of MHL is equal to the ligand bound in the complex so the quantity $[H_3L^{2-}]$ was computed through the stoichiometric concentration of ligand and the released metal concentration. The data obtained enabled us to evaluate the thermodynamic stability constant, extrapolated to zero ionic strength, knowing the correction factor arising from the ionic strength, with α_L and α_M, calculated as a function of pH. The data obtained by the ion exchange procedures were compared with the values obtained with a spectrophotometric method [6-8].

RESULTS AND DISCUSSION

Retention of Copper

The uptake of copper as a function of pH or ligand to metal ratio was evaluated for each dye and 2M HNO_3 was employed as the stripping agent after confirming its capability to completely strip the metal ion immobilized on the resin. For the experiments at different pHs the ligand-to-metal ratio was kept constant at 100:1, and at the different concentrations of ligands the pH was kept equal to 9, in order to ensure optimum conditions for the invariant parameter. The results showed that in all cases except C2R and NY the uptake efficiency increases with increasing pH, and ligand concentration. The recovery yields for C2R decrease at pH higher than 7.0. NY did not allow immobilization of copper on the resin at any pH studied, probably because of the electron-withdrawing effect of the two nitro-groups, which decrease the electron density on the chelating center, thus lowering its tendency to give co-ordinative bonds. The steric effect of the NO_2 group in the ortho position to the chelating center could hinder complexation. Table 1 summarizes the pH ranges and ligand/metal ratio of quantitative recovery for each dye.

Previous investigations [9-11] on ligands of similar structures showed that the complexation of the metal ion occurs through the azo-center and the oxygen atoms in the position ortho to it. In this case, such atoms belong to hydroxide groups, but they could also be part of carboxy moieties. The ligands which allow a quantitative recovery in wider ranges of pH and excess of reagents are those with two OH-groups in the ortho position with respect to the azo moiety, that is AVN, CA, PL, PCB. For O2, OG and AR4, which have only one o-OH group, the uptake of copper is complete only at higher pH and ligand concentrations. This behaviour can be explained with the stability of the corresponding complexes, which is higher for oo'-dihydroxyazo dyes than for monohydroxy ones [12]: this probably happens because the second hydroxyl group co-operates in the chelation of the metal ion. A complete uptake of copper was never achieved with AO8 and AR8: both ligands have a methyl group ortho to the azo-group, which causes a steric hindrance to chelate formation, thus reducing the amount of metal ion complexed.

TABLE 1
pH and ligand/metal (L/M) ratio ranges for quantitative recovery of copper.

LIGAND	pH range	L/M ratio range
AO8	-	-
O2	7 - 9	50 - 100
OG	7 - 9	40 - 100
AR4	6 - 9	80 - 100
AR8	-	-
C2R	7	-
AVN	2.5 - 9	10 - 100
CA	4 - 9	10 - 100
PCB	4 - 9	10 - 100
PL	4 - 9	10 - 100
NY	-	-

Preconcentration

Six of the eleven azo-dyes, namely O2, OG, AVN, CA, PCB, PL were tested for the recovery of other trace metal ions, namely Cd(II), Co(II), Cr(III), Fe(III), Mn(II), Ni(II), Pb(II). The recovery efficiency in metal preconcentration was evaluated in the presence of different ionic strengths, surfactants and ligands able to compete with the proposed chelating agent and simulating the matrix of real samples. Among the ligands chosen Plasmocorinth showed the best results and the good recoveries obtained (see Table 2) show the suitability of the method for the analysis of natural and seawater samples.

Natural Waters

The method was applied to the analysis of Cd, Cu, Fe, Mn, and Ni in Antarctic Lake Water (Carezza Lake). The sample, Antarctic expedition summer 1988/89, was frozen soon after collection and kept at -20°C until before the analysis. The determinations were performed on 1.0 L samples after acidification with ultrapure HNO_3 to pH = 2 and next submitted to the procedure developed. Standard additions, for Cd, Cu, Ni, Mn, and Fe, and external calibration were used. Table 3 collects the results obtained together with the detection limits of the method defined as 3 times absolute blank.

TABLE 2

Effect of Interfering Agents on Percent Cu(II), Mn(II), and Ni(II) Recovery (100 µg /L).

Interferent type	Recovery, %		
	Cu	Ni	Mn
no interferent	99.7 ± 0.6	99.1 ± 1.1	96.8 ± 0.8
NaCl 0.5	98.5 ± 0.6	97.5 ± 1.7	97.6 ± 0.7
NaCl 1.0 M	101.6 ± 0.5	102.4 ± 5.3	104.1 ± 0.5
[NTA]:[L] = 1:5	97.4 ± 1.6	98.7 ± 1.9	97.7 ± 2.4
[NTA]:[L] = 1:1	101.7 ± 2.6	100.1 ± 2.6	99.8 ± 2.2
[NTA]:[L] = 2:1	98.6 ± 1.0	100.1 ± 0.9	97.0 ± 0.6
Brij 10 mg/L	100.1 ± 1.4	100.5 ± 0.9	98.9 ± 1.3
Brij 100 mg/L	104.5 ± 3.2	103.2 ± 1.7	101.9 ± 3.2
Ca 20, Mg 2 mg/L	101.8 ± 0.4	101.8 ± 1.0	101.9 ± 1.8

Code : NTA = Nitrilotriacetic acid; Brij 35 = Polyoxyethylene 23 Lauryl Ether. Mean and relative standard deviations for three measurements.

TABLE 3

Values of Detection Limits (D.L., µg/L) obtained for the described procedure and analytical data (µg/L) for the analysis of Carezza Lake Water (Antarctica).

	D.L.	Plasmocorinth B[b]	GF Zeeman AAS[c]
Cd	< 0.09	< 0.09	0.07
Cu	0.30	1.02 ± 0.09	1.26 ± 0.50
Fe	0.72	48.4 ± 4.9	48.8 ± 4.4
Ni	0.04	0.53 ± 0.03	0.57 ± 0.09
Mn	0.03	3.64 ± 0.04	4.20 ± 0.83

[c] Reference [13]. Mean and relative standard deviations for three measurements.

The data reported have been corrected for the overall recovery yield obtained for each element in the operating conditions adopted and compared with those found with a different method [13]. The accuracy of the values obtained seems to be acceptable and the precision of analysis of Lake Water is good, except for Cd present in concentration below the detection limit of the method.

Evaluation of Stability Constants

Tables 4-5 collect the average values and relative standard deviations obtained for the computed stability constants for the metal complexes of the ligand Plasmocorinth; as can be seen the results obtained from the several methods are in good agreement.

One feature to be considered when dealing with separation of different species on ion-exchangers is a modification of equilibria involved in complex formation. In some cases the ligand when bound to a resin shows different behaviour with respect to the metal complexation in solution [14].

TABLE 4

Stability constants evaluated with the ion-exchange method (Ionic Strength extrapolated to 0; T = 25 °C).

	$\log K_{cond}$	$-\log \alpha_L$	$-\log \alpha_M$	f	$\log K$
Cu	4.83 ± 0.10	13.58	0	0.13	18.54
Ni	4.71 ± 0.15	8.80	0	0.19	13.70
Pb	3.94 ± 0.11	5.53	0.01	0.23	9.71
Cd	4.19 ± 0.05	5.77	0	0.23	101.9
Mn	3.87 ± 0.04	5.42	0	0.23	9.52

TABLE 5

Stability constants evaluated with the spectrophotometric method (Ionic Strength extrapolated to 0; T = 25 °C).

	$\log K_{cond}$	$-\log \alpha_L$	$-\log \alpha_M$	f	$\log K$
Cu	4.98 ± 0.32	13.58	0	0.13	18.69
Ni	4.82 ± 0.51	8.78	0	0.19	13.79
Pb	-	-	-	-	-
Cd	5.25 ± 0.21	4.81	0	0.23	10.30
Mn	3.92 ± 0.51	5.34	0	0.23	9.49

Mean and relative standard deviations for three measurements in all cases.

Additional experiments were performed in order to check whether the retention of free ligand on the resin does affect the complexation equilibria. The results were in good agreement with the previous ones, and this fact confirms that no modification of equilibrium occurs during the retention of free species onto the resin. The pH seems to be the only limiting factor of this ion-exchange procedure to all metal ions. In fact for metal ions such as Fe(III) exhibiting high values of stability constants, one is forced to choose low pH values (2 or less) in order to obtain the balance equilibrium needed between complexed and free metal ion, pH_{eq}. In these conditions the efficiency of the exchanger is reduced and the complete separation of free and complexed metal ion is not feasible.

CONCLUSIONS

Ten out of eleven of the azo-dyes examined allow the uptake of copper traces onto an ion exchange resin and its recovery with an acidic eluent. A general trend was observed: oo'-dihydroxy- substituted ligands gave quantitative yields in broader ranges of acidities and ligand concentrations than the mono-substituted ones, owing to a more efficient complexation of copper. The technique examined has the advantages of a good precision, relative freedom from interferences and the possibility of determining several analytes with a single enrichment procedure. The method can be used not only for analytical applications, but also for the removal of trace metal ions from aqueous solutions, either for the purification of reagents, or the treatment of effluents. Finally, the proposed ion exchange procedure seems very useful in evaluating stability constants for metal ions complexes and studies are in progress for similar ligands.

REFERENCES

[1] Sarzanini, C., Porta V. and Mentasti, E., New J. of Chem., **13**, 463, (1989).
[2] Abollino, O., Mentasti, E., Sarzanini, C., Porta, V. and Kirschenbaum, L.J., Analyst, **116**, 1167 (1992).
[3] Ohzeki, K., Toki, C., Ishida, R. and Saitoh, T., Analyst, **112**, 1689 R82, (1987).
[4] Anderson, R.G. and Nickless, G., Analyst, **92**, 207, (1967).
[5] Brush, J.S., Anal. Chem., **33**, 6 (1960).
[6] Coates, E. and Rigg, B., Trans. Far. Soc., **57**, 1088, (1961).
[7] Coates, E. and Rigg, B., Trans. Far. Soc., **58**, 88 (1962).
[8] Coates, E. and Rigg, B., Trans. Far. Soc., **58**, 2058 (1962).
[9] Drew, H.D.K. and Landquist, J.K., J. Chem. Soc., 292, (1938).
[10] Drew, H.D.K. and Fairbairn, J., Chem. Soc., 823, (1939).
[11] Beech, W.F. and Drew, H.D.K., J. Chem. Soc., J. Chem. Soc., 608, (1940).
[12] Snavely, F.A and Fernelius, W.C, Science, **117**, 15, (1953).
[13] Mentasti, E., Porta, V., Abollino, O. and Sarzanini, C., Ann. Chim. (Rome), **81**, 343 (1991).
[14] Shriadah, M.M.A. and Ohzeki, K., Analyst, **111**, 197 (1986).

Acknowledgements. The financial support from Ministero dell'Università e della Ricerca Scientifica e Tecnologica (MURST, Rome) and from the Italian National Research Council (CNR, Rome) are kindly acknowledged.

SELECTIVE AND REVERSIBLE SORPTION
OF TARGET ANIONS BY LIGAND EXCHANGE

YUEWEI ZHU, ARUP K. SENGUPTA and ANURADHA RAMANA
Environmental Engineering Program
Fritz Engineering Lab 13
Lehigh University
Bethlehem, PA 18015
U.S.A.

ABSTRACT

Polymeric sorbents, which can selectively remove target contaminants and be regenerated efficiently, are highly desirable for large-scale commercial applications. In this regard, selective sorption of contaminating inorganic and organic anions such as, selenites, arsenates, cyanides, phthalates, oxalates and others from the background of competing sulfate and chloride ions is a challenging separation problem yet to be resolved. Using copper-loaded specialty chelating polymers and a novel regeneration technique, we have been successful in overcoming commonly encountered difficulties.

INTRODUCTION

Background and Related Studies

LIGAND EXCHANGE was conceptualized and formally introduced by Helfferich (1, 2) in 1961. Cu(II)- or Ni(II)-loaded weak-acid cation exchange resins were used by Helfferich to enhance the sorption of various ligands through relatively strong Lewis acid-base interactions. In such ligand-exchange processes, the water molecules (weak ligands) present at the coordination spheres of immobilized Cu(II) and Ni(II) in the cation-exchange resins are replaced by relatively strong ligands, such as ammonia or ethylenediamine. The following provides a typical ligand exchange reaction with ammonia where "M" represents a divalent metal ion like Ni(II) or Cu(II) with strong Lewis acid properties:

$$\overline{(RCOO^-)_2\, M^{2+}\, (H_2O)}_n + NH_3 = \overline{(RCOO^-)_2\, M^{2+}\, (NH_3)}_n + n\, H_2O \qquad (1)$$

The overbars denote the exchanger phase and R represents the polymer matrix. During the last thirty years, a great deal of work has been done to apply the concept of ligand exchange in the areas of analytical chemistry, separation technology and pollution control processes (3, 4 and many others) with varying amount of success.

Shortcomings of the Conventional Ligand-Exchange Process

Investigations in ligand exchange have, however, been The overbars denote the exchanger phase and R represents the polymer matrix. During the last thirty years, a great deal of work has been done to apply the concept of ligand exchange in the areas of analytical chemistry, separation technology and limited to primarily non-ionized (uncharged) ligands, namely, various amines and ammonia derivatives. In reality, most of the inorganic and organic ligands are anionic, such as cyanide, selenite, sulfide, acetate, oxalate, phthalate, phenolate and many others including naturally occurring humates and fulvates. Ligand-exchange process using polymeric cation exchangers as metal hosts, as depicted in eqn. 1, are unable to sorb any anionic ligand i.e.,

$$\overline{(RCOO^-)_2\, M^{2+}\, (H_2O)}_n + \begin{cases} CN^- \\ SeO_3^{2-} \\ HS^- \\ CH_3COO^- \end{cases} \longrightarrow \text{No Reaction} \qquad (2)$$

The metal-loaded weak-acid cation exchange resins are electrically neutral and do not have any anion exchange capacity and also, the negatively charged fixed co-ions (carboxylates in this case) of the polymer will not allow uptake of any anions according to the Donnan co-ion exclusion principle. Thus, in spite of being strong ligands, the anions in eqn. 2 cannot displace water molecules (much weaker ligands) from the coordination spheres of the metal ions (Lewis acids).

Premises of the Present Study

It is recognized that, in order to sorb the anionic ligands selectively, the polymeric substrate (metal hosts) upon metal loading must possess fixed positive charges i.e., it should act as an anion exchanger. Obviously, such functional polymers should have high preference towards metal ions so that the metals do not bleed or bleed only negligibly during the ligand-exchange process. In essence, the ligand-exchange reactions in such cases involve ion exchange (IX) accompanied by metal-ligand or Lewis acid-base (LAB) interactions. The overall free-energy change may be presented as follows:

$$\Delta G^{\circ}_{Overall} = \Delta G^{\circ}_{IX} + \Delta G^{\circ}_{LAB} \tag{3}$$

where subscripts IX and LAB represent ion exchange and Lewis acid-base interactions, respectively.

If the overall equilibrium constant is $K_{overall}$, then

$$-RT \ln K_{Overall} = -RT \ln K_{IX} - RT \ln K_{LAB}$$

$$K_{Overall} = K_{IX} K_{LAB} \tag{4}$$

Thus, all other conditions remaining identical, an anion with stronger ligand or Lewis base characteristic is expected to show much higher selectivity towards this sorbent.

The primary objectives of this short communication are two-fold: first, to present a specialty chelating polymers as metal-hosts for ligand-exchange processes involving anions; and second, to present convincing experimental results exhibiting high sorption affinities of some anionic ligands of interest towards these metal-loaded sorbents. From an application viewpoint, the process looks quite promising in areas where selective sorptions of trace concentrations of target anionic ligands (namely, selenites, arsenates, phthalates, oxalates and natural organic matters etc.) are desirable from the background of high concentrations of competing chloride and sulfate ions.

EXPERIMENTS: MATERIALS AND PROCEDURE

This study primarily includes experimental results of a chelating exchanger from Dow Chemical Co., namely, DOW 2N (or XFS 43084). For comparison, two other synthetic exchangers, namely, IRC-718 (a chelating exchanger) and IRA-900 (a strong-base exchanger), both from Rohm and Haas Co, are also included in the study. All three sorbents are macroporous with polystyrene matrix and divinylbenzene crosslinking but they possess different functional groups. The names of the manufacturers, chemical compositions of the functional groups, and other salient properties of these polymeric ion exchangers have been provided elsewhere (5, 6). All of them were received in spherical bead forms; the average particle size after screening was 500 ± 50 microns. Detailed methodologies and analytical procedures used in this study have been discussed elsewhere (5, 6).

EXPERIMENTAL RESULTS

Sorption of Se(IV) Oxy-anions i.e., Selenite

Selenite or Se(IV) oxy-anions are sometimes present as trace contaminants under reducing conditions of groundwater and their selective removals by traditional strong-base anion exchangers (SBA) are particularly difficult in the presence of competing sulfate ions. Since sulfate is: first, preferred over selenite; and second, often present in water at concentrations several orders of magnitude greater than selenite, SBA are not viable sorbents for selenite removal. Figure 1 shows effluent histories of Se(IV) and sulfate anions during the fixed-bed column run with strong-base IRA-900 (Rohm and Haas Co., Pennsylvania) where Se(IV) was present in trace concentrations(2.0 mg/L) in the presence of much higher concentrations of competing sulfate and chloride ions. As anticipated, Se(IV) breakthrough occurred quite early, at less than fifty bed volumes. Suflate, on the contrary, appeared in the effluent much later (after 350 bed volumes).

Figure 2 shows the effluent histories of selenite and sulfate during a column run using copper-loaded DOW 2N or DOW 2N-Cu. Note that, in contrast to IRA-900 column run in Figure 1, Se(IV) or selenite breakthrough occurred long after sulfate confirming the suitability of this new sorbent for selective removal of Se(IV) oxy-anions in the presence of high concentrations of competing sulfate and chloride ions.

Sorption Behaviors of Oxalate

Oxalate is a divalent organic anion at around neutral pH. Figure 3 shows effluent histories of oxalate for four different sorbents, namely, activated carbon, IRA-900, IRC 718-Cu and DOW 2N-Cu in the presence of competing sulfate and chloride anions. Influent composition, pH and hydrodynamic conditions were identical for the four column runs and provided in Figure 3. Note that, compared to other three sorbents, oxalate

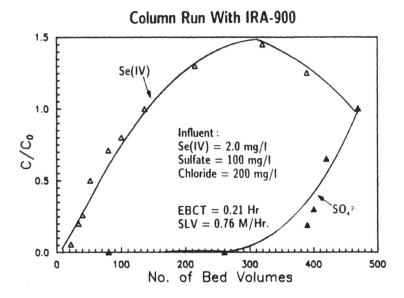

Figure 1.   Effluent histories of selenite and sulfate during a column run
with strong-base anion exchanger (IRA-900). (EBCT = Empty bed
contact time; SLV = Superficial liquid-phase velocity).

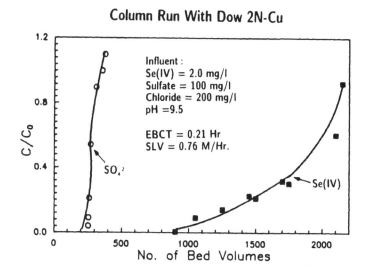

Figure 2.   Effluent histories of selenite and sulfate during a column run
with newly identified DOW 2N-Cu. (EBCT = Empty bed contact
time; SLV = Superficial liquid-phase velocity).

breakthrough for DOW 2N-Cu occurred much later and after 3000 bed volumes indicating its extremely high sorption affinity toward oxalate as compared to sulfate and chloride. In contrast, activated carbon (Filtrasorb 300, Calgon Corpn., Pittsburgh, PA) and IRC 718-Cu did not offer practically any oxalate removal capacity while the strong-base anion exchanger (IRA-900) was completely exhausted with less than 500 bed volumes. No commercially available sorbent, according to information in the open literature, offers such high oxalate selectivity as DOW 2N-Cu. Other mono- and di-carboxyltes (both aliphatic and aromatic) can be removed, concentrated and probably recycled from very dilute solutions using DOw 2N-Cu or similar sorbents. Small amount of virgin IRC-718 at the bottom of the column was sufficient to avoid any bleeding of copper into the treated water during the entire column run which lasted almost a month.

Regeneration of Oxalate-Loaded Column

The column in Figure 3 was saturated with the influent and subsequently, three equal portions (1.6 ml each) of the exhausted DOW 2N-Cu were taken out and regenerated separately under identical hydrodynamic conditions i.e., same empty bed contact time (EBCT) and superficial liquid velocity (SLV). The three different regenerants were: (4% NaCl + 1% NaOH); 4% Na_2Co_3; and 4% ammonia. Figure 4 shows the desorption profiles of oxalate for the three regenerations.

Both sodium chloride and sodium carbonate were capable of desorbing/eluting oxalate from DOW 2N-Cu and of them, sodium carbonate was more efficient than sodium chloride. Mass balance calculations for a separate set of experiments showed almost complete recovery (over 90%) with both regenerants. During regeneration with both sodium chloride and sodium carbonate, elution or bleeding of copper was practically negligible due to DOW 2N's extremely high affinity toward Cu(II).

Compared to sodium chloride and sodium carbonate regenerations, ammonia regeneration was distinctly more efficient; less than three bed volumes of 4% ammonia was sufficient for complete regeneration. However, copper(II) was also desorbed along with oxalate. In order to capture and reuse copper(II), a weak-acid cation column was placed in series with

293

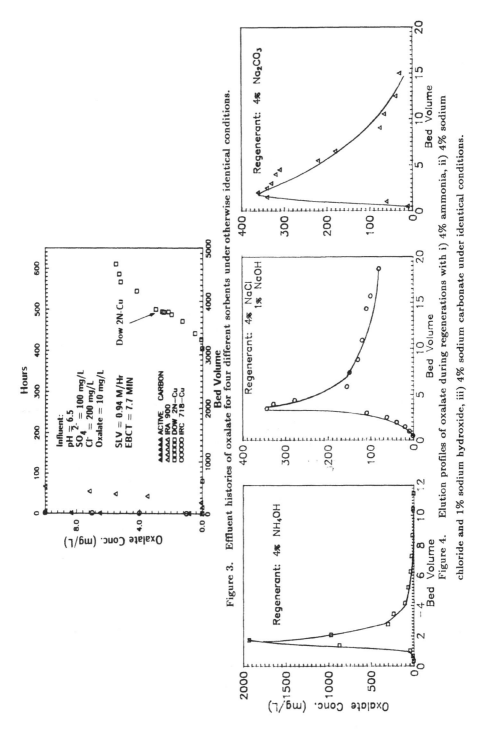

Figure 3. Effluent histories of oxalate for four different sorbents under otherwise identical conditions.

Figure 4. Elution profiles of oxalate during regenerations with i) 4% ammonia, ii) 4% sodium
chloride and 1% sodium hydroxide, iii) 4% sodium carbonate under identical conditions.

DOW 2N-Cu. This regeneration process, although operationally somewhat complex, is attractive when complete desorption or product recovery from a dilute stream are desirable.

CONCLUSIONS

This study identifies an obscure but commercially available chelating polymer which, when loaded with Cu(II) ions, exhibit very high affinity toward the anionic ligands (Lewis bases). Major conclusions of the study can be summarized as follows:

- Copper-loaded DOW 2N or XFS 43084 offers very high selectively towards anionic ligands such as, selenites, arsenates, oxalates, phthalates over competing sulfate and chloride ions. Mechanistically, the sorption of ligands may be viewed as an anion exchange process accompanied by Lewis acid-base interaction.

- Since DOW 2N has very high affinity toward Cu(II), the bleeding of dissolved copper during the lengthy column runs was negligible and that too was completely arrested by the virgin chelating exchanger provided at the bottom of the column.

- Upon exhaustion, the bed can be regenerated with brine or sodium carbonate or ammonia. Ammonia regeneration, although operationally more complex, provided almost stoichiometric efficiency with complete desorption of organic ligands.

REFERENCES

1. Helfferich, F., Nature, (1961), 189, 1001.
2. Helfferich, F., J. Am. Chem. Soc., (1962), 84, 3237.
3. Dobbs, R. A. et al., AIChE Symp. Series, (1975), Vol. 75, No. 152, 157.
4. Hernandez, C. M., de and Walton, H. F., Anal. Chem., (1972), 46, 6, 890.
5. Sengupta, A. K. et al., Environ. Sci. Technol., (1991), 25, 3, 481.
6. Ramana, A., M.S. Thesis: A new class of sorbents for selective removal of As(V) and Se(IV) oxy-anions. (1990), Civil Engineering Department, Lehigh University, Pennsylvania, U.S.A.

THE TREATMENT OF A BOARD MILL EFFLUENT USING A COMBINATION OF UF, IX AND RO SYSTEMS

D HEBDEN
F A HAWKINS
Hebden & Associates
265 Persimmon Street, Malvern
Johannesburg 2094, Republic of South Africa

ABSTRACT

Secondary effluents from a Board mill are heavily contaminated with fibre, lignosulphonates and sodium. The disposal
of these effluents without the subsequential contamination of
adjacent water courses, has proved to be a severe problem.

The process to be described has produced clear white, low-
conductivity water for re-use in the mill. A combination of
ultrafiltration, ion exchange and reverse osmosis systems has
proved successful at lab and pilot plant levels. The design
for a 2 500 m³/day plant has been completed.

INTRODUCTION

In designing a process system to treat the effluent arising
at a typical Board mill, a brief description of the effluent
streams, the present treatment (together with its inability
to conform to discharge limits) is given. The following
effluent streams arise from the mill process :

PULP EFFLUENT

This effluent is highly contaminated with fibre (suspended
solids), sodium, COD (chemical oxygen demand) and TDS (total
dissolved solids). It has the highest lignosulphonic
acid/lignosulphate content of any of the streams.

MACHINE WATER

This effluent is also high in fibre, whilst the TDS, sodium
and COD are about 35% that of the pulp effluent above.

These two streams are blended in an approximate 1:1 ratio
which discharges from plant as :

SECONDARY EFFLUENT

From a treatment point of view it has been found by experiment that the combination of the above two effluents exhibits all the worst characteristics of either. Secondary effluent is pumped to the 'storage dams' through which it passes over a period of 2 to 4 months,(depending on the degree of by-passing). The dams are open and ambient temperatures range from 14ºC to 30ºC. The dam effluent is pumped to the 'irrigation fields' as :

IRRIGATION EFFLUENT

Due to considerable enzymatic and bacterial (anaerobic) action in the ponds, the quality of this effluent is much improved, as is shown in the Table 1 (based on three months data).

TABLE 1 :   CONCENTRATION REDUCTIONS AFTER PASSAGE THROUGH THE 'STORAGE PONDS'

Conductivity	19,4%
Total dissolved solids	53,2%
Suspended solids	96,3%
Chemical oxygen demand	63,3%
Sodium	3,1%
Total hardness	0,5%

It was fairly obvious that the irrigation effluent was more amenable to treatment than the others.

PROCESS DESIGN

In the laboratory many systems were tried, but were found wanting. It soon became apparent that only the precipitation systems and membrane systems had any chance of success on secondary effluent. The best precipitation system tried was as follows :

FERRIC SULPHATE & LIME

This system had a certain amount of success on pulp effluent and secondary effluent. Eventually the results were found to be erratic due to the variance in the amount of lignosulphates in the feed which would have required adjustment to the chemical dosage and altered the pH relationships. This would have made control of the plant very difficult. These systems also suffered from the fact that large quantities of precipitate ensued, which resulted in large amounts of sludge for disposal.

LABORATORY SCALE ULTRAFILTRATION (UF)

Small scale UF trials showed definite promise as a preparatory process for the removal of the bulk of the lignosulphonate fraction. The permeate however was yellow and the removal of this colour proved difficult. As it had been decided to use a reverse osmosis (RO) plant for the final removal of sodium, the removal of all organic constituents from the RO feed was essential if long RO membrane life times were to be ensured.

UF PERMEATE TREATMENT

After much testwork, it was found that passing the UF permeate through a protonated, weak anion IX column removed the colour and other organics. Further treatment with activated carbon (AC) removed odour and some further organics, even though the amount removed was small. Finally, the AC product was filtered through diatomaceous earth (DE) to provide feed for the RO plant.

PILOT PLANT

The pilot plant was designed to a nominal 1 m³ throughput, a factor of 100 down on the anticipated requirement for the main plant. The pilot plant comprised :

PRETREATMENT FOR THE UF SYSTEM

Originally a dispersed air flotation (DAF) plant and an electroflotation plant were used to remove fibre, however the `irrigation effluent' feedstock requires the same plants to remove `frothing chemicals' of as yet unknown nature. It was found by observation, that two different species of frothers are removed by the two plants. The product of these plants is filtered and passed to the UF plant.

ULTRAFILTRATION PLANT

The ultrafiltration (UF) plant was originally equipped with Osmonics Type HP09 membranes. These membranes were polysulphone, spiral wrapped, 4" diameter, with "45 thou" spacers and a molecular weight "cut-off" of 1 500 Daltons. The UF plant performed well on both `machine water' and pulp effluents, but it proved exceedingly difficult to pretreat the feed - essentially to remove the fibre load. Microscopic fibre fragments passed through both float plants and filters, eventually collecting in the feed channels (spacers) of the membranes. This problem was not solved. On "irrigation effluent" however, no such problem occurred since 96% of the suspended solids had been removed, and long filter runs were achieved. The membranes produced a straw-coloured permeate which proved to be excellent feed stock for the ion exchange system, however the flux per membrane was

relatively low.   In order to improve this aspect (and reduce
the cost of the main UF plant) Osmonics HP05 membranes were
installed.   These membranes were also made of polysulphone,
spiral-wrapped, "45 thou" spacers but with a molecular weight
"cut-off" of 5 000 Daltons.   The fluxes achieved were
considerably improved, but the quality of the permeate
decreased.   This in turn put a greater load on the anionic
column of the IX plant, and, it is thought, to some extent,
poisoned it.   Osmonics produced a HG19 membrane which had
the same molecular weight cut-off as the HP09, but modified
by increasing the pore density to give much the same flux as
the HG05.   The plant is operated for 22 hours, cleaned with
EDTA, detergent and NaOH (for pH adjustment to pH=11,2) for
two hours and is then put back on stream.

ION EXCHANGE/ACTIVATED CARBON PLANT (IX/AC)

Since gas was generated in the removal of colour by the anion
IX column, the IX columns have to be operated in upflow and
much effort was required to make this system work.   The IX
resin was a weak base macroporous resin comprising a highly
porous tertiary amine anion with a styrene divinyl benzene
matrix.   The IX column is regenerated with NaOH, the eluant
is passed to the mill digester plant for use in the digestion
process.

The AC plant serves essentially to remove odour, but in fact
it removes a small amount of phenols and cresols.   Note that
these chemicals are not removed significantly in the RO
process.

The cation exchange plant removes calcium.   This is to
prevent scaling as with high recoveries on the RO plant, the
calcium sulphate (gypsum) content will increase to as much as
twenty times the inlet concentration.

REVERSE OSMOSIS PLANT (RO)

This plant is required to reduce the sodium content of the
effluent ;   the extent of removal is in excess of 92%.   The
removal of sodium by any other means is virtually impossible
at the levels appearing in the Board mill effluents.   No
problems were experienced with the operation of this section,
and bearing in mind the high quality of the feed, a fairly
long 'life' of the membrane is expected (i.e. in excess of 3
years).

PLANT PERFORMANCE

HP09 MEMBRANE

The performance characteristics of these membranes are shown
in Figures 1 and 2.   The short runs (20 hours) were
generally caused by feed shortages.   The data are

FIGURE 1.   HP09 MEMBRANES – SERIES 1.

PROCESS PERFORMANCE

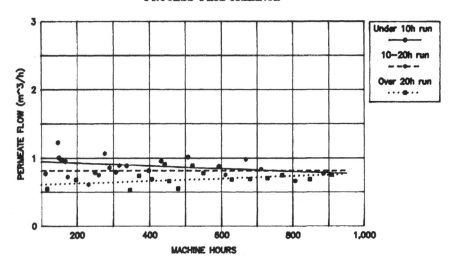

FIGURE 2.   HP09 MEMBRANES – SERIES 2

PROCESS PERFORMANCE

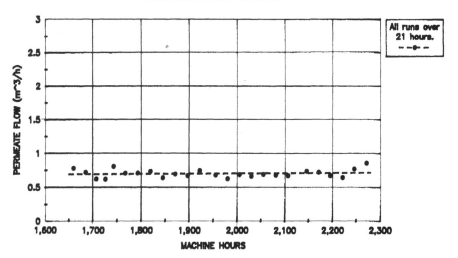

FIGURE 3.   HG05 MEMBRANES

STANDARD RUNS

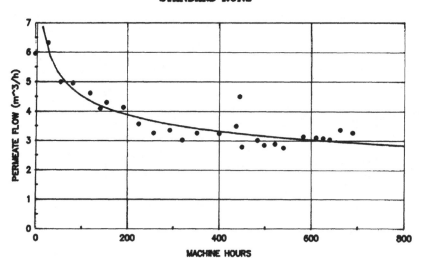

FIGURE 4.   HG05 MEMBRANES.

PROCESS PERFORMANCE

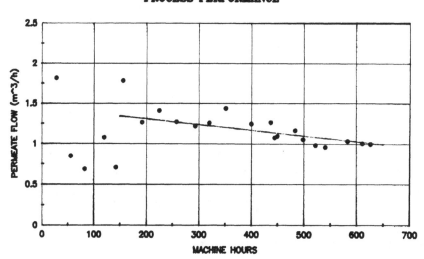

compensated for temperature. It is clear that for runs of over 20 h and the 21 h runs of the second series (Fig. 2) the membranes gave increased performance. The average permeate production over a 24 h run was 0,74 m³. The number of bed volumes passed through the anion IX column during these runs was about 120. Note that 200 bed volumes were achieved in the laboratory, the difference is attributed to inefficient regeneration procedures.

HG05 MEMBRANES

As can be seen from Figures 3 and 4, these membranes exhibited intra-pore fouling, and the quality of the permeate was much inferior to that obtained from HP09 membranes. It can be clearly seen that these membranes are not the correct types for this application. Note that these membranes had larger pore sizes than the HP09 membranes. The IX performance also deteriorated when HP05 membrane permeate was fed to the column, dropping to 28 bed volumes. The capacity of the IX column increased again(up to 120 bed volumes) after the HP09 membranes were replaced.

RO PLANT

The RO plant performed according to expectations if the feed was restricted to an efficient IX plant product.

AN ION EXCHANGE EXTRACTION OF VALUABLE COMPONENTS OUT OF WASTE WATERS

M.R. TANASHEVA, M.S. KAZYMBETOVA, S.B. GUNTER,
D.A. SMAGULOVA, R.S. MAKHATOVA, U.D. DZUSIPBEKOV

Kazakh State University, Chemical Department,
Vinogradov str.,95 A, Alma Ata 480012, Kazakhstan

ABSTRACT

The results of investigating sorptional properties of inorganic sorbents: phospho-gypsum, boro-gypsum, polygalite and serpentine are discussed and sorption of phosphate-ions from industrial waste waters is described. There is the possibility of obtaining valuable secondary raw materials by ion-exchange sorption of HPO_4^{2-} on inorganic sorbents. The importance of waste utilization in chemical industry connected with more strict demands on ecological purity of the environment gives rise to the necessity of studying new efficient methods.

INTRODUCTION

The main purpose of this work was to obtain sorbents from non-toxic chemical industry waste material for the extraction of low concentrations of phosphate and borate ions contained in industrial waste waters. In the neighbourhood of phosphorus/boron-processing plants solid waste such as phospho-gypsum, phospho-slag, boron-gypsum and boro-slag is accumulated. In general, these wastes are sulphates, carbonates, silicates of calcium, magnesium, iron and aluminium.

The possibility of isomorphic replacement of sulphate by phosphate due to similar ionic radii has been shown in previous work (1-3). Polygalite, boro-gypsum, serpentine and phospho-gypsum were therefore studied (Table 1).

Table 1 Content of basic components and
composition of sorbents used

Sorbent	Content of basic components wt%						
	P_2O_5	B_2O_3	CaO	MgO	SO_3	SiO_2	Cl
Phospho-gypsum	1.14	--	33.5	--	48.2	--	--
Boro-gypsum	--	4.41	27.8	2.43	26.4	--	0.3
Polygalite $K_2SO_4.MgSO_4$ $2CaSO_4.2H_2O$	--	--	19.0	7.2	52.2	1.0	--
Serpentine $3MgO.2SiO_2.2H_2O$	--	--	13.3	30.8	--	48.5	--

The sorbents generally contain sulphate salts of alkaline and alkaline-earth metals with the exception of serpentine. The sorbents were ground and 100, 80, 70 and 30 mesh (BSS) fractions were taken. The sorption capacity was determined using 80 mesh particles. The phosphorus-containing waste waters are the wastes of phosphorus-working plants of Kazakhstan. They are not utilized because of the small content of phosphorus (1400-2450 mg/L of P_2O_5). Boron-containing waste waters (B_2O_3 in the range 2930 mg/L to 4460 mg/L) also are accumulated near the boron-working plants. The sorbent (1-10g) was contacted with phosphorus-containing waste water (PWW), boron-containing waste water (BWW) or their mixture.

Under equilibrium conditions the solid and liquid phases were separated and analyzed. We determined the content of basic components in each phase and used IR and X-ray spectroscopy for the solid phase.

The influence of various factors (P_2O_5 and B_2O_3 concentration, S:L phase ratio, pH change) on the degree of extraction of phosphorus and boron was studied. The dependence of P_2O_5 content in the solid phase on solution pH is given in Figure 1.

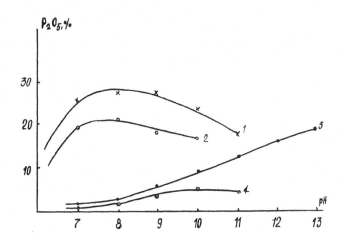

Figure 1 P_2O_5 content in solid phase:
dependence on pH: 1-phospho-gypsum,
2-boro-gypsum, 3-polygalite, 4-serpentine

The required solution acidity was obtained by adding concentrated solutions of caustic soda. The capacity of sorbents studied decreases in the sequence:

phospho-gypsum - boro-gypsum = polygalite - serpentine.
The maximum phosphate sorption depends significantly on solution pH.

In the next series of experiments the sorption of phosphate ions out of the PWWs was studied under the optimum solution pH for each sorbent, but P_2O_5 content in solution was varying. The experimental data are given in Figure 2.

The influence of the P_2O_5 initial concentration on the sorption of phosphate depends on the sorbent nature. In particular, for phospho-gypsum, this dependence has an extreme character with a peak in the range of 400-500 mg/L of B_2O_5. The P_2O_5 concentration change in WW is less noticeable for boro-gypsum and polygalite. The content of phosphate-ions in the solid phase does not change significantly when P_2O_5 concentration increases. This is found for polygalite also: P_2O_5 content in the solid phase is small (3.0-5.1%) and remains through all the range of concentrations studied.

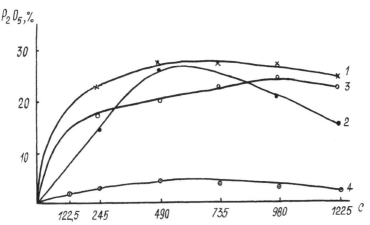

Figure 2 P_2O_5 content in solid phase of waste water concentration: 1-phospho-gypsum, 2-boro-gypsum, 3-polygalite, 4-serpentine

Sorptional capacity was determined according to the ratio of solid and liquid phases. It was shown that the optimal S:L ratio for the systems with phospho-gypsum is 1:200, boro-gypsum - 1:200, polygalite - 1:500 and serpentine - 1:200. The data obtained permits us to suppose that precipitation of phosphate-ions takes place due to the exchange of SO_4^{2-} ions of the sorbent by HPO_4^{2-} -ions of the waste water solution.

To confirm the suggestion that exchange of the phosphate-ions has an ion-exchange mechanism the well-known equation using the distribution coefficient of the analyzed component and concentration of hydrogen ions was used. To characterize the precipitation of the phosphate-ions in the solid phase the 10gK/pH dependence was studied (Figure 3).

Observing this isotherm we emphasize that it has the character generally found in the alkaline range. This indicates that the law of mass action is applicable. It permits us to suggest that under the above experimental conditions there is an equivalent exchange of ions. Moreover, we emphasize that serpentine has no sulphate-ions. It is therefore clear why there is no sorption of phosphate-ions on this sorbent.

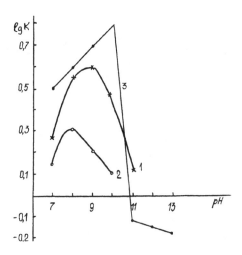

Figure 3 Isotherm for distribution of the hydrogen-ion

Experiments were made on equilibrium sorption of phosphate-ions and desorption of sulphate-ions. In other words the equivalence of ion exchange was tested for HPO_4^{2-} (in solid phase) and SO_4^{2-} (in liquid phase). Chemical analysis of the solid sedimentary material and the liquid phase has shown that under defined conditions equivalent quantities of ions were sorbed and desorbed.

Summarising the data obtained here and from references (1-3) we contend that the conversion of a sorbent into di-calcium-phosphate proceeds in all systems, i.e. the sulphatic part of the sorbent converts into the appropriate phosphate.

$$CaSO_4 + HPO_4^{2-} \quad -- \quad CaHPO_4 + SO_4^{2-} \qquad (1)$$

When phospho-gypsum is used as a sorbent, the solid phase is represented mainly as $CaHPO_4.2H_2O$ with insignificant admixtures of, for example, anhydrous $Ca(H_2PO_4)_2$ and gypsum.

When boro-gypsum sorbs phosphate-ions there are three crystalline phases in the solid phase: di-calcium-phosphate, calcium-meta-borate and double-water gypsum, and the conversion of calcium and magnesium sulphates into the corresponding phosphates takes place. When polygalite is a sorbent, $CaHPO_4$ and $MgHPO_4$ with a small admixture of $CaSO_4$ are produced. The roentgengrams of some specimens are in Figure 4.

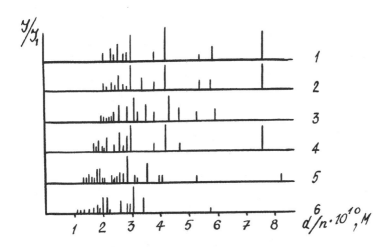

Figure 4 Dot-and dash diagrams of systems:
1-CaO:PG=1:3, 2-CaO:BG-1:2, 3-CaO:Polygalite-1:1,
4-standard specimens:CaHPO4.2H2O, 5-MgHPO4.3H2O, 6-
Ca(BO2)2.H2)

The maximum sorbent capacity is obtained generally for alkaline solutions. The necessity to create a particular solution pH is not desirable and hinders the devising of a general scheme of complex waste utilization and WW purification methods. In this connection the next series of experiments were devoted to the study of sorptional properties of complex sorbents. These sorbents were produced by adding alkaline materials (oxides and hydroxides of alkaline and alkaline-earth metals) to the sorbents (phospho-gypsum, boro-gypsum and polygalite). This obviates the necessity to adjust the solution pH by addition of caustic soda.

The dependence of P_2O_5 content in the solid phase on the alkaline addition (CaO) is given in Figure 5. The condition under which the maximum P_3O_5 content is reached depends on the properties of a sorbent. When CaO:PG ratio is equal to 1:3, nearly 30% of P_2O_5 is contained in the solid phase. For boro-gypsum and polygalite the optimal relations are 1:2 and 1:1

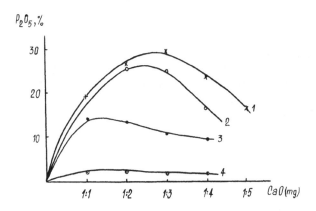

Figure 5 P2O5 content in solid phase
on CaO sorbent relation:
1-CaO:BG, 2-CaO:PG, 3-CaO:Polygalite, 4-CaO:Serpentine

To determine the possibility of mother liquor utilization experiments were made with usage of the ion-exchange properties of the sorbents under investigation. This opened new possibilities of solving the problems of direct waste utilization. The experimental data are given in Table 2. These data show the dependence of P_2O_5 and boron content in the sedimentaries on the PWW:BWW correlation and on medium pH. The sorption process was carried out in the same way: the sorbent was flooded by different PWW+BWW mixtures, the sedimentaries obtained were separated and analyzed by IR spectroscopy and RP analysis. It is interesting to note that the maximum sorptional capacity of phosphate-ions is within the former limits of pH whereas borate-ions are best sorbed out of slightly acid and neutral media. The P_2O_5 content in the sedimentaries are high (18.0-20.5%), the boron concentration changes from 0.40%. Almost all the phosphatic part of phosphatic fertilizers with boron is represented by citrate-soluble form which is easily assimilable by plants. As a result there were devised ways to produce pure di-calcium-phosphate (4) and di-calcium-phosphate with boron, the basis of phospho-gypsum and boro-gypsum correspondingly, and the complex phosphate-magnesium-containing fertilizer, re-working polygalite.

Table 2 Composition of sedimentaries on pH
and PWW and BWW correlation

Correlation	Solution pH		Composition of sedimentaries mass portion, %		
PWW:BWW	Initial	Equilibrium	P_2O_5 Complete	P_2O_5 Absorbed	B
20:1	8.50	7.60	18.80	9.50	0.14
15:1	8.48	7.10	19.15	8.30	0.12
10:1	9.18	8.94	20.52	19.31	0.24
9:1	9.16	8.76	20.85	19.40	0.25
8:1	9.50	9.17	19.85	18.75	0.22
5:1	8.50	7.96	14.80	8.70	0.31
1:1	10.80	9.70	12.54	6.60	0.40

REFERENCES

1. Dahlgren, S.E., Phosphate Substitution in Calcium Sulphatic in Phosphoric Acid Manufacture, Brit. Chem. Eng., 1965, N10, N11, pp.776-777.
2. US Patent N 3353908, Process for the Manufacture of Di-calcium-phosphate, Off. Gas., 1967, N844, p.1.
3. Osborn, G., Synthetic Ion Exchangers, M.Miz, 1964, p.505.
4. Kazymbetova, M., Investigation of Phospho-gypsum and Phosphorus-containing Waste Water Utilization, Bachelor of Chemistry Science, Alma-Ata, 1983, p.8.
5. Author certificate N4708995/26, 30.10.90, Method of Obtaining Phosphoric Fertilizer with Boron, Tanasheva, M., Kazymbetova, M., Nurachmetov, N.

INORGANIC SORBENTS FOR AQUEOUS EFFLUENT TREATMENT

E.W.HOOPER

AEA Technology, Radwaste Treatment Division,
B353, Harwell Laboratory, Harwell, OX11 0RA, UK

ABSTRACT

The Harwell Laboratory is engaged in the design and development of processes to reduce the radioactivity present in low and intermediate level waste streams to a very low level. Amongst the processes being examined are precipitation, crossflow filtration and ion-exchange using inorganic sorbents; these processes are used singly or in combination depending on the composition of the waste and the extent of decontamination required.

INTRODUCTION

Many aqueous waste streams arising at nuclear plants are contaminated with low levels of radioactive nuclides and, when necessary, these are treated by conventional filtration and ion-exchange prior to discharge to the environment. Soluble species cannot be filtered directly but if additives or "seeds" which absorb the species are added then even soluble radionuclides can be dealt with.

The use of finely divided sorbents in conjunction with ultrafiltration offers a number of advantages over conventional ion-exchange processes:

(a) ultrafiltration provides good solid/liquid separation, enabling the removal of colloidal material,

(b) smaller quantities of floc and/or ion-exchange materials are required, leading to smaller volumes of secondary wastes,

(c) they provide the opportunity to "tailor" a process to remove specific radionuclides.

(d)   the application of ion exchangers as additives, rather than as conventional packed beds, results in a wider range of sorbents being available and their preparation is frequently simpler and cheaper. The higher surface area of the finely divided form, compared to granular materials, results in improved sorption kinetics.

In the past few years a number of novel sorbent materials, both organic and inorganic, have appeared on the market - some claiming to achieve very large decontamination factors for metal ions, including those having radioactive isotopes. A unified and standardized testing programme making use of available expertise is necessary to provide a fair and meaningful comparison.

In November 1988, as part of its commitment to minimize discharge of radioactive material into the environment, representatives of the United Kingdom nuclear industry agreed to form the Novel Absorbent Evaluation Club. The Club was formed to assess absorber materials and to undertake the necessary work to identify the extent and rate of adsorption of radionuclides by such materials from a set of typical reference waste streams. The administration of the Club and the experimental programme are both undertaken by AEA Technology at their Harwell Laboratory.

This paper collects together the information obtained at Harwell from a number of investigations involving seeded ultrafiltration and also summarises the work undertaken for the Novel Absorbent Evaluation Club.

EXPERIMENTAL

The ultrafiltration tests were carried out in small rig consisting of a solution reservoir, a pump and a 140 mm long x 6mm internal diameter (inner surface area 22 cm^2) vertically mounted ultrafiltration membrane with a nominal pore size of 2 nm (molecular weight cut off 20,000). The operating pressure was 20 psi (1.3 bar) and permeation rates were 100 mL h^{-1} (~1.0 m^3/m^2/d).

The tests carried out for the Novel Absorbent Evaluation Club consisted of batch contact experiments in which 1mL portions of the conditioned sorbents were contacted with 50mL of the reference waste stream and agitated in a thermostatted shaker bath at 20°C. Samples of liquid were removed after 1,2,4,6 and 24 h contact, centrifuged and then analysed for radionuclide content.

SORBENT MATERIALS

Some of the sorbent materials used in this work were prepared at Harwell, others were obtained from commercial sources or prepared elsewhere. The sorbents prepared at Harwell were produced by the following methods:-

a) Sodium Nickel Hexacyanoferrate(II) (NHCF): Equal volumes of 0.094M $Na_4Fe(CN)_6$ and 0.113M $Ni(NO_3)_2.6H_2O$ solutions were mixed well and a portion of the resulting slurry was added to the waste simulant to provide the required $Fe(CN)_6^{4-}$ concentration.

b) Manganese Dioxide (MnO₂): 20mL of an aqueous solution of potassium permanganate (14.4g $KMnO_4$ per litre) were adjusted to pH11 using sodium hydroxide solution. 400mg of solid sodium dithionite were then slowly added over 15-20 min whilst maintaining the pH at 11 ± 0.5. The slurry was centrifuged and then washed with water adjusted to the pH of the solution to be treated.

A number of experiments were carried out with a sample of MnO_2 prepared by an alternative route. 1mL of a manganese nitrate solution (93.8g $Mn(NO_3)_2.6H_2O$ per litre) was added to 10mL of a potassium permanganate solution (3.45g $KMnO_4$ per litre). The slurry was centrifuged and then washed with water adjusted to the pH of the solution to be treated.

c) Hydrous Titanium Oxide (HTiO): 10mL of a 15% w/v titanium sulphate solution were treated with 1M sodium hydroxide solution until no further precipitate formed. The precipitate was centrifuged, washed with water to remove excess alkali and then washed with water adjusted to the pH of the solution to be treated.

d) Zirconium Phosphate (ZrP): 0.1M orthophosphoric acid was added dropwise to a solution of 1.12g of zirconium nitrate $[Zr(NO_3)_4]$ in 100mL of water. The precipitate of $Zr(HPO_4)_2$ was filtered off and washed with water.

e) <u>Ferric Hydroxide:</u> For most experiments this was prepared in situ by addition of a solution of ferric nitrate to the waste solution followed by adjustment of the pH to 7. In other tests the iron was added as ferrous sulphate and the ferric hydroxide precipitated by bubbling with oxygen.

f) <u>Other sorbent materials</u> were supplied by:

i) Duratek Corporation (USA) - cation exchangers -
Durasil D70 - functionalised carbon
Durasil D230 - functionalised inorganic oxide

ii) Toray Industries (Japan) - fibres of
TIN 100H - Polystyrene/polyethylene - sulphonic acid
TIN 200A -   "   "   - trimethyl ammonium
TIN 600H -   "   "   - iminodiacetic acid

iii) Universal Chemicals - Synthetic alumino silicates R2W and R2U

iv) STM1-RAN - an unspecified mix of several inorganic sorbents

v) University of Reading (Dr M J Hudson)
Goethite surface modified with quaternary amine
Layered double oxide (Mg + Aℓ)
Silica gel modified with diethylenetriamine
Goethite (unmodified)
Modified clinoptilolite

Clinoptilolite, AW500 and IRN 77/78L mixed resin were also tested for comparison of the test results with those obtained with sorbents currently in use.

For the batch contact experiments the sorbents were washed with water adjusted to pH9 or 4 using sodium hydroxide or nitric acid solution respectively. The washing was continued until the pH of the washings was constant.

Gamma spectrometry was used for the determination of the γ-emitting radionuclides. Samples containing ^{99}Tc, ^{90}Sr, ^{239}Pu + ^{241}Am, were analysed for total beta or total alpha using a liquid scintillation counter.

WASTE STREAM COMPOSITIONS

A number of waste stream simulants were used in this work. In some studies, water spiked with the radionuclides and then pH adjusted with either nitric acid or sodium hydroxide was used. In the experiments using sorbent mixtures, a simulated PWR waste containing boron was spiked with radionuclides.

In the test programme of the Novel Absorbent Evaluation Club three reference waste streams were used, known as NAEC S1, S2 and S3. With some absorbers a fourth waste stream was also used to examine the performance in the presence of the complexing agents EDTA and citric acid. All waste streams were 0.05M in sodium nitrate and were adjusted to pH 9 or 4 using sodium hydroxide or nitric acid solutions.

Waste NAEC/S1 contained ^{137}Cs, ^{60}Co, ^{65}Zn, ^{51}Cr, ^{59}Fe, ^{54}Mn, ^{125}Sb, ^{106}Ru, ^{203}Hg and ^{109}Cd each at the 100Bq/mL level. The pH was 9 or 4.

NAEC/S2 contained ^{99}Tc as TcO_4 at the 100 Bq/mL level. The pH was 9.

NAEC/S3 contained ^{239}Pu at 2Bq/mL, ^{241}Am at 1 Bq/mL and ^{90}Sr at 5 Bq/mL. The pH was 9.

NAEC/S4 was NAEC/S1 plus 0.25g/L EDTA and 0.15g/L citrate. The pH was 4.

RESULTS AND DISCUSSION

All of the sorbents tested in this study enhanced the removal of some of the fourteen main radionuclides considered when used individually and different sorbents enhanced the removal of different radionuclides. A summary of the main enhancements in removal is presented in Table 1. Only very small increases in activity removal were achieved with ^{51}Cr and ^{125}Sb so there was only a modest overall Decontamination Factor (DF) for total $\beta\gamma$ activity. ^{239}Pu is insoluble at the pH investigated and reasonable DFs were achieved by ultrafiltration alone.

TABLE 1

Summary of Most Significant Enhancements in Activity Removal Using Seed Materials

Radionuclide	Most suitable absorber(s)
^{51}Cr	None
^{54}Mn	MnO_2, HTiO
^{59}Fe	$Fe(OH)_3$, MnO_2
^{60}Co	HTiO, MnO_2*
^{63}Ni	NHCF, MnO_2, HTiO
^{65}Zn	High DF by UF alone
^{90}Sr	Generally low DF, but some removal with MnO_2*
^{95}Zr	$Fe(OH)_3$
^{106}Ru	MnO_2, HTiO, $Fe(OH)_3$
^{110m}Ag	High DF by UF alone
^{125}Sb	None
^{137}Cs	NHCF
^{139}Ce	Small DF in absence of seeds, some enhancement with $Fe(OH)_3$
^{239}Pu	Small DF in absence of seeds, some enhancement with $Fe(OH)_3$

* Prepared by Mn^{2+} + MnO_4
Prepared by aerial oxidation of Fe^{2-}

Mixtures of nickel hexacyanoferrate(II) with MnO_2, $HTiO_2$ and $Fe(OH)_3$ gave results equivalent to the sum of those obtained when the sorbents were used singly. This was not true however when mixtures of NHCF, MnO_2 and HTiO were used; similar behaviour was found with mixtures of MnO_2 and HTiO.

The results summarised in Table 1 were all obtained at a solution pH of 7. From results obtained at different pH values it is apparent that solution pH has a marked effect on the decontamination achieved.

The results of the tests on materials supplied to the Novel Absorbent Evaluation Club have been published [1-5] and are summarised in Table 2. The table gives only the decontamination factors obtained after 24 hours contact with the test solution. Data also exists for contact times of 1, 2, 4 and 6 hours contact and is included in the final reports. In the table, a > sign indicates that the nuclide was removed to below the limit of detection of the analytical method used.

TABLE 2

Decontamination Factors after 24h contact.

	pH	51Cr	54Mn	60Co	65Zn	106Ru	109Cd	125Sb	137Cs	203Hg	99Tc	90Sr	Tot. α
Clinoptilolite	9	11.4	6.0	9.1	25.6	> 50	8.5	1.1	63	1.8	1.0	19.0	14.9
AW500	9	12.2	49.6	37.6	> 60	> 60	> 100	1.1	66	2.7	1.0	7.1	8.7
IRN 77/78	7	–	2.5	2.8	> 5	–	–	9.1	2.0	–	–	2.5	3.0
Magnesium Electron Absorber													
Zirconium Phosphate	4	> 10	55.9	5.0	34.2	8.3	45.9	5.5 p	86	11.5	1.04	100	59
" "	4	> 3	>100	2.9	> 50	7.3	11.2	4.1	75.6	2.0	–	–	–
" "	4*	4.8	1.0	1.0	1.0	1.2	1.0	1.0	68.6	1.1	–	–	–
Zirconium Hydroxide	9	> 10	>170	15.1	>100	76.1	60.2	> 6	1.1	56.7	1.08	5.0	> 20
Recherche Appliquee du Nord Absorbers													
RAN P577	9	12.5	17.5	74.7	> 60	8.5	> 10	1.0	25.5	1.0	1.06	1.5	> 3
University of Reading Absorbers													
A336	9	> 25	128	1078	> 60	> 22	> 10	> 20	1.0	3.4	6.9	3.2	> 6
A336	4	> 3	1.5	2.9	23	> 16	>2.3	> 24	1.0	4.6	–	–	–
A336	4*	> 8.7	1.0	1.0	1.0	4.1	1.3	27.8	1.0	4.5	–	–	–
DDB	9	> 50	>128	647	> 47	> 20	> 10	> 20	1.0	3.8	3.7	3.0	> 6
DDB	4	> 3	11.8	26.7	87.8	> 16	>19.9	> 20	1.0	3.9	–	–	–
DDB	4*	> 4.5	1.0	1.0	1.0	2.9	>1.0	> 33	1.0	1.1	–	–	–
Goethite	9	> 24	>160	360.6	> 60	> 20	> 14	> 30	1.1	2.7	1.1	1.3	> 3
Layered Hydroxide	9	> 10	>238	265	> 70	> 18	> 11	1.8	1.0	1.5	1.2	7.8	21.1
Modified SiO2	9	> 10	>238	2.4	> 70	12.3	> 11	1.9	1.1	7.6	1.1	10.2	4.1
Clino Bis	9	> 10	>238	1.7	> 70	12.9	> 11	2.0	41.6	45.5	1.2	2.6	7.3
Duratek Absorbers													
D70	9	–	–	17.5	–	2.8	–	–	1.0	–	–	–	115
D70	7	2.8	3.8	14.7	–	–	–	1.2	1.1	–	–	–	–
D230	12	–	–	–	–	1.7	–	1.0	6.0	–	–	–	9.3
Universal Chemical Absorbers													
R2U	9	–	–	2.5	–	1.5	–	–	22.9	–	–	35	3.5
R2W	9	–	–	2.7	–	2.0	–	–	31.1	–	–	17.5	6.5
Toray Absorbers													
TIN 100	9	–	–	4.8	–	2.9	–	–	9.6	–	–	4.4	1.0
TIN 200	9	–	–	15.6	–	7.8	–	–	1.4	–	–	3.5	34.0
TIN 600	9	–	–	3.0	–	1.8	–	–	1.2	–	–	5.8	3.5

CONCLUSIONS

The use of finely divided inorganic sorbent materials in combination with ultrafiltration, so-called 'seeded ultrafiltration', can reduce the amount of radioactivity in aqueous effluents to very low levels. The correct choice of seeds is essential if a variety of different species are to be treated simultaneously. In effect, seeded ultrafiltration offers a way of tailoring ion exchange to the treatment of complex mixtures of chemicals. Seeded UF also does away with the need for the material to be granular in form and of sufficient strength that it could be used in a conventional ion exchange bed (often a problem with inorganic absorbers), and in principle allows a wider range of absorbers to be consiu.

ACKNOWLEDGEMENTS

Some of the results presented in this paper have been obtained from work undertaken for Nuclear Electric plc, the UK Department of the Environment and for the Novel Absorbent Evaluation Club. Approval to publish the data obtained on behalf of these organisations is gratefully acknowledged.

REFERENCES

(1) Hooper, E.W., Holwell, G.J. Moreton, A.D.
     Investigation of Durasil Absorbers for the Removal of Radionuclides
     from a range of Reference Waste Streams. Report AERE-G5408, AEA
     Technology Harwell (1990).

(2) Hooper, E.W., Moreton, A.D.
     Investigation of Universal Chemicals Absorbers for the Removal of
     Radionuclides from a Reference Aqueous Waste Stream. Report
     AERE-G5491, AEA Technology Harwell (1990).

(3) Hooper, E.W., Moreton, A.D.
     Investigation of Toray Industries (Japan), IONEX Ion Exchange Fibres
     for the removal of Radionuclides from a Reference Aqueous Waste
     Stream. Report AERE-G5492, AEA Technology Harwell (1990).

(4) Hooper, E.W., Moreton, A.D.
     Investigation of Absorber Prepared at Reading University for the
     Removal of Radionuclides from Aqueous Waste Streams. Report
     AEA-D&R-0244 (1991).

(5) Hooper, E.W., Moreton, A.D.
     Investigation of Magnesium Elektron Absorbers for the Removal of
     Radionuclides from Aqueous Waste Streams. Report AEA-D&R-0245
     (1991).

SURFACE CHARGE AND ION SORPTION PROPERTIES INFLUENCING THE FOULING AND FLOW CHARACTERISTICS OF CERAMIC MEMBRANES

STEPHEN GALLAGHER and RUSSELL PATERSON
Colloid & Membrane Research Group,
Chemistry Dept.,
University of Glasgow, U.K.

JOCELYN ETIENNE, ANDRE LARBOT and LOUIS COT
Ecole Nationale Supérieure de Chimie de Montpellier,
France

ABSTRACT

Ion exchange and sorption characteristics of a range of calcined oxides which constitute the active layers of ceramic
membranes have been determined using new automated titration
methods employing ion selective electrodes. The effects of pH
and salt concentration on exchange capacities have been evaluated and explained mechanistically. These were shown to be
consistent with the exchange mechanisms of non-calcined oxides
determined previously (1-6). The sign of the surface charge
and the exchange capacities for each counterion may be altered
in a predictive way, by changing the pH and the activity of
the sorbing ion. The effects of altering pH and salt concentration on the flow characteristics of whole ceramic membranes
have been determined. Studies on the minimisation of fouling
by charged substrates on ceramic membranes have been
initiated, with promising results.

INTRODUCTION

Surface charge and ion sorption properties are major contribu-
tory factors to fouling by polar or ionic substrates, since
surface sorption is often the first step in fouling leading to
pore blocking within the membrane structure. By altering the
surface charge density and changing its polarity, the opportunity exists for minimising fouling and optimising separation
processes. A fundamental understanding of the surface charge
and sorptive properties within the pores of the active layers

of ceramic membranes is therefore essential to their efficient
use.

This research was performed on the calcined oxide active
layers of ceramic membranes, as prepared by the sol-gel process (7,8). From the gels used to prepare the membrane active
layers, calcined oxides in powder form were obtained. These
powders are identical to those of the active layers of the
membrane. Aqueous dispersions of these materials were examined
titrimetrically.

EXPERIMENTAL

A fully automated computer-controlled system was designed and
constructed, enabling the performance of multi-ionic titrations with a maximum of six syringes and the monitoring of the
solution activity of up to four different ions simultaneously,
using ion-selective electrodes (ISEs) (9). Titrations were
made over predetermined ranges of pH or pX (by X-ion ISE)
using acids, bases or salts as required. At each titration
point the equilibrium activities of hydrogen (pH) and those of
a variety of ions (chloride, sodium, potassium, calcium) were
determined using the relevant ion selective electrodes. From
these data, the uptake and/or release of the monitored ions by
the exchanger were determined from observed concentration
changes in solution and calculated by mass balance. Uptake of
ions by the exchanger, expressed as milliequivalents (meq) per
gram of exchanger, is taken as positive and release as negative.

A traditional problem in the determination of pH or other
p(ion) in colloidal solution, particularly in dilute solutions
with highly charged colloidal particles, is the problem of the
suspension effect, due to the potential of the electrical
double layer surrounding charged particles (the suspension
potential, Ψ). In the region of the double layer, the activity
of counterions exceeds those of coions due to the influence of
the electrical potential near the surface. From the condition
of uniform electrochemical potential, it is shown that

$$p(ion)_{(\infty)} = p(ion)_{(s)} - \frac{z_i F \Psi_{sp}}{2.303 RT} \tag{1}$$

It is not possible to detect a suspension effect in a single
electrode titration. However, it has been shown (10) that it
is possible to determine and correct for suspension potentials

if all ions which participate in exchange processes are monitored by ISEs. These corrections were employed in the current research.

RESULTS AND DISCUSSION

Previous research (1-6) on the (non-calcined) oxide precursors of ceramic membranes showed that the pH dependent equilibria for exchange processes on monoclinic zirconia (with NaCl as the electrolyte) may be represented by equations 2 and 3.

$$M - OH_{pore} + H_{aq}^+ + Cl_{aq}^- \rightleftharpoons M - OH_{2\,pore}^+ + Cl_{pore}^- \tag{2}$$

$$M - OH_{pore} + OH_{aq}^- + Na_{aq}^+ \rightleftharpoons M - O_{pore}^- + Na_{pore}^+ + H_2O \tag{3}$$

The magnitude and sign of the surface charge is sensitive to both pH and the activity of the counterion in solution. The mechanisms of equations 2 and 3 were validated by the observations that anion capacity is a single valued function of pA (= pH + pCl) and cation capacity is a single valued function of pB (= pNa + pOH), below and above the zero point charge (zpc) respectively.

Titration of a dispersion of active layer calcined monoclinic zirconia gave similar results to those obtained previously (4,6), with capacities an order of magnitude lower (11). The dual effects of pH and salt concentration on the exchange capacity were seen, with zpc between pH 7 and 8. Monofunctional relationships between pA and anionic capacity and between pB and cationic capacity were observed. The exchange mechanisms (equations 2 and 3) for the oxide precursors of ceramic membranes therefore apply equally to calcined oxides. Figure 1 demonstrates the stoichiometry of the ion exchange processes. Below the zpc there is stoichiometric uptake of H⁺ and Cl⁻ ions (negative capacity), with no cation uptake. Above the zpc there is cation uptake only (positive capacity), with a stoichiometric uptake of Na⁺ and release of H⁺.

Similar studies were performed on a range of active layer calcined oxides (γ alumina, silica, titania). All were amphoteric, being positively charged below the zpc and negatively charged above the zpc, with the exception of silica, which showed cation exchange behaviour throughout the pH range studied (Table 1). The mechanisms of equations 2 and 3 applied in all cases, with the observance of the dependency of anion exchange on pA and cation exchange on pB.

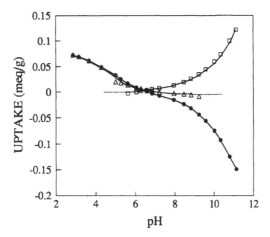

Figure 1. Ion uptakes for H^+ (●), Na^+ (□) and Cl^- (△) on titration of a dispersion of calcined monoclinic zirconia with HCl and NaOH. Data were obtained by mass balance calculations from ISE solution activity measurements, after correction for suspension potential.

Table 1 Anion capacities (with Cl^- as the sorbing ion) and cation capacities (with Na^+ as the sorbing ion) of active layer calcined oxides, as a function of total acid pA (= pH + pCl) and total base pB (= pNa + pOH) respectively.

Oxide	zpc	pA						pB					
		5	6	7	8	9	10	5	6	7	8	9	10
ZrO_2	7-8		-0.074	-0.044	-0.018	-0.003	0	0.147	0.085	0.042	0.019	0.005	0
γAl_2O_3	8-9			-0.207	-0.104	-0.03	0			0.080	0.043	0	0
SiO_2	<3	0	0	0	0	0	0				0.067	0.035	0.020
TiO_2	6-7	-0.025	-0.020	-0.013	-0.005	0	0	0.053	0.036	0.021	0.010	0	0
		Anion capacity (meq/g)						Cation capacity (meq/g)					

Ion sorption studies were extended to include a wider range of ions to which the membranes may be routinely exposed. For indifferent electrolytes (such as KCl), similar results to those obtained with NaCl electrolyte were observed. Specific sorption was observed when the oxides were exposed to multivalent ions (Ca^{2+}, SO_4^{2-}). In these cases considerably higher capacities were observed. Parallel research on the electrophoretic mobility of calcined oxides in the presence of Ca^{2+} and SO_4^{2-} confirmed that specific sorption was occurring.

The exchange characteristics of calcined oxides can therefore be determined precisely, if the pH and salt concentration in the solution phase are known. Figure 2 shows the ion exchange capacity of monoclinic zirconia at varying pH and NaCl concentrations, calculated from pA and pB relationships. From the exchange capacity and the specific surface area, the surface charge may be calculated.

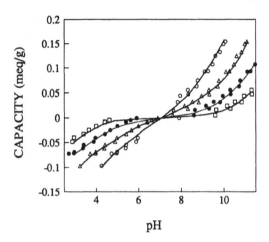

Figure 2. pH dependency of ion exchange capacity of calcined monoclinic zirconia at pNa (and pCl) = 4 (□); = 3 (●); = 2 (△); = 1 (○).

PRACTICAL APPLICATIONS

This research has led to a fundamental understanding of the operational factors which govern ceramic membrane function. The magnitude and sign of the surface charge may be manipulated by small changes in pH or salt concentration. The final stage of this research was concerned with the practical applications of this fundamental knowledge, with a view to optimising operating conditions and minimising membrane fouling.

The effects of varying the pH and salt concentration of feed solution on the filtration characteristics of ceramic membranes were investigated. The results discussed refer to an ultrafiltration membrane with an 4.5μm zirconia active layer on a 1 mm microporous alumina disc. Fluxes were measured at constant pressure of 2 bar and constant temperature of 25°C. Filtrations were performed until constant fluxes were reached

and permeate pH was similar to retentate pH. Figure 3 shows
the influence of pH on fluxes across the zirconia membrane, at
constant NaCl concentration (0.0051 M). Maximum flux was
observed at the zpc of zirconia (pH 7-8), where interactions
between the solution and the electrical double layer are at a
minimum. When the pH of the feed solution was decreased (surface charge increased), there was a gradual decrease in the
flux (relative to the maximum flux), with a flux reduction of
almost 10% at pH 3.5. Similarly, by raising the pH, the relative flux decreased with increasing surface charge.

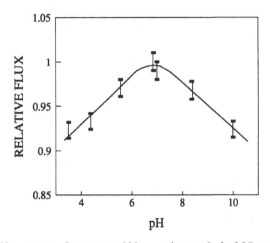

Figure 3. Influence of pH on filtration of 0.005M NaCl solutions, for a 4.5 μm thick zirconia ultrafiltration membrane.
Fluxes are relative to the highest observed. Error bars
correspond to fluxes at ±1%.

An increase in the ionic strength leads to a collapse of the
double layer inside the pore, which reduces the interactions
between the solution and the charged walls of the membrane.
Higher fluxes were observed at high ionic strength (NaCl
0.1M), at equivalent pH (Figure 4). At high ionic strength,
the effect of altering the surface charge on flux across the
membrane is considerably less than with solutions of low ionic
strength.

Studies on the effects of manipulation of surface charge (by
changing pH and salt concentration) on membrane fouling by
charged substrates were initiated. Preliminary studies were
performed on the sorption of amino acids on active layer calcined oxides. Qualitative results showed that sorption

occurred only when the sign of the charge on the amino acid and on the oxide surface were opposite. An increase in the oxide surface charge caused an increase in the sorption of an oppositely charged amino acid. In this way it is possible to minimise amino acid sorption by selection of suitable pH and salt concentration (based on pA and pB relationships).

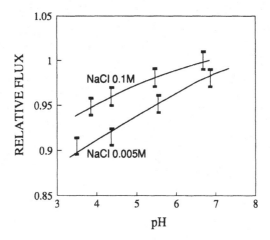

Figure 4. Influence of pH and ionic strength on measured fluxes for NaCl filtration (zirconia membrane).

CONCLUSIONS

The factors influencing the ion exchange and sorptive properties of the oxide active layers of ceramic membranes have been isolated and their influence explained in mechanistic terms. These mechanisms were shown to be the same as for those previously determined on non-calcined microcrystalline oxides, with lower capacities observed.

By changing the parameters influencing the sorption characteristics of oxide active layers (solution pH, salt concentration, nature of sorbing ion), it has been demonstrated that it is possible to alter the membrane surface charge density and reverse the polarity in a predictable way.

Studies on the effect of manipulation of surface charge on the filtration characteristics and on membrane fouling were initiated, with encouraging results. The effects of pH and salt concentration on fluxes and on membrane fouling by charged organic substrates were shown to be as predicted by the relationships determined in this research. By appropriate

selection of the membrane active layer material and manipulation of the operational parameters, it should be possible to
optimise conditions for separation processes across UF/RO
membranes. Similarly, optimisation of operating conditions and
membrane surface modification raise the possibility of reducing membrane fouling by charged substrates such as proteins,
thus improving efficiency and increasing the membrane
lifetime.

NOMENCLATURE

F	Faraday's constant	$C\ mol^{-1}$
$pH_{(x)}$	pH at distance x (0 - ∞) from colloid surface	
R	Gas constant	$J\ mol^{-1}\ K^{-1}$
T	Temperature	K
z_i	valency of ion i	
Ψ	electrical potential	V
Ψ_{sp}	suspension potential	V

REFERENCES

(1) Paterson R., Rahman H., J. Coll. Int. Sci., 1983, **94** 60
(2) Paterson R., Rahman H., J. Coll. Int. Sci., 1984, **97** 423
(3) Paterson R., Rahman H., J. Coll. Int. Sci., 1985, **98** 494
(4) Paterson R., Rahman H., J. Coll. Int. Sci., 1985, **103** 106
(5) Paterson R., Smith A., J. Coll. Int. Sci., 1988, **124** 581
(6) Paterson R., 1st Int. Conference on Inorganic Membranes,
    Montpellier, July 1989, 127
(7) Cot L., 1st Int. Conference on Inorganic Membranes, Mont-
    pellier, July 1989, 17
(8) Larbot A., Julbe A., Randon J., Guizard C., Cot L., 1st
    Int. Conference on Inorganic Membranes, Montpellier, July
    1989, 31
(9) Gallagher S., Paterson R., Chapter 5 in: Improvement of
    Practical Countermeasures against Nuclear Contamination in
    the Urban Environment (ed. A. Cremers, R. Paterson, J.
    Roed, F.J. Sandalls), Commission of the European Commu-
    nities, EUR 12555 EN, in press
(10 Etienne J., Gallagher S., Paterson R., in preparation
(11)Gallagher S., Paterson R., Etienne J., Larbot A., Cot L.,
    2nd Int. Conference on Inorganic Membranes, Montpellier,
    July 1991, in press

ADSORPTION AND DESORPTION BEHAVIOUR OF ARSENIC COMPOUNDS ON VARIOUS INORGANIC ION EXCHANGERS

Mitsuo ABE, Yasuo TANAKA and Masamichi TSUJI
Department of Chemistry, Faculty of Science,
Tokyo Institute of Technology,
2-12-1, Ookayama, Meguro-ku, Tokyo 152,JAPAN

ABSTRACT

Adsorption and desorption behaviour of arsenic compounds were studied on six inorganic ion exchangers; hydrous oxides of Al, Ti(IV), Zr(IV), Mn(IV) and Fe(III) and hydrotalcite by batch and column experiments. Among these inorganic ion exchangers, amorphous hydrous titanium dioxide showed high selectivity towards As(III), As(V) anions and cacodylic acid in 0.01M(M=mol dm^{-3}) NaCl solution.

INTRODUCTION

Synthetic inorganic ion exchangers have been reported to show an excellent selectivity towards some metal ions[1-7]. Particularly, oxides and hydrous oxides of multivalent metals have a relatively large ion-exchange capacity and an excellent ion-exchange selectivity towards certain cations[6-16]. These inorganic ion-exchange materials have found applications for the selective separation of cations in the field of hydrometallurgy, radioanalytical chemistry, environmental chemistry, and biochemistry.

Organometallic compounds and oxides of heteroelements, e.g., As, Sn, Se, Te are widely used in industry, biomedical fields, agriculture and marine product industry. Recently, environmental pollution by these materials has occurred.

The present paper describes selectivity data for cacodylic acid (dimethylarsinic acid, DMAA) along with As(III) and As(V) acids on various ion exchangers.

EXPERIMENTAL

Ion-exchange Materials

The inorganic ion exchangers(a-e) were synthesized according to methods reported earlier.

a) Hydrous titanium dioxide,HTDO (in amorphous form, $TiO_2 \cdot 2.3H_2O(15)$,

b) Hydrous manganese(IV) oxide, (cryptomelane type): $MnO_2 \cdot 0.45H_2O(9)$

c) Iron(III) oxide hydroxide (β-type):$FeOOH \cdot 0.2HCl \cdot 0.2H_2O(2)$. d) Hydrous zirconium oxide (amorphous form): $ZrO_2 \cdot 1.76H_2O(18)$

e) Hydrous aluminum oxide (boehmite-type): $Al_2O_3 \cdot 1.2H_2O(2)$.

These exchangers were identified by thermal studies (thermogravimetric and differential thermal analysis, TG and DTA) and an X-ray diffraction with Ni-filtered Cu-Kα radiation, except hydrous manganese(IV) oxide in which case Mn-filtered Fe-Kα radiation was used.

f) Hydrotalcite(KYOWAAD 500) $Mg_6Al_2(OH)_6CO_3 \cdot 4H_2O$: This was supplied by Kyowa Chemical Industry. Ion exchangers resins were used for comparison.

g) Strong-acid type cation exchange resin, Dowex 50W-X8 (H^+ form,100-200 mesh).

h) Strong-base type anion exchange resin, Dowex 2-X8 (Cl^- form, 100-200 mesh).

All other chemicals used were of analytical grade and supplied by Wako Pure Chemical Ind. Ltd.

Selectivity Study by the Batch Method

Samples of 0.250 g or 0.100 g of exchanger were equilibrated with 25 cm^3 or 10cm^3 of solutions containing various concentrations of arsenic compounds with ionic strength of 0.1 or 0.01 with respect to (NaCl and/or HCl). The equilibrium pH and arsenic concentration were determined after equilibration at 30°C.

Sorption-desorption Study by the Column Method

The breakthrough capacity was determined by injecting into a column an arsenic compound solution of pH4 (2×10^{-2} M as arsenic) with the ionic strength adjusted to 0.1 (NaCl + HCl) or (NaCl + NaOH). After saturation, the column was washed by a small amount of 0.1 M NaCl solution of pH4, and then sorbed arsenic species was eluted by 0.1 M NaOH solution. The adsorbed or desorbed amounts were calculated from the areas of breakthrough or elution curves.

RESULTS AND DISCUSSION

Inorganic Ion-exchange Materials

The chemical composition and the crystal data were in good agreement with those reported previously(2,9,14,18).

Adsorption Properties of Arsenic Compounds on Organic Ion Exchange Resins

A low uptake of arsenic compounds was observed on both of the cation and anion exchangers at pH>7 (Fig.1)(20), though increased uptake of cacodylic acid was found at pH< 4 on Dowex 50W-X8. Therefore, these organic ion exchangers will not be effective for separation of arsenic compounds in an aqueous environmental and industrial waste solution containing a large amount of other ions.

Selectivity Study for Arsenic Compounds on Inorganic Ion Exchangers.

Uptake for three arsenic compounds was studied on inorganic ion exchangers (Fig.2)(21). The equilibrium of adsorption of arsenic compunds was attained within 5 days on these exchangers. A commercially available hydrotalcite and synthetic ferric oxide hydroxide known to be an anion exchanger showed a low selectivity for cacodylic acid, though they showed a high selectivity towards arsenic and arsenious acids in low concentration ($<10^{-3}$ M). Hydrous aluminum oxide and hydrous zirconium dioxide had a relatively small capacity for cacodylic acid (<0.2 mol/Kg). Hydrous titanium dioxide (HTDO) and Hydrous manganese(IV) oxide showed a relatively large capacity (>0.3 mol/Kg) at the pH range studied. Especially, the HTDO showed a relatively large uptake with 1.0mol/Kg for cacodylic acid at pH<5.5 ,and 1.8 mol/Kg for arsenious acid at pH around 8. Thus, the selectivity of arsenic compounds was studied in more details in the following experiment.

Ion-exchange Selectivity for Arsenic Compounds

The plot of log Kd vs pH for the sample showed a similar slope for three arsenic compounds(Fig.3)(21): -0.92 for As(III) at pH>9.5, -0.85 for As(V) at pH>7.5 and -0.96 for cacodylic acid at pH>6. Each plot of the results of the reverse reaction can be plotted on the curve of the forward reaction. In these pH ranges and concentrations, the predominant chemical species is mononuclear $As(OH)_4^-$, $AsO_3(OH)^{2-}$, and $(CH_3)_2AsOO^-$. Therefore, the ion-exchange for $As(OH)_4^-$ and $(CH_3)_2AsOO^-$ is "ideal" and reversible for these conditions. However, the slope for $AsO_3(OH)^{2-}$ was different from the expected value at lower pH. A lower concentration of $AsO_3(OH)^{2-}$ may be needed for the "ideal" exchange reaction.

The Breakthough and Elution Curves of the Arsenic Compounds on HTDO

The breakthrough capacity on the sample **a** was found to be 0.91, 0.90 and 2.28 mol/Kg for cacodylic, As(V) and As (III) acids respectively (Fig.4a, c, e)(21). Those for cacodylic and As(V) acids are of sigmoidal shape with a steep rise in the breakpoint. That for As (III) showed a slow increase in the concentration of the effluent, which may be attributed to a relatively low rate of particle diffusion. The predominant chemical species are monomeric $(CH_3)_2AsO(OH)$, $AsO_2(OH)_2^-$ and $As(OH)_3$ in the solution of <0.1 M at pH4(19). However, As(III) is polymerized at higher concentrations. Within the HTDO, the concentration of As(III) will attain 2-3 M, and so As(III) is possibly exchanged in the polymerized species. It may be responsible for the slow uptake. These adsorbed arsenic compounds could be eluted by 0.1 M NaOH with a yield of 85%, 73% and 51% for cacodylic, As(V) and As(III) acids respectively (Fig.4b, d, f). The small fraction of elution for As(III) is attributed to the high selectivity, as shown in the isotherm. A higher concentration of NaOH solution is necessary for the complete elution of adsorbed As(III).

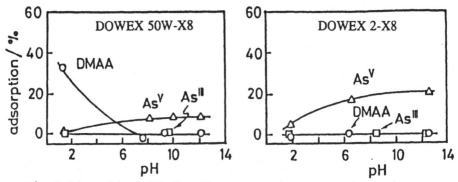

Fig.1.Adsorption (%) for three arsenic compounds on ion
exchange resins

Exchanger, 0.10g. Vol. of solution, 10 cm³.Temp.,
30°C.

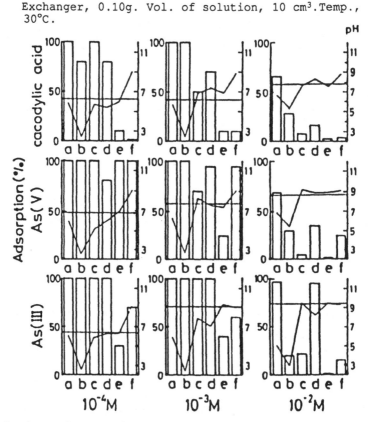

Fig.2.Adsorption profile for three arsenic compounds on
inorganic ion exchangers (a-f).

Exchanger, 0.10g. Vol. of solution, 10 cm³. Temp.
,30°C. Equlibration time,5 days. Horizontal line
shows the initial pH. The equilibrium pH was
connected by the zigzag line.

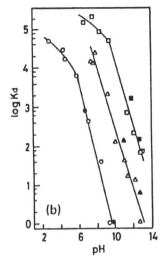

Fig.3.Ion exchange ideality and reversibility of adsorption of arsenic compounds on sample **a**(HTDO).

Initial concn., 10^{-4}M. Vol. of solution, 10 cm^3. Exchanger, 0.10g. Temp. 30°C. Ionic strength, 0.10. Cacodylic acid, Δ and $\blacktriangle$; As(V),$\bigcirc$ and$\bullet$; As(III),$\square$ and $\blacksquare$.Black symbol shows reverse reaction.

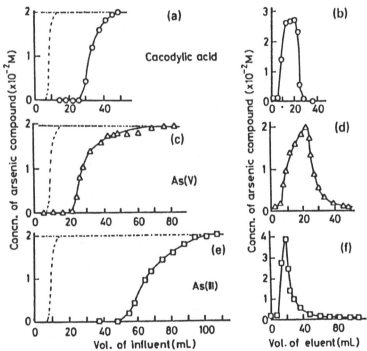

Fig.4.Breakthrough (left) and elution (right) curves for arsenic compounds.
Dotted line, blank curves. Column, 2.5 cm x 0.5 cm diam.(HTDO, 0.5g, 100-200 mesh size). Concn. of influent,0.02 M (pH4 by HCl). Ionic strength, 0.1 with NaCl. Flow rate, 0.05 cm^3 min^{-1} for cacodylic acid, 0.2 cm^3 min^{-1} for As(V) and As(III) acids.

332

Chromatographic Separation of Cacodylic Acid on HTDO
The adsorption features of the sample **a** described above will
enable a separation of cacodylic acid from aqueous solutions.
The chemical composition of the waste solution supplied by
Kyoto University is given in Table 1.

Table 1 Waste solution containing cacodylic acid

Element	Cr	As	Pb	Cu	Zn	Fe	PO_4^{3-}
g/m³	13	1300	30	28	38	180	5000

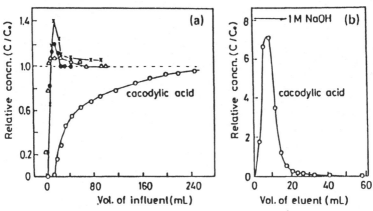

Fig.5.Breakthrough (left) and elution (right) curves
    for real waste.
    Column: 10 cm x 0.5 cm diam. (HTDO:2.0g, 100-200
    mesh size). Flow rate, 0.1cm³ min-1. O;Cacodylic
    acid,X; Os, ●; Fe,Δ; Cu.

It contains cacodylic acid along with other heavy metals,
osmium compound and a small amount of sucrose and aldehydes.
The color of waste solution was black, due to presence of
Os(VI).Cacodylic acid adsorbed was 0.560 mol/Kg (Fig.5a)(21)
and quantitatively eluted (0.566 mol/Kg) with 1 M NaOH as an
eluant after washing with a small amount of water(Fig.5b)[21].
The effluent was pale orange and contains almost no osmium
compound and other heavy metals. Heavy metals other than zinc
were not adsorbed, probably because of complexation with
organics.

Thus, the sample **a** was demonstrated to be effective for
the separation of cacodylic acid from real waste. After
elution with 1 M NaOH, adsorbed Na⁺ can be eluted and the column
can be used repeatedly.

ACKNOWLEDGMENT

A part of the present work was supported by the Grant-in-Aid for Scientific Research No.01602027 from the Ministry of Education, Science and Culture.

REFERENCES

1. Amphlett, C. B., Inorganic Ion Exchangers, Elsevier, Amsterdam,.1964
2. Abe, M. and Ito, T., Nippon Kagaku Zasshi, 1965,**86**, 818,1259
3. Vesely, V. and Pekarek, V., Talanta, 1972, **19**, 219
4. Pekarek, V. and Vesely, V., Talanta, 1972, **19**, 1245
5. Abe, M., Bunseki Kagaku, 1974, **23**, 1254, 1561
6. Abe, M., Oxides and Hydrous Oxides of Multivalent Metals as Inorganic Ion Exchangers.In:Inorganic Ion Exchange Materials, Clearfield, A(ed), CRC press, Boca Raton, 1982, pp161-273
7. Volkhin, V. V., Leonteva, G. V. and Onorin, S. A. Neorg.Mater., 1973, **9**, 1041
8. Abe, M. and Tsuji, M., Chem.Lett., 1983, 1561
9. Tsuji, M. and Abe, M., Solvent Extr. Ion Exch., 1984, **2**,253
10. Tsuji, M. and Abe, M., RADIOISOTOPES., 1984, **33**,218
11. Tsuji, M. and Abe, M., Bull. Chem. Soc. Jpn., 1985, **58**, 1109
12. Abe, M., Chitrakar, R., Tsuji, M. and Fukumoto, K.,Solvent Extr. Ion Exch., 1985,**3**, 149
13. Ooi, K., Miyai, Y., Katoh, S., Maeda, H. and Abe, M., Chem. Lett., 1988,989
14. Chitrakar, R. and Abe, M., Mater. Res. Bull., 1988, **23**, 1231
15. Abe, M., Tsuji, M., Qureshi, S. P. and Uchikoshi, H. Chromatographia, 1980, **13**, 626
16. Abe, M., Tsuji, M., Kimura, M., Bull. Chem. Soc. Jpn., 1981, **54**,130
17. Craig, P J, Organometallic Compounds in the Environment, John Wiley & Sons, New York, 1986
18. Sugita, M., Tsuji, M. and Abe, M., Bull.Chem.Soc.Jpn., 1990, **63**, 559
19. Baes, C F and Mesmer, R E The Hydrolysis of Cations, John Wiley and Sons, New York, 1976, p366

INORGANIC MEMBRANES IN ION SEPARATION BY DIALYSIS

D.A. WHITE, R. LABAYRU,[*] H. OBERMOLLER [*] and E. ZAKI
Department of Chemical Engineering, Imperial College,
London SW7 2BY.
[*] Chilean Nuclear Commission, Amunategui 95,
Santiago, Chile.

ABSTRACT

The present paper describes work in which electro-dialysis
has been carried out using a barium phospho-silicate
membrane. A simple perspex cell was used to do experiments
in which it was shown that kinetics are faster with
inorganic than organic membranes. Theoretical work is
described to predict pH and ion contents of the cell during
dialysis runs and supported by actual data obtained in the
cell. Future membrane developments are discussed.

INTRODUCTION

In a recent series of publication we have reported some
initial experiments on the manufacture and use of inorganic
membranes. (1-3). These membranes made by the precipitation
of an insoluble salt precipitate such as barium phosphate
have been known for a number of years but have not been
studied very much. The literature on dialysis using organic
membranes is more extensive (4-6). Inorganic membranes are
very fragile in nature and break up if they are allowed to
dry. The innovation in our work has been the

production of membranes that can be dried and are stable at
ambient conditions. A recipe for the manufacture of a
barium phospho-silicate membrane is given elsewhere (2).

In recent work it has been shown that barium phospho-silicate membranes have pore sizes of around 15nm and are stable in strong alkalis and in the presence of high intensive γ - radiation, which would destroy organic ion exchange membranes. The membranes have been used to remove strontium and shown to be suitable for alkaline production of hydrogen peroxide.

Electrodialysis is a separation technique that is suitable for treatment of aqueous streams of high ionic strength. Ions migrating across the membrane carry the current and cause separation of ionic species to occur. Inorganic membranes are not specific to cations or anions but can transfer both freely when an electric current passes through them. Cations pass into the cathode chamber and anions in the opposite direction. Inorganic membranes also have very marked electro-osmotic properties unlike their organic counterparts.

COMPARISON OF ORGANIC AND INORGANIC MEMBRANES

The experimental work described here was carried out in a simple laboratory perspex cell described in more detail elsewhere (1). Two platinum electrodes were used for the dialysis runs. The cathode and anode cells had the same volume of 200mL and the membranes used had a surface area of 9.6 cm^2. Some simple experiments are described and the results are given in Figs 1 and 2. The runs were carried out starting with a 0.1 M solution of potassium nitrate in both cells. In the first series of experiments the initial variation of cell current with applied voltage was studied.

It was found that the current versus voltage relationship is substantially the same for both the organic membrane (a cationic one supplied by BDH) and the inorganic

336

barium phosphate membrane. Very little current passes through the membranes until the applied voltage has exceeded 2 V. In the next series of experiments a continuous dialysis run was carried out with both membranes at a constant cell voltage of 10 volts for 5 h. In Fig. 1 the variation of cell current with time is given for both membranes at a constant applied voltage of 10 V. After about an hour the current passed when the cell is fitted with an inorganic membrane exceeds that of the run with the organic membrane.

The reason for this is connected with the extent of mass transfer. Fig. 2 gives the variation of the potassium content in the cathode chamber for the runs with the two membranes. To check reproducibility two runs were done for the organic membrane. The rate of transfer of potassium ions is much more rapid for the inorganic membrane and at this end of this run, the solution in the cathode chamber would be largely KOH and that in the anode is largely HNO_3.

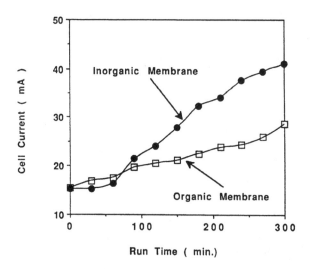

Figure 1. Cell current variation at constant cell voltage

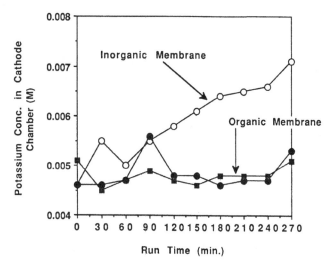

Figure 2   Comparison of dialysis performance for organic &
inorganic membranes

PROPERTIES OF INORGANIC MEMBRANES

1.   pH Variations.   Consider a cell with the electrode
chambers separated by an inorganic membrane.   The volume of
the anode chamber is V_A and the cathode chamber V_c. H^+ ions
are produced at the anode and $O\bar{H}$ at the cathode.   Since the
solutions in the cell are electrically neutral the quantity
of H^+ (in keq) in the anode chamber must be equal to the $O\bar{H}$
in the cathode side.   If this quantity is denoted by P, .then
the concentration of H^+ in the anode chamber is P/V_A.   The
anode pH is

$$pH_A = -\log_{10}\left[\frac{P}{V_A} \right] \tag{1}$$

The concentration of $O\bar{H}$ ions in the cathode chamber is P/V_c
so the hydrogen ion content is 10^{-14} divided by this
concentration.   The cathode pH is

$$pH_C = 14 + \log_{10}\left[\frac{P}{V_c} \right] \tag{2}$$

Hence, $$pH_A + pH_C = 14 + \log_{10}\left[V_A/V_c \right] \tag{3}$$

If VA = Vc the sum of the compartment pH's is exactly 14. In table 1 some actual experimental data is given for a cell where VA and Vc was the same. Both comparments were initially filled with a 1 kg/m³ solution of strontium nitrate.

Table 1 Dialysis Run pH Data

Time (min)	pH_C	pH_A	$pHc + pH_A$
0	11.37	2.63	14.00
15	11.79	2.19	13.98
30	11.96	2.03	13.99
60	12.10	1.87	13.97
90	12.27	1.72	13.99
120	12.37	1.66	13.99
240	12.61	1.44	14.05

The data agrees with equation (5) and it should be noted that pH_A (the anode pH) always decreases with time to an assymptotic value. If it should rise it denotes that the membrane has failed as a result of being attacked by acid or other causes.

2. Ion migration across the Membrane

The transfer of cation from the cathode is given by

$$- V_A \frac{dC}{dt} = \eta_C \frac{I}{z_C F} \tag{4}$$

Where I is the cell current η_C the cation current efficiency and F Faraday's constant. The analogous expression for anion transport is

$$- V_C \frac{dA}{dt} = \eta_A \frac{I}{z_A F} \tag{5}$$

Following classical electrochemical theory η_C and η_A may be equated to the following

$$\eta_C = \theta_C C / (\theta_C C + \theta_A A + \theta_p P)$$

and $\eta_A = \theta_A A / (\theta_C C + \theta_A A + \theta_p P)$ (6)

Here θ_A, θ_C and θ_p are the transport numbers for the passage of A, C and the current flux due to the OH^- H^+ couple respectively. If however, the values for η_C and η_A are substituted in equation (7) and (8) and if (7) is divided by (8) the following simple equation is obtained

$$\frac{V_A \, dC}{\theta_C \, C} = \frac{V_C \, dA}{\theta_A \, A}$$

or

$$\frac{dA}{A} = \frac{V_A \, \theta_A}{V_C \, \theta_C} \frac{dc}{C} \tag{7}$$

This equation integrates with the boundary condition that A = A_0 when C = C_0 to give

$$\frac{A}{A_0} = \left[\frac{C}{C_0} \right]^\lambda \tag{8}$$

where $\lambda = V_A \, \theta_A / V_C \, \theta_C$

To demonstrate the validity of this expression a run was done with $A_0 = C_0 = 0.01$ and $V_A = V_C$ using a solution of potassium iodide and a barium phosphate membrane. Samples were taken from the cell at intervals and analysed for potassium by flame emission spectrophotometry and for iodide by visible spectrometry. The results are plotted as a graph of $\ln (A/A_0)$ as a function of $\ln (C/C_0)$ and a reasonable straight line fit obtained as shown in Fig. 3.

In a similar manner it is possible to study the simultaneous separation of two cations C and B. B has a transport number of θ_B and a valency z_B. From the reasoning employed and expression similar to (10) can be obtained for migration out of the anode chamber

$$\frac{dB}{dC} = \frac{\theta_B}{z_B} \frac{z_A}{\theta_A} \frac{B}{C} = m \frac{B}{C} \tag{9}$$

340

Where $m = \theta_B z_A / z_B \theta_A$. This finally gives

$$(B/B_o) = (C/C_o)^m \qquad (10)$$

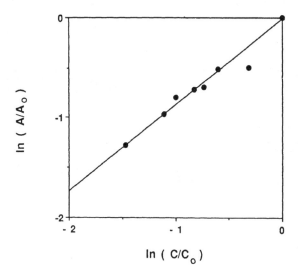

Figure 3.   Determination of Transport Number Ratio

CONCLUSIONS

In the preceeding pages some of the simpler aspects of dialysis using inorganic membranes have been considered. If the transport numbers for all the ionic species can be determined it would be possible to model the behaviour of a batch experiment on a computer where a known voltage is applied across the membrane. The ionic fluxes would be used to estimate the concentration of species in the chamber. From this the electrolyte conductivities could be calculated and hence the resistance of the cell. In this way current flux can be determined with time at a fixed voltage. Another future area of application is to extent theoretical considerations to continuous flow plant.

NOMENCLATURE

A	=	Anion concentration in cathode chamber	(keq/m^3)
B	=	Cation concentration in anode chamber	(keq/m^3)
C	=	Cation concentration in anode chamber	(keq/m^3)
F	=	Faradays Constant	(coulomb/keq)
H	=	H^+ concentration	(keq/m^3)
I	=	Cell current	(Amps)
m	=	$(\theta_B z_A / z_B \theta_A)$	eq (12)
P	=	Total ionic flux	(keq)
pH_A	=	Anode chamber pH	
pH_C	=	Cathode chamber pH	
V_A	=	Anode chamber volume	(m^3)
V_C	=	Cathode chamber volume	(m^3)
z_A, z_B, z_C	=	Valencies of ions A, B, and C	
$\eta_A \ \eta_C$	=	Current efficiencies of ions A and C	
$\theta_A, \theta_B, \theta_C$	=	Transport numbers of ions A, B and C	
θ_p	=	Transport numbers of H^+ OH^- couple	
λ	=	$V_A \theta_A / V_C \theta_C$ in equation 11	
o	=	Subscripts for initial concentrations.	

REFERENCES

(1). Labayru, R. and White, D.A., Trans. I. Chem. E., 1991, 69, 30-34.
(2) Labayru, R., Nucleotécnica, 1991, 20, 15-24.
(3) White, D.A., Labayru, R., Obermoller, H.R. and Deryde, X.,
Proc. XII Int. Symp. Desalination and Water Re-use, Malta 1991, 4, 177-188.
(4) Hann, R.E., Eyres, R. and Cottier, D., (Flett, D. Ed.), SCI London, 1983, 31-38.
(5) Fabiani, R., Int. Conf. Sep. Sci. Technol., (King Ed.), 1985, 2, 449-458.
(6) Fabiani, R. and De Francesco, M., Membrane and Membrane Processes. (Plenum Press, New York), 1986, 263-272.

HYDROLYSIS AND TRACE Cs$^+$ EXCHANGE IN Na- AND K-MORDENITE

RISTO HARJULA[1], JUKKA LEHTO[1], JANNECO H. POTHUIS[1],
ALAN DYER[2] AND RODNEY P. TOWNSEND[3]
1. University of Helsinki, Department of Radiochemistry,
Unioninkatu 35, 00170 Helsinki, Finland.
2. University of Salford, Department of Chemistry,
Salford M5 4WT, UK
3. Unilever Research, Port Sunlight Laboratory,
Bebington, Wirral, Merseyside L63 3JW, UK

ABSTRACT

Hydrolysis and trace Cs$^+$ exchange in Na- and K-mordenite were studied in Na-or K-nitrate solutions. The zeolites hydrolyse through hydronium exchange, which brings OH$^-$ and Na$^+$ (Na-mordenite) or K$^+$ (K-mordenite) into the solution. In addition, Si and Al dissolve slightly from the zeolites. Equations were derived for the prediction of equilibrium pH, Na$^+$- and K$^+$- concentrations, and the distribution coefficient of trace Cs$^+$ for the zeolites. The predictions for the distribution coefficients failed in dilute salt solutions due to strong decrease of selectivities for trace Cs$^+$ ions.

INTRODUCTION

Zeolites are promising ion exchangers for aquatic pollution control. At present they are mainly used in detergency and nuclear waste treatment, but they have a large potential for many other applications. Mordenite is effective for the removal of ^{137}Cs from nuclear waste solutions [1].

Ion exchange equilibria in zeolites may be interfered by the concurrent hydrolysis, for instance by converting the zeolites partially to H$_3$O$^+$-form [2], as metal ions in zeolite are exchanged for H$_3$O$^+$ from water. The resulting metal hydroxide in the solution, too, is likely to affect other ion exchange equilibria, especially at low ionic concentrations.

EXPERIMENTAL

Na-mordenite (Laposil 3000, Laporte Adsorbents, UK) was used in the experiments as supplied. The K-form of mordenite was prepared by successive equilibrations in 0.1 M KNO_3. The unit cell compositions of the zeolites were [3]:

$Na_{7.8}[(SiO_2)_{8.0}(AlO_2)_{46.0}] \cdot 39\ H_2O$ and
$K_{7.8}[(SiO_2)_{8.0}(AlO_2)_{46.9}] \cdot 38\ H_2O$.

Samples of Na-mordenite (NaM) and K-mordenite (KM) were equilibrated in $NaNO_3$ or KNO_3 with varying concentrations of salts at a constant solution to zeolite ratio (V/m) of 100 $cm^3 \cdot g^{-1}$. Into the salt solutions, $2 \cdot 10^{-7} - 5 \cdot 10^{-8}$ mol/l of Cs^+ with ^{134}Cs tracer (Amersham Int., UK) was added. After one weeks' contact time, which was found sufficient for the attainment of all ion exchange equilibria in the solutions, the zeolites were separated from the solution by centrifugation. The solutions were analysed for ^{134}Cs, Na^+, K^+, pH, total Si, total Al and total CO_2 [3].

THEORY

The corrected selectivity coefficient $K_G^{Cs/M}$ for the exchange of Cs^+ and an univalent cation M^+ in a zeolite can be written as [3]:

$$K_G^{Cs/M} = \frac{[\overline{Cs^+}][M^+]}{[Cs^+][\overline{M^+}]}\ \Gamma \tag{1}$$

where $[Cs^+]$ and $[M^+]$ are the concentrations of the ions in the solution $(mol \cdot dm^{-3})$, $[\overline{Cs^+}]$ and $[\overline{M^+}]$ are the concentrations in the zeolite $(mol \cdot kg^{-1})$. Γ is the solution phase activity correction $(\Gamma = \gamma_M/\gamma_{Cs})$. Trace ion exchange equilibrium may decsribed by the distribution coefficient K_D. For trace Cs^+ in this case, $K_D = [\overline{Cs^+}]/[Cs^+]$, and can be calculated from [3]:

$$logK_D = log(K_G^{Cs/M}[\overline{M^+}]) - log[M^+] - log\Gamma \tag{2}$$

For trace Cs^+ exchange $[M^+]_i \gg [Cs^+]_i$ (subscripts i denote initial concentrations). From this it follows that $[M^+]=[M^+]_i$.

However, concurrent hydrolysis of the zeolite may bring considerable amounts of M^+ into the solution. Zeolites hydrolyse through H_3O^+ exchange (assuming that they are not in H_3O^+-form). The net hydrolysis reaction is represented by equation:

$$\overline{M^+} + 2\ H_2O \longleftrightarrow \overline{H_3O^+} + OH^- + M^+ \tag{3}$$

Thus, in case of concurrent hydrolysis $[M^+] = [M^+]_i + [M^+]_x$, where $[M^+]_x$ is the concentration of M^+ resulting from the H_3O^+ exchange, and at low $[M^+]_i$, it may effectively control the equilibrium concentration of M^+.

The true corrected selectivity coefficient $K_G^{H/M}$ of H_3O^+/M^+ exchange is defined as

$$K_G^{H/M} = \frac{[\overline{H_3O^+}]\,[M^+]}{[H_3^+O]\,[\overline{M^+}]}\ \Gamma \tag{4}$$

The apparent selectivity coefficients $K_M^{H/M}$ and $K_H^{H/M}$, determined independently from the measurements of $[M^+]_x$ or $[H_3O^+]$ alone, respectively, can be written as [4]:

$$K_M^{H/M} = \frac{[M^+]_x^2 (V/m)\,([M^+]_i + [M^+]_x)}{k_w\,[\overline{M^+}]}\ \Gamma \tag{5}$$

and $\quad K_H^{H/M} = \dfrac{(k_w/[H_3O^+])\,(V/m)\,(k_w/[H_3O^+] + [M^+]_i)}{[H_3O^+]\,[\overline{M^+}]}\ \Gamma \tag{6}$

assuming that $[M^+]_x \approx [OH^-]$. This holds in initially neutral, or nearly neutral, solutions that contain no weak acids that could consume OH^- produced by hydrolysis. It is also required that $[M^+]_x \gg [H_3O^+]_i$. When these conditions are met, measurements of selectivity coefficient by Eqs. 5 and 6 yield the true corrected selectivity coefficient $K_G^{H/M}$. If, however, OH^--consuming species are present in the solution, $K_M^{H/M}$ and $K_H^{H/M}$ are only apparent selectivity coefficients [4], and in this case , the calculation of $[H_3O^+]$ and $[M^+]$ using the true selectivity coefficient $K_G^{H/M}$ would be complicated. By using the apparent selectivity coefficients, presence of OH^--consuming species can be ignored, and this assumption greatly simplifies the calculations. For the calculation of $[M^+]$ and $[H_3O^+]$ at different $[M^+]_i$ and V/m, Eqs. 5 and 6 are solved for $[M^+]_x$ and $[H_3O^+]$, respectively.

In nearly neutral salt solutions only very minor conversion of the zeolites to H_3O^+-form is likely to occur at V/m = 100 $cm^3 \cdot g^{-1}$ used in this study. For instance for NaM and KM in water, solution to zeolite ratios of ca. 2000-3000 $cm^3 \cdot g^{-1}$ would be needed to achieve 5 % conversion [3]. Thus the zeolites are expected remain essentially in pure M^+- form and this means that the selectivity coefficients of H_3O^+ and Cs^+ exchange should be constant as a function of $[\overline{H_3O^+}]$, $[\overline{Cs^+}]$ and $[M^+]$, provided that salt imbibition is negligible [5,6].

RESULTS AND DISCUSSION

Hydrolysis and Dissolution of the Zeolites

Hydrolysis brought considerable amounts of Na^+ and K^+ in to the solutions (Fig. 1). The pH of the solutions decreased with increasing concentration of M^+ due to the competing effect of M^+ on H_3O^+ exchange. Significant amounts of Si and Al were found in the solutions, resulting from the dissolution of the zeolites. Concentrations of total Si in the solutions were almost constant, being ca. $2 \cdot 10^{-4}$ M for NaM and ca. $1 \cdot 10^{-4}$ M for KM. Concentrations of total Al in the solutions for NaM and KM varied with pH and were in the range of $1 \cdot 10^{-5}$–$5 \cdot 10^{-7}$ M. The concentrations of carbonate ions (as CO_2) were below the detection limit $(1 \cdot 10^{-4}M)$ of the CO_2-sensing electrode used in the measurements. However, based on the earlier hydrolysis results for zeolite NaX [4], it was assumed that carbonate ions were present in the solutions. Total CO_2 concentration of about $5 \cdot 10^{-5}$ M would be sufficient for the validity of electroneutrality condition:

$$[Na^+]_x = [OH^-] + [HCO_3^-] + 2[CO_3^{2-}] + [Al(OH)_4^-] + [SiO(OH)_3^-]$$
$$+ 2[SiO_2(OH)_2^{2-}] \tag{7}$$

which was found valid for NaX equilibrated in deionised water. NaX was found to preadsorb the CO_2 from air prior to the equilibrations.

Calculation of pH and [M$^+$]

The presence of silicate and carbonate ions in the solutions meant that the calculation of the equilibrium pH and $[M^+]$ from the true selectivity coefficient $K_G^{H/M}$ would be complicated. Therefore, apparent selectivity coefficients $K_H^{H/M}$ and $K_M^{H/M}$ were used, as appropriate, for the calculation pH and $[M^+]$. Constant values, determined in 0.1 M $NaNO_3$ and KNO_3, were used for the selectivity coefficients in the calculation of $[Na^+]_x$ ($K_{Na}^{H/Na} = 1.39 \cdot 10^4$) and $[K^+]_x$ ($K_K^{H/K} = 4.02 \cdot 10^3$). There was a good agreement between calculated and observed $[M^+]$ for both Na^+ and K^+ (Fig. 1). Calculated from the measured $[Na^+]_x$ and $[K^+]_x$, the maximum conversions of NaM and KM to H_3O^+-form were 0.83 % and 0.55 %, respectively.

For NaM, the equilibrium pH of the solutions was calculated using a constant $K_H^{H/Na} = 48.9$. The agreement of the calculated pH's with the observed ones was only moderate (Fig. 1). For KM, the calculation of pH deviated even more from observations. It is assumed that CO_2 uptake from air during the pH-measurements is a major reason for this. In addition, silicate and carbonate ions released into the solution from the zeolites may have interfered with the predictions. The prediction of pH was found to work very well over a wide range of conditions for zeolites NaX and NaY [4]. Hydrolysis of these zeolites produced much higher pH into the solutions and the interference of silicate and carbonate ions was thus less effective.

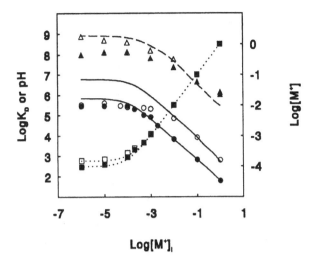

Figure 1. Equilibrium concentration [M$^+$] of the macro-ion M$^+$ (squares), pH (triangles) and distribution coefficient K of trace Cs$^+$ (circles) as a function of initial macro-ion concentration [M$^+$]$_i$ for Na-mordenite (open symbols, M$^+$= Na$^+$) and for K-mordenite (filled symbols, M$^+$= K$^+$). Solid, dotted and broken lines are the calculated values for K$_D$, [M] and pH, respectively.

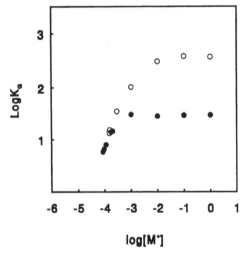

Figure 2. True corrected selectivity coefficients K$_G^{Cs/M}$ for trace Cs$^+$ exchange as a function of equilibrium macro-ion concentration [M$^+$] for Na-mordenite (open symbols, M$^+$= Na$^+$) and for K-mordenite (filled symbols, M$^+$= K$^+$).

Calculation of Trace Cs$^+$ Equilibrium.

In the calculation of K_D's (Eq. 3) constant selectivity coefficients, $K_G^{Cs/Na} = 380$ and $K_G^{Cs/K} = 28.8$, were used for NaM

and KM. These were determined in 0.1 M NaNO$_3$ and KNO$_3$ The predictions were very good when $[Na]_i >$ ca. 10^{-2} M and when $[K^+]_i >$ ca. 10^{-3} M (Fig. 1). Below these concentrations, the observed K_D-values levelled off at higher concentrations than the calculated ones and for NaM, the deviation became large at low sodium concentrations.

In dilute solutions, the presumed constancy of the $K_G^{Cs/M}$-values used in the calculation of K_D's (Eq.5) was no longer valid. The observed K_D-values remained constant though $[M^+]$ in the solutions decreased. This meant that $K_G^{Cs/M}$'s for the zeolites strongly decreased with decreasing $[M^+]$, in contradiction to present theory (5,6), (Fig 2). As the zeolites remained in practically pure Na$^+$- and K$^+$-forms in the equilibrations, the changes in the selectivity coefficients cannot arise from the changes of zeolite compositions.

In our earlier study selectivities were found to decrease with decreasing concentration of the macro-ion for trace Ca^{2+} and H$_3$O$^+$ exchange in zeolites NaX and NaY [4,7].The phenomenon is not restricted to trace ion exchange, but operates over the whole composition range of the exchanging ions in zeolites when the total concentration of the ions is lowered below certain limits [8,9]. It has also been shown that electrolyte imbibition or possible suspended colloidal particles arising from partial dissolution of the zeolites cannot be responsible for this phenomenon (9).

CONCLUSIONS

Hydrolysis of zeolites, ie. H$_3$O$^+$-exchange from water, controls the solution composition in dilute solutions. The solution compositon arising from hydrolysis can be predicted using apparent selectivity coefficients of H$_3$O$^+$- exchange, but dissolution of zeolites and presence of carbonate ions in the solution may interfere with the predictions.

In dilute solutions, decrease of selectivity at constant zeolite composition with decreasing solution concentration has been observed in several earlier studies. This phenomenon is inconsistent with current theories of ion exchange. This study gives further indication that the phenomenon may well be common in zeolites.

ACKNOWLEDGEMENTS

This study has been supported by the Royal Society European Programme with the Academy of Finland. Support has also been recieved from Unilever Research (UK), Imatran Voima Foundation (Finland) and Finnish Cultural Fund. All these sponsors are thankfully acknowledged.

REFERENCES

1. Harjula, R. and Lehto, J., Nucl. Chem. Waste Manag., 1986, **6**, 133.
2. Franklin, K.R., Townsend, R.P., Whelan, S.J. and Adams, C.J., Proc. 7th Int. Conf. Zeolites, Tokyo, 1986, Elsevier, Amsterdam (1986).
3. Harjula, R., Lehto, J., Pothuis, J.H., Dyer, A. and Townsend, R.P., submitted in J. Chem. Soc., Faraday Trans.
4. Harjula, R., Lehto, J., Dyer, A. and Townsend, R.P., submitted in J. Chem. Soc., Faraday Trans.
5. Kraus, K.A., "Ion Exchange", in "Trace Analysis", Yoe, J.H. and Koch, H.J., Eds., pp. 34-102, John Wiley & Sons, New York (1957).
6. Barrer, R.M. and Klinowski, J., J. Chem. Soc., Faraday Trans, 1974, **70**, 2180.
7. Harjula, R., Dyer, A. and R.P.Townsend, submitted in J. Chem. Soc., Faraday Trans.
8. Franklin, K.R. and Townsend, R.P., J.Chem.Soc., Faraday Trans., 1988, **84**, 2755.
9. Dyer, A., Harjula, R., Pearson, S. and Townsend, R.P., submitted in J. Chem. Soc., Faraday Trans.

INORGANIC ANION EXCHANGERS FOR 99-Tc AND 125-I UPTAKE

A. DYER AND M. JAMIL

Department of Chemistry and Applied Chemistry

University of Salford, Salford M5 4WT

ABSTRACT

Earlier work (1) reported investigations on 21 inorganic materials suggested for technetium (99-Tc), sulphur (35-S), iodine (125-I) isotope removal from simulated aqueous nuclear wastes by anion exchange. This work has been extended to a systematic study of five of the most promising of the original anion exchangers. These materials are lead sulphide, mercarbide and the oxides of zirconium (Zirox), iron (Ferrox) and aluminium. The work presented here details anion kinetics and equilibria and leach tests of exchanger/cement composites. The ion pairs studied were NO_3/TcO_4, NO_3^-/SO_4^{2-}, NO_3^-/I^-. The perrhenate ion was used as carrier for $99\text{-}TcO_4^-$.

INTRODUCTION

Long lived isotopes of both technetium (99-Tc, $t_{1/2} = 2.13 \times 10^5$ y) and iodine (129-I, $t_{1/2} = 1.57 \times 10^7$ y) are created by fission in nuclear reactors, and both elements are used as isotopes in nuclear medicine. 35-S species also occur in nuclear wastes, albeit of short half-life.

Generally the preferred method of treatment of aqueous nuclear effluents is ion exchange and anion species usually are taken up onto organic resins.Although organic resins can be encapsulated in say cement waste forms clearly organic resins loaded with high isotope concentrations will have a limited radiation stability.In organican ion exchangers would bepreferable and an earlier paper (1) described a survey of 21 candidate inorganic substances with anion capacities. Five materials showed reasonable capacities for $99\text{-}TcO_4^-/ReO_4^-$, $125\text{-}I^-$, and $35\text{-}SO_4^=$ uptake and are the subject of the further studies described here. They are a (i) mercarbide polymer, (ii) lead sulphide, (iii) hydrated zirconium oxide (Zirox), (iv) hydrated ferric oxide (Ferrox) and (v) hydrated aluminium oxide (Camag alumina).

EXPERIMENTAL

Materials: Zirox,Ferrox and lead sulphide were supplied as 0.1-0.5 mm granules by Recherche Applique du Nord, Hautmont, France. Alumina was obtained as 63-150 μm particles from Fisons, Loughborough, UK and the mercarbide polymer $((CHg_3O)_n^{n+}$

$(NO_3^-)_n)$ was synthesised by the method of Weiss and Weiss(2).

Ammonium perrhenate $(NH_4Re\overline{O_4})$ (carrier for pertechnate) and all the other chemicals used were at least GPR grade and all solutions were in deionised water.

The isotopes were supplied by Amersham International, Amersham, UK as follows (i) 125-I as NaI in dilute-NaOH, (ii) 35-S as Na_2SO_4 in pH 6.8 solution and (iii) 99-Tc as solid NH_4TcO_4.

METHODS

Radiochemical analyses were by (a) liquid scintillation counting (99-Tc, 35-S) of stable microemulsions of composition 1 part aqueous isotopic solution, 3 parts Triton-X 100 and 6 parts toluene based scintillant, or (b) γ-ray spectrometry (125-I).

Ion exchange isotherms were constructed from analyses of solutions equilibrated with 0.1 g aliquots of exchanger in nitrate form. The equilibrating solutions (20 mL) were 0.1

352

Fig.1- Anion Exchange Isotherms on Hydrous Oxides

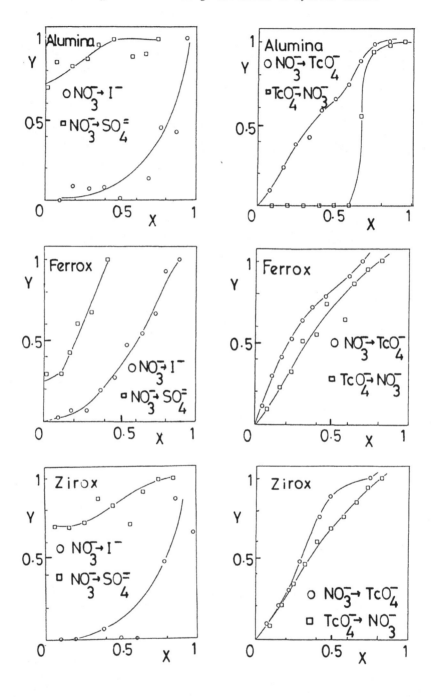

Fig. 2 - Anion Exchange Isotherms on Mercarbide and PbS

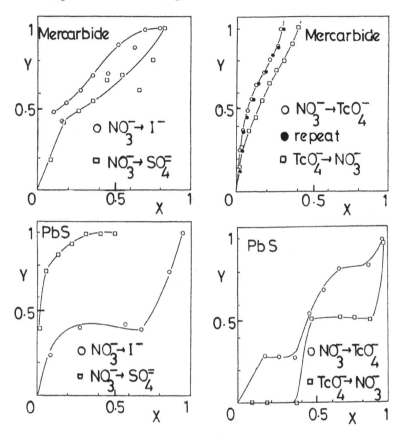

Fig. 3 - Arrhenius Plots for TcO_4^- Diffusion

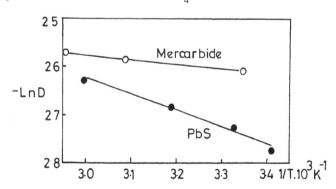

isonormal solutions of nitrate and the ingoing anion ($99-TcO_4^-$ /ReO_4^-, $125-I^-$ and $35-SO_4^{2-}$). The isotherms obtained are in Figs. 1 and 2.

Kinetic analyses were made in which 0.05 g of exchanger (nitrate form) were contacted with 5 mL of isotopically labelled solution. Sampling at different time intervals enabled diffusion profiles to be produced. In three cases (uptake of $99-TcO_4^-$ onto PbS, mercarbide, $99-TcO_4^-$ and Zirox) it was possible to extend the kinetic analysis over a temperature range (293 → 338 K) and derive energies of activation (E) (Fig.3).

Composites were made with Portland Cement (OPC) and Camsave BFS. OPC composites were of 4 g exchanger to 36 g OPC with water (solid ratio maintained at 0.40). OPC/BFS composites were made at the same water/solid ratio from 4 g exchanger, 18 g OPC and 18 g BFS. The exception was mercarbide when 2 g were encapsulated in 38 g OPC or in 19 g OPC plus 19 g BFS. Samples were left for 24 h and cured in polypropylene cubes for 24 days at 100% relative humidity. Leaching experiments were carried out at 25°C using deionised water (DW) and synthetic sea (SSW) and ground waters (SGW).

Elucidation of kinetic and leaching properties used computer programs based upon the Carman-Haul equation(3) for diffusion from a fixed amount of solid into a fixed amount of solution. In all cases kinetic and leach parameters were obtained from good fits to computer predicted profiles. Some examples of these from the leaching experiments are in Figs.4,5. (Note - no composite/leaching experiments for PbS or involving $35-SO_4^-$ were carried out because of the known deleterious effects of S containing moieties on cement).

DISCUSSION

Isotherm Data:

PbS: The unusual nature of the isotherms in Fig. 2 meant that the process studied was unlikely to be true ion exchange. The lack of water, or OH groups, in PbS (TGA) ruled out the possibility of exchange via OH^-. A crude selectivity series can

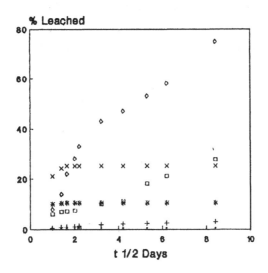

FIG 4a LEACHING OF Tc 99
M- MERCARBIDE
Z-ZIROX,F-FERROX

% Leached

t 1/2 Days

· M/SW OPC/BFS	+ M/SW OPC	* Z/GW OPC/BFS
□ Z/GW OPC	× F/SW OPC/BFS	◇ F/SW OPC

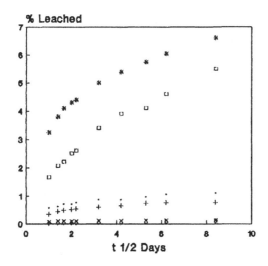

FIG 4b LEACHING OF I-125
M- MERCARBIDE
Z-ZIROX,F-FERROX

% Leached

t 1/2 Days

· M/GW OPC/BFS	+ M/GW OPC	* Z/DW OPC/BFS
□ Z/DW OPC	× F/GW OPC/BFS	◇ F/GW OPC

be suggested as I > SO_4^{2-} > 99-TcO_4^- ≥ NO_3.

Mercarbide: NO_3^- → 99-TcO_4^-/ReO_4^- exchange was a reversible process of reasonable selectivity for 99-TcO_4^-. Reversibilities of NO_3^- → SO_4^{2-} and NO_3^- → I^- were less convincing (not shown). The selectivity series observed was; I^- > TcO_4^- >> NO_3^- ≥ SO_4^{2-}.

Hydrated oxide: All three hydrated oxides showed similar selectivity trends, viz.SO_4^{2-} >> NO_3^- ≥ TcO_4^- > I^-. This was similar to those of Hallaba[4] and Dyer and Malik[5] for hydrated zirconia and titania respectively, but unlike that of Amphlett[6] for hydrated zirconia. Not all reversibilities were checked but it was possible that the NO_3^- → 99-TcO_4^-/ReO_4^- exchange on Ferrox was a true anion exchange.
The selectivities were in accord with the K_d results reported earlier[1]. The slight hysteresis observed in NO_3^- → 99-TcO_4^-/ReO_4^- systems may well be an artefact of the presence of the perrhenate carrier.

Kinetics: No literature values exist for comparison — clearly 99-TcO_4^- experienced a lower energy barrier (E = 17 $kJmol^{-1}$) in entering the mercarbide than PbS (30 $kJmol^{-1}$). The process of 99-TcO_4^- uptake onto Zirox was judged to be a two stage process.

Leaching studies: The leach rates from OPC encapsulates were low. Those for 99-TcO_4^- were in the range 1.10^{-15} → 9.10^{-17} $m^2 s^{-1}$ and those for I^- fell between 1.10^{-15} and 9.10^{-16} $m^2 s^{-1}$.

Hydrous oxide composites: Leach rates were obtained for 99-TcO_4^- from Ferrox, Zirox and alumina in OPC. In all cases the order of leach rates was SSW > DW ≥ SGW 125-I^- leach rates from Ferrox and Zirox followed the same pattern (alumina had too low a capacity for 125-I^- to warrant study). Additions of BFS to the composite was very successful if 99-TcO_4^- was being removed — virtually complete inhibition was seen (Fig. 4). Surprisingly BFS addition increased leach rate when 125-I^- was encapsulated (up to x 100 in the case of Zirox for all three leachants).

Mercarbide: Here 99-TcO_4^- leaching from mercarbide/OPC followed the pattern DW > SSW > SGW and again was completely inhibited by BFS addition. Leach curves for 125-I$^-$ also showed good retention with the trend SSW > SGW > DW, but, as before, BFS addition was ineffective.

CONCLUSIONS

All the materials studied show promise for the removal of 99-TcO$_4^-$ and 125-I$^-$ from aqueous environments. Containment in cement composites was possible and BFS greatly improved the leach resistance of 99-TcO$_4^-$ exchanges encapsulated in OPC.

ACKNOWLEDGEMENTS

Gift of materials from Recherche Applique du Nord, Hautmont, France were much appreciated as was financial support from H.M.I.P. (for M.J.) throughout this week. The views expressed herein do not necessarily reflect UK government policy.

NOMENCLATURE

D   diffusion coefficient m^{2}s^{-1}

E   energy of activation kJmol^{-1}

K_d distribution coefficient L K$_g^{-1}$

t   time

X   equivalent fraction of anion in solution.

Y   equivalent fraction of anion in exchanger.

REFERENCES

1. Dyer, A., and Jamil, M. in "Ion Exchange for Industry" Ed. M. Streat, Ellis Horwood, 1988, Pg. 494.

2. Weiss, A. and Weiss, A., Z. Anorg. Allgem. Chem., 1955 **282**, 324.

3. Carman, P.C., and Haul, R.A.W., Proc. Roy. Soc. London 1954, A<u>222</u>, 109.

4. Hallaba, E., Ind. J. Chem., 1973 **11**, 580.

5. Dyer, A and Malik, S.A., J. Inorg. Nucl. Chem., 1981 **43**, 2975.

6. Amphlett, C.B., McDonald, L.A., and Redman, M.J., J. Inorg.Nucl. Chem., 1958 **6**, 236.

THE EFFECTS OF EXTERNAL GAMMA RADIATION AND OF *IN SITU* ALPHA PARTICLES ON FIVE STRONG-BASE ANION EXCHANGE RESINS

S. FREDRIC MARSH
Nuclear Materials Process Technology Group,
Mail Stop E501, Los Alamos National Laboratory,
Los Alamos, New Mexico 87545, USA

ABSTRACT

The effects of external gamma radiation and of *in situ* alpha particles were measured on a recently available macroporous, strong-base, polyvinylpryidine anion exchange resin and on four strong-base polystyrene anion exchange resins. Each resin was irradiated in 7 M nitric acid to 1-10 megaGray of gamma radiation from external ^{60}Co, or to 5-15 megaGray of *in situ* alpha particles from sorbed ^{238}Pu. Each irradiated resin was measured for changes in dry weight, wet volume, weak-base and strong-base chloride exchange capacities, and exchange capacities for Pu(IV) from nitric acid. Alpha-induced resin damage was significantly less than that caused by an equivalent dose of gamma radiation. The polyvinylpyridine resin showed the greatest resistance to damage from both gamma radiation and alpha particles.

INTRODUCTION

Anion exchange in nitric acid is the major aqueous process used to recover and purify plutonium from a variety of impure nuclear materials. Although this system is quite selective for Pu(IV), the sorption kinetics for Pu(IV) are particularly slow(1).

The Plutonium Facility at Los Alamos National Laboratory has an on-going program whose objective is to develop anion

exchange resins that provide improved sorption kinetics and
increased safety for processing Pu(IV) in nitric acid. The
best anion exchange resin previously available, in terms of
performance, chemical resistance, and radiation stability,
was Permutit™ SK, a vinylpyridine polymer.

Because Permutit™ SK is no longer manufactured, we initiated
a collaborative effort with Reilly Industries, Inc., Indian-
apolis, Indiana, to develop a polyvinylpyridine resin having
similar beneficial properties. The resulting resin, Reillex™
HPQ, is a macroporous copolymer of 1-methyl-4-vinylpyridine
and divinylbenzene. Its functional group is compared with
that of conventional polystyrene resin in Fig. 1.

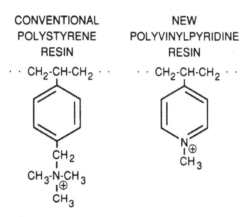

Fig. 1. Functional groups of conventional strong-base
polystyrene and the new polyvinylpyridine resins.

First, we evaluated the performance and chemical resistance
of Reillex™ HPQ in nitric acid(2). Next, we measured the
effects of external gamma radiation on this resin and on four
strong-base polystyrene resins: Dowex™ 1x4, a gel-type resin;
and three macroporous resins: Dow™ MSA-1, Amberlite™ IRA-
900, and Lewatit™ MP-500-FK(3). (The gamma-irradiation
portion of this investigation was done at Harwell Laboratory,
Oxfordshire, UK, during 1989-1990, when the author was
attached to the Chemistry Division as a Visiting Scientist.)

Finally, we measured the effects of *in situ* alpha particles on these five resins and one other resin, Lewatit^{TM} UMP-950(4).

PROCEDURES AND RESULTS

Effects of Nitric Acid

Vinylpyridine resins are expected to be more resistant to attack by nitric acid than polystyrene resins because polystyrene is susceptible to electrophilic aromatic substitution, whereas the electronegativity of nitrogen in the pyridine ring makes polyvinylpyridine highly resistant to such substitution.

To evaluate the chemical resistance of Reillex^{TM} HPQ resin under worst-case conditions, we subjected a portion of this resin to boiling concentrated nitric acid, under reflux, for 3 hours. During this treatment, a small amount of NO_2 was the only indication of any reaction. After air-drying, the resin weight was essentially unchanged from its initial weight, although its volume in water had increased by nearly 10%.

Because the resin damage appeared to have been minor, we next compared the capacities of treated and untreated resin for Pu(IV) from 7 M nitric acid (the plutonium processing medium). These measurements demonstrated that, instead of a decreased capacity, the capacity of acid-boiled Reillex^{TM} HPQ resin *increase*d by nearly 20%(2).

Reillex^{TM} HPQ resin consists of 1-methyl-4-vinylpyridine crosslinked with approximately 30% divinylbenzene (DVB). The most vulnerable portion of the polymer, therefore, is the DVB crosslinking groups. Oxidative cleavage of DVB during nitric acid refluxing would relax the resin network structure by lowering the crosslinking density and allow the resin beads to expand. Such resin swelling, evidenced by the observed increase in wet resin volume, would make the deeply buried exchange sites more accessible, which would explain the increased resin capacity and faster sorption kinetics.

Effects of Radiation

Because Reillex™ HPQ resin also was expected to be more resistant to radiolytic degradation, we tested the radiation stability of Reillex™ HPQ and of various other anion exchange resins previously used to process plutonium.

Each resin was irradiated in 7 M nitric acid to total absorbed doses of 1-10 MGray of external gamma radiation from ^{60}Co, or to 5-15 MGray of *in situ* alpha particles from sorbed ^{238}Pu. All irradiated resins were measured for changes in dry weight, wet volume, chloride exchange capacity, and Pu(IV) capacity.

Weight losses for the various resins as a result of gamma and alpha irradiations are presented in Tables I and II. (No values are presented for Lewatit™ MP-500-FK at the highest gamma doses in Table I because at those levels this resin was almost completely destroyed.) Note that all resins survived alpha-particle irradiation (Table II) better than gamma radiation, and that 97% of the initial weight of Reillex™ HPQ resin remained even after the highest alpha dose of 14.3 MGray.

TABLE I. Resin Weight Loss from Gamma Irradiations
(post-irradiation wt./pre-irradiation wt.)

Nominal Exposure (MGray)	Dowex™ 1x4	Dow™ MSA-1	Amberlite™ IRA-900	Lewatit™ MP-500-FK	Reillex™ HPQ
0	1.000	1.000	1.000	1.000	1.000
1.0	0.996	0.994	0.992	0.978	0.997
2.0	0.973	0.952	0.968	0.881	0.990
3.6	0.937	0.884	0.933	0.710	0.997
5.5	0.824	0.750	0.830	0.351	0.898
7.0	0.796	0.674	0.753	0.059	0.842
8.2	0.814	0.641	0.702	---	0.834
10.0	0.762	0.483	0.536	---	0.683

TABLE II. Resin Weight Loss from Alpha-Particle Irradiations
(post-irradiation wt./pre-irradiation wt.)

Nominal Exposure (MGray)	Dowex™ 1x4	Dow™ MSA-1	Amberlite™ IRA-900	Lewatit™ MP-500-FK	Reillex™ HPQ	Lewatit™ UMP-950
0	1.000	1.000	1.000	1.000	1.000	1.000
4.9	0.958	0.964	0.953	0.902	0.992	0.929
8.2	0.931	0.920	0.937	0.838	0.995	0.875
11.1	0.898	0.890	0.905	0.770	0.992	0.832
14.3	0.880	0.835	0.871	0.680	0.970	0.771

Capacities of these same resins for Pu(IV) from 7 M nitric
acid during a 15-minute dynamic contact are presented in
Tables III and IV. These Pu(IV) capacity data confirm that
the resin damage caused by alpha particles is significantly
less than that caused by a comparable dose of gamma radiation.

TABLE III. Sorption of Pu(IV) on Gamma-Irradiated Resins from
7 M Nitric Acid During 15-Minute Dynamic Batch Contacts
(mg Pu/g pre-irradiated resin)

Nominal Exposure (MGray)	Dowex™ 1x4	Dow™ MSA-1	Amberlite™ IRA-900	Lewatit™ MP-500-FK	Reillex™ HPQ
0	85	150	143	273	237
1.0	80	146	142	234	219
2.0	75	143	120	194	228
3.6	68	109	93	137	222
5.5	37	58	75	---	204
7.0	45	47	73	---	126
8.0	38	28	54	---	123
10.0	38	18	12	---	87

TABLE IV. Sorption of Pu(IV) on Alpha-Irradiated Resins from
7 M Nitric Acid During 15-Minute Dynamic Batch Contacts
(mg Pu/g pre-irradiated resin)

Nominal Exposure (MGray)	Dowex™ 1x4	Dow™ MSA-1	Amberlite™ IRA-900	Lewatit™ MP-500-FK	Reillex™ HPQ	Lewatit™ UMP-950
0	85	150	143	273	237	210
4.9	52	120	103	220	210	168
8.2	50	123	89	216	208	148
11.1	43	95	93	174	193	130
14.3	35	54	64	153	212·	107

Much of the radiolytic resin damage appears to be caused by
secondary reactions between the resin and radiolysis products
of nitric acid, rather than by primary processes. We therefore attribute the measured differences in resin damage for
comparable doses of gamma and alpha irradiation to differences
in the magnitudes of these secondary reactions.

Specifically, during both gamma and alpha irradiations, the
resin containers always included a measured excess of nitric
acid above the resin. Alpha particles, whose range in
solution is only about 30 micrometers, could interact with
only a very small volume of nitric acid, whereas gamma rays
from external ^{60}Co would irradiate the entire volume of nitric
acid present.

A previous Soviet study(5) showed that polyvinylpyridine resin
irradiated while immersed in 7 M nitric acid sustained much
greater damage than resin subjected to an equivalent dose
while immersed in water. These data, shown in Table V,
support our contention that the radiolysis products of nitric
acid constitute a major mechanism for resin damage.

TABLE V. Damage to Resin Irradiated in Various Media[*]

Irradiation Medium	Resin Weight Loss	Resin Capacity Loss
Vacuum	0	7.1%
Water	0	13.7%
7 M Nitric Acid	35.8%	58.0%

[*]Published Soviet study(5) using nitrate-form polyvinylpyridine resin irradiated to 11.2 MGray with accelerated electrons in all cases.

CONCLUSIONS

1. Reillex[TM] HPQ nearly always retained more of its initial weight and more of its initial exchange capacity than the other resins evaluated in our study. (Reillex[TM] HPQ resin previously has been shown(2) to offer greater resistance to attack by nitric acid.)

2. External gamma radiation causes substantially more resin damage than does an equivalent does of *in situ* alpha particles. We attribute this difference to secondary reactions involving highly reactive radiolysis products of nitric acid, which are more extensively produced during gamma irradiations.

REFERENCES

1. Ryan, J. L., "Ion Exchange Reactions," *Gmelin Handbuch der Anorganishen Chemie*, Band 21, Teil D2, Chapter 7 (Springer-Verlag, Berlin - Heidelberg - New York, 1975), p. 422.

2. Marsh, S. F., "Reillex[TM] HPQ: A New, Macroporous Polyvinylpyridine Resin for Separating Plutonium Using Nitrate Anion Exchange." *Solvent Extr. Ion Exch.*, 1989, 7(5), pp. 889-908.

3. Marsh, S. F., "The Effects of Ionizing Radiation on Reillex™ HPQ, a New Macroporous Polyvinylpyridine Resin, and on Four Conventional Polystyrene Anion Exchange Resins," Los Alamos National Laboratory report LA-11912 (November 1990).

4. Marsh, S. F., "The Effects of *In Situ* Alpha-Particle Irradiations on Six Strong-Base Anion Exchange Resins," Los Alamos National Laboratory report LA-12055 (April 1991).

5. Kiseleva, E. D., Khasanova, V. M., and Chmutov, K. V., "Chemical Resistance of KU-2 Resin in Various Ionic Forms to the Actions of Radiations," *Z. Prikl. Khim.*, 1960 45(5), p. 1196.

ION—EXCHANGE FIBRES FOR THE RECOVERY OF GOLD CYANIDE

MARTHA H. KOTZE* and F.L.D. CLOETE
Dept. of Metallurgical Eng., University of Stellenbosch,
Stellenbosch 7600, South Africa
(*present address : MINTEK, Private Bag X3015,
Randburg 2125, South Africa)

ABSTRACT

Fibrous ion exchangers have been synthesised using polypropylene staple as a cheap, stable and robust base. Styrene was graft copolymerised onto polypropylene fibres as a starting point. Various anionic groups were then fixed on this material to find the most effective ion exchanger for the selective recovery of gold in the presence of other metal cyanides. The gold-loading capacities and selectivities of functionalised fibres with strong base or guanidine functional groups compared favourably with a conventional anion exchange resin. The physical properties of the fibres show that operation in packed beds is feasible.

INTRODUCTION

Most of the total world production of about 2000 tons per annum of gold is produced by leaching milled ore with dilute alkaline cyanide solution. Recovery of gold and silver from the resulting mixture of metal cyanide complexes has been done in past years using precipitation from clear solutions with zinc powder, but most new plants use activated carbon to recover gold and silver directly from the leach slurry.

There have been several applications of ion exchangers in the recovery of precious metals from leach solutions during the last few decades: The application of anionic resins developed by Wells at the NCRL (Teddington) in the 1950s is described by Davison et al.(1) while work at MINTEK (Randburg) is covered in recent articles by Fleming (2) and Green (3)˙ and co-authors. The costs of regenerating activated carbon at temperatures above 600 °C are one of the reasons to consider ion exchange as an alternative.

Work on the development of fibrous ion exchangers arose from the possibility of obtaining faster overall reaction rates, due to the smaller diameter of fibres (30-60 μm), which could lead to smaller inventories of ion exchanger in plants. Fibrous ion exchangers have been made in small quantities by several workers since 1955 and more recent developments were reported by Soldatov (4). It is known that both Diaprosim (8) and Courtaulds (9) have made trial batches of their own ion-exchange fibres to test the market.

The first stage of the project consisted of a preliminary evaluation of ion-exchangers based on several readily available fibres. Cellulose, nylon and poly-acrylonitrile were modified to induce anion exchange properties, but these all showed inferior loading for gold when compared with conventional resins. Polybenzimidazole-based anion exchange fibres gave promising results (5), but only functioned up to a pH-value of 8, while the pH of gold leach solutions is above 10.

PREPARATION OF FIBRES

The skeleton of the ion exchanger was prepared from polypropylene staple fibres and purified styrene in methanol, as given in detail by Kotze (6). The mixture was irradiated with gamma rays using a Co-60 source at a dose rate of 0.157 Mrad/h to a dosage of 0.47 Mrad. The excess styrene and homopolymer were extracted with benzene using Soxhlet equipment and the mass gain of the dried fibres determined. The increase in mass was expressed as a percentage of the original polypropylene, e.g. Pp55 is a fibre containing 55 g of graft copolymerised styrene per 100 g of polypropylene.

Functional groups were fixed on the polypropylene-styrene fibres first by reaction with chloromethylmethylether. Various amines were then reacted with the fibres. Strongly basic fibres were made using trimethylamine(TMA), triethylamine(TEA), triethanolamine(TEOA) among others including a group code named SBF. The only promising weakly basic group was guanidine(GUAN), since it remained functional at pH-values above 10. Some values for dimethylamine(DMA) fibres at pH 7 are included.

The amount of functional groups, measured as ion exchange capacity, which could be fixed to the skeleton was proportional to the grafting yield of styrene as well as to the chlorine content of the fibres prior to reaction with the amine. However an anomalous effect was found when the capacity for gold was considered - this was highest when the ion exchange capacity of the fibre was about 1/3 of the maximum attainable. Thus the maximum gold loading does not correspond with the maximum ion exchange capacity and milder reaction conditions can be used in preparing the fibres, which would have less effect on the mechanical properties.

LOADING AND ELUTION OF GOLD

Fibres with various grafted styrene contents and anionic groups were equilibrated with excess synthetic leach solutions. Two leach solutions were made up to correspond to a poor and a rich solution in relation to industrial practice. After equilibrium was attained, the fibres were washed, dried and analysed for seven metals as listed in Table 1, which is a summary of the most important results from (6).

The loading and selectivity for gold increased significantly from trimethylamine and triethylamine, to SBF. The triethanolamine fibre showed reasonable selectivity for gold but gave a very poor loading.

The best fibre for adsorption was a 54% graft copolymer of styrene on polypropylene with strongly basic functionality, code named SBF. The selectivity and loading for gold were both substantially better than for IRA400, a Rohm and Haas strong base resin which has also been used for the recovery of gold.

TABLE 1: CAPACITIES OF POLYPROPYLENE FIBRES AND SELECTIVITIES FOR GOLD AND OTHER METALS FROM CYANIDE LEACH SOLUTIONS

Fibre	IEX Cap. (mol/kg)	Au	Ag	Zn	Ni	Co	Fe	Cu	Au/Σmet (-)
		(Metals as g metal/kg dry fibre)							
STRONG BASE		(rich leach soln.)							
Pp116/TMA	1.94	25.1	4.6	8.0	8.6	9.6	7.5	6.2	0.36
Pp116/TEA	0.81	22.5	1.0	15.0	4.7	3.0	0.5	0.5	0.48
Pp116/SBF	0.30	45.0	1.4	2.1	2.2	0.2	0.2	0.7	0.84
Pp116/TEOA	0.11	9.2	0.4	0.5	1.0	0.2	0.2	0.5	0.77
		(poor leach soln.)							
Pp54/SBF	0.26	64.1	0.32	2.8	4.1	0.07	0.67	2.0	0.87
IRA400 res.	3.8	14.9	13.6	22.8	4.7	21.7	9.3	0.2	0.17
WEAK BASE		(rich leach soln.)							
Pp116/DMA	-	29.6	0.9	9.0	8.2	7.0	4.3	2.5	0.48
Pp116/GUAN	1.5	42.8	3.2	1.8	6.1	0.2	0.2	2.1	0.76
		(poor leach soln.)							
Pp54/GUAN	0.9	32.7	0.44	0.1	0.6	0.05	0.1	1.1	0.93
GUAN resin	1.7	27.5	0.2	4.0	7.9	1.7	5.1	0.8	0.58

Leach solutions Ph 10, 400% molar excess of cyanide, except pH 7 for Pp116/DMA

| (mg metal/L) | | 19 | 9.6 | 5.7 | 5.7 | 5.2 | 4.1 | 5.8 | rich |
| (mg metal/L) | | 5 | 0.5 | 2 | 5 | 1 | 10 | 10 | poor |

Guanidine was the only weak base fibre which could still adsorb in solutions of pH 10-11 and showed excellent selectivity, but slightly lower capacity for gold than SBF. The weakly basic dimethylamine fibre showed acceptable loading and selectivity, but since it functioned only at pH up to a value of 7, could not be considered for industrial use.

The elution properties of the two most promising fibres, namely SBF and guanidine, were tested by pre-loading to the values given in Table 2 and then eluting the column at 7.5 to 8.5 bed volumes per hour for 50 bed volumes(6).

Eluants containing various anions were tested on the strong base SBF fibre, namely zinc cyanide, ammonium thiocyanate and acidic thiourea. Results with several concentrations of zinc cyanide at various temperatures failed to remove more than 20% of the gold. This was quite unexpected in comparison with the efficient elution of IRA400 resin in this way.

The best conditions for elution with ammonium thiocyanate were a pH value of 7, a temperature of 60 °C and a concentration of 2M. This was in contrast to the behaviour of IRA400 resin which elutes readily at ambient temperature (6). The thiocyanate anion must be stripped from the resin with ferric nitrate or sulphate before adsorption. Most gold was removed but not all the copper, nickel and iron.

Thiourea at 60 °C, acidified with sulphuric acid, was shown to be the most useful eluant for SBF fibre, since no separate regeneration process is required after elution. Virtually all the gold as well as most of the other metals were removed. This eluant will also be cheaper and less complicated to use from the electrowinning point of view than thiocyanate.

TABLE 2 : ELUTION OF GOLD AND OTHER METALS FROM MODIFIED POLYPROPYLENE FIBRES

Eluant	Efficiency of elution after 50 bed vol.(%)						
	Au	Ag	Zn	Ni	Co	Fe	Cu
STRONG BASE (Pp247/SBF)							
2M NH$_4$thiocyanate,pH 7,60°C	96	90	89	24	-	63	0
1M thiourea,0.1M H$_2$SO$_4$,60°C	99+	94	96	99+	-	91	88
WEAK BASE (Pp54/GUAN)							
0.1M NaOH,10% EtOH,60°C	97	77+	0	69	-	67	93+
0.1M NaOH,20% EtOH,40°C	96	49	18	60	-	77	86
0.1M NaOH,20% EtOH,60°C	99	61	0	67	-	69	89+
0.1M NaOH,30% EtOH,60°C	99+	77+	9	87	-	77	93+
Loaded fibres before elution (metals as g metal/kg dry fibre)							
STRONG BASE (Pp247/SBF)	41	0.6	1.0	1.4	<0.1	0.5	0.3
WEAK BASE (Pp54/GUAN)	33	0.4	0.7	3	<0.1	0.5	1.6

Elution of weak base fibres is in principle much simpler than for strong bases since deprotonation of the functional group with alkali to release the anion only is required. The simplest eluant, sodium hydroxide, was studied first, but this could only elute gold from guanidine resin at temperatures of this 80 °C using 1M solution. Such conditions could be expected to break the resin down and hydrolyse the guanidine groups in practice. Other eluants tested included mixtures of sodium hydroxide and cyanide, sodium hydroxide and ethanol or methanol, and zinc cyanide.

The most effective eluant for weak base fibres appeared to be 0.1 M sodium hydroxide in 30% ethanol at 60 °C. The low concentration of sodium hydroxide and a temperature of 60 °C should not cause breakdown of the fibre. The only danger could arise from the flammability of the solution. Nickel, zinc and iron were not fully eluted and might have to be removed periodically. We do not know whether these metals will build up any further at this stage.

The two promising fibres, Pp54/SBF and Pp54/GUAN, were subjected to five cycles of loading and elution and analysed after each cycle. Metals which were tending to accumulate were zinc and iron on the guanidine fibre and zinc, nickel and cobalt on the SBF fibre. No deterioration of the fibres was observed in these preliminary tests.

PHYSICAL PROPERTIES OF FIBRES

Some of the physical properties of the fibres which affect their behaviour in a process are listed in Table 3, from (6).

TABLE 3 : PHYSICAL PROPERTIES OF ION EXCHANGE FIBRES

	Type of fibre based on polypropylene		
Physical property	Pp54/GUAN	Pp54/SBF	Pp247/SBF
diameter (μm)	39.2	39.2	53.5
density, dry (kg/m^3)	965	998	876
swelling (% H_2O in moist fibre)	36	39	46
theoretical exchange capacity (eq/kg)	0.9	0.26	0.5

Diameter of untreated polypropylene 24 μm

The diameter of the polypropylene fibres was increased by the graft copolymerisation, as shown in Table 3. It was noted that no further increase in diameter took place during chloromethylation and amination. The diameters of fibres are much smaller than conventional resins and thus exhibit a much larger surface for reaction, typically about 100 m^2/kg, which is about ten times greater than for a standard grade resin. If one considers a resin bed of 1 m^3 with a voidage of 0.35, the absolute volume of resin would be about 0.65 m^3 and the surface area of the beads about 7000 m^2. A fibre bed of 1 m^3 with a voidage of 0.77 would contain 0.23 m^3 fibres with a surface area of 23 000 m^2.

The pressure drop across a bed of fibres was measured in terms of the voidage and superficial velocity through the bed. These data were correlated adequately by the Carman equation (7). The value of the modified Reynolds number indicated that flow was just turbulent. Typical values of pressure drop per unit depth of bed at a superficial flow rate of 60 m^3/m^2.h were 120 kPa/m and 38 kPa/m for voidages of 0.77 and 0.86 respectively.

Measurements showed that the rate of loading of gold was much faster for strong base than for weak base fibres. Comparative values (6) of reaction half-time for various ion exchangers were : Pp54/SBF, 8 min.; Pp54/GUAN, 22 min.; Imidazoline resin, 200 min.

CONCLUSIONS

The studies summarised in this paper show that ion exchange fibres with very selective affinity for gold cyanide can be synthesised from polypropylene fibre onto which styrene is graft copolymerised, followed by functionalisation. Either strongly or weakly basic fibres can be used but different elution processes must be employed.

The results indicate that guanidine (weak base) and SBF (strong base) are the most promising functional groups. Both types were shown to remain stable through a small number of cycles. Longer-term trials are required to study the stability and the build up of poisons.

Preliminary measurements show that operation of packed beds of fibres in columns is feasible at flow velocities comparable to conventional resins. The overall kinetics for the adsorption of gold were much faster for fibres than for resins, as anticipated.

NOMENCLATURE

Au/Σmet	selectivity for gold, expressed as the capacity for gold to that for total metals adsorbed
DMA	functional group based on dimethylamine
GUAN	functional group based on guanidine
Pp54	polypropylene fibre graft copolymerised with 54 g styrene per 100 g original polypropylene
SBF	code name for a strong base functional group
TEA	functional group based on triethylamine
TMA	functional group based on trimethylamine
TEOA	functional group based on triethanolamine

ACKNOWLEDGEMENTS

The financial and technical assistance given by Mintek, which made the project possible, is gratefully acknowledged. Dr.B.R. Green made many helpful suggestions.

REFERENCES

(1) Davison,J. et al., Trans.Instn.Min.Met., 1960-61, 70, 247-290.

(2) Fleming,C.A. and Cromberge,G., J.S.Afr.Inst.Min.Metall., 1984, 84, 125-137.

(3) Green,B.R., Schwellnus,A.H. and Kotze,M.H., Int.Conf.on Gold, 1986, Johannesburg, 321-333.

(4) Soldatov,V.S., Shunkevich,A.A. and Sergeev,G.I.,Reactive Polymers, 1988, 7, 159-172.

(5) Kotze,M.H., Green,B.R. and Ellis,P., Reactive Polymers, 1991, 14, 129-141.

(6) Kotze,M.H., MSc Thesis, Univ. Stellenbosch, 1991.

(7) Carman,P.C., Flow of Gases Through Porous Media, Butterworths, 1956.

(8) Nolan,J.D., Dia-prosim, Pontyclun, Personal commun.

(9) Kaye,A., Courtaulds Research, Coventry, Personal commun.

MEDIUM BASE POLYAMINE ION EXCHANGE RESINS FOR THE EXTRACTION OF GOLD FROM CYANIDE SOLUTIONS

W.I. HARRIS, J.R. STAHLBUSH, W. C. PIKE
Liquid Separations Research, 1604 Building
The Dow Chemical Company
Midland, Michigan, USA, 48674

ABSTRACT

Medium base polyamine anion exchange resin based on 1,3-diaminopropane or 2,4-diamino-2-methylpentane and chloromethylated/cross-linked polystyrene were evaluated for gold cyanide recovery. Equilibrium and kinetic data are presented. The resins are compared against activated carbon by calculating minimum flow rates, minimum number of stages, and kinetic limitations for a plant processing 100 tons per hour of gold ore.

INTRODUCTION

Much of the gold in the world today is recovered by the carbon-in-pulp process. The steps in this process are: grinding the ore to a fine powder in a water slurry (pulp), cyanide leaching the gold from the pulp, adsorption of the gold cyanide onto activated carbon, followed by hot caustic elution and electrowinning to gold metal. The activated carbon must undergo periodic thermal regeneration and tends to foul with organics and calcium. Ion exchange resins have long been considered as a solid absorbent to replace activated carbon. Some of the advantages cited for resins are: higher equilibrium capacity, easier elution of gold (weak base anion exchange resins), no thermal regeneration, greater fouling resistance, and synthetic rather than natural product. Possible disadvantages are: lower selectivity for gold over base metal cyanides, smaller particle size, and lower density (1).

The major reasons ion exchange resins have not been widely accepted are: strong base resins require expensive eluants such as thiourea, thiocyanide, zinc cyanide, and/or polar organic solvents, while weak base resins are ineffective for loading gold at the normal leaching pH of 10-11. Recently a series of medium base resins were prepared from polyamines, ie. 1,3-diaminopropane or 2,4-diamino-2-methylpentane, and chloromethylated/crosslinked polystyrene (2,3). These resins were shown to have: good selectivity for gold, high capacity at leaching pHs, and to be readily eluted with sodium hydroxide solution.

The polyamine resins have now been evaluated against carbon for equilibrium gold capacity and kinetics of loading. These data were used to calculate the minimum adsorbent flow rates, equilibrium stages, and kinetic limitations for a mine processing 100 tons/hour of ore. The resin-in-pulp process was shown to compare favorably with a carbon-in-pulp process.

EXPERIMENTAL

The preparation of the resins has been described in detail elsewhere (3). The macroporous copolymer matrix was suspension-polymerized from styrene monomer and divinylbenzene cross-linker. The copolymer was chloromethylated and aminated with either 1,3-diaminopropane (DAP) or 2,4-diamino-2-methylpentane (DAMP). The carbon was a coconut base carbon, G210AS, available from North American Carbon Inc. Resin and carbon properties are presented in Table 1.

Stock Solution: To study gold equilibrium and kinetics a stock solution was prepared by dissolving 1.56 g of gold chloride hydrate (50% gold by weight) and sodium cyanide in 1L of de-ionized water.

TABLE 1, PROPERTIES OF RESINS TESTED FOR KINETICS

	PDA	DAMP	CARBON
Dry Weight Cap., meq/g	7.48	5.13	NA
Wet Volume Cap.,meq/mL	2.62	1.78	NA
% Strong base	0.4	0.7	NA
%Water	50.4	47.5	NA
Density, g/mL	1.06	1.04	1.32,(1.58 Dry)
Vol. Avg. Dia. μm	567	620	1830

Equilibrium Gold Loading: Ten mL of gold stock solution were added to 250 mL glass bottles containing 0.1, 0.3, 0.5, 0.7, or 0.9 g of resin pre-conditioned to the loading pH, and 150 g of de-ionized water. Lesser amounts of carbon, 0.025, 0.05, 0.075, 0.15, 0.20, or 0.25 g were used because of its higher equilibrium capacity. The pH of each bottle was adjusted and placed on a shaker. The pH was monitored and adjusted daily until it stabilized for at least 48 h. Once equilibrated , 15 mL of solution were withdrawn from each bottle and analyzed for gold by atomic absorption.

TABLE 2, GOLD EQUILIBRIUM DATA at pH 10

Solution*	DAMP*	Solution	Carbon	Solution	DAP
11.1	57300	30.0	103000	25.9	47400
2.0	25300	21.9	80700	4.7	29800
0.7	16100	12.4	74000	2.0	18900
0.5	11300	1.3	49500	0.8	13800
0.2	7160	0.3	37400	0.7	10800
0.1	5300	0.1	29900		

*Solution conc. in μg/mL and solid conc. in mg/kg

Kinetics: The experimental method was conducted as follows: a) 1.5 g of resin or carbon were weighted into a 250 mL glass bottle and filled with de-ionized water. The pH was initially adjusted to kinetic loading pH of 10-10.5. The pH was re-adjusted until stabilized for at least 48 hours. Once stabilized, the water was decanted from the resin. b) Into a 3.8 L glass bottle was added 2.5 L of an aqueous 10 μg/mL gold cyanide solution made from stock solution. The pH was adjusted to the desired level. c) A 25 mL aliquot was withdrawn before the resin or carbon was added to the 3.8 L bottle. The resin or carbon was added to the bottle and it was placed on a bottle roller. The bottle was removed from the roller and 25 ml aliquots withdrawn at elapsed times of 15, 30, 60, 90, 180, 300 and 1440 min. d) The aliquots were analyzed for gold by atomic absorption. The gold concentration on the resin or carbon was calculated from a mass balance.

DISCUSSION

Equilibrium Gold Loading: The equilibrium gold loading curves are shown in Figure 1. In synthetic solutions the carbon equilibrium capacity is clearly superior to the medium base resins. In real mine solutions containing base metals and foulants the equilibrium loading may be significantly affected. Future work needs to be conducted on real mine streams.

FIGURE 1: GOLD EQUILIBRIUM

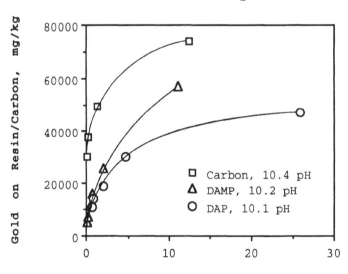

Gold in Solution
micro-g/mL

Kinetics: At a typical gold concentration, <10 μg/mL, the loading of gold onto either carbon or resin is expected to be under film-diffusion control. For low conversions the film-diffusion controlled batch loading (4) can be expressed by the following function:

$$\ln (C_b/C_{bo}) = -(4n\pi r^2 Dt)/(V\delta)$$

Where:
δ=boundary layer thickness C_b=bulk concentration
C_{bo}=initial bulk conc.   D=diffusion coefficient
n=number of particles   r=particle radius
t=time   V=solution volume

TABLE 3, KINETIC LOADING DATA

	PDA	DAMP	CARBON
g Resin	1.485	1.516	1.518 (Dry)
mL Solution	2500	2500	2500
mL Aliquot	25.0	25.0	25.0
pH	10.2	10.2	10-10.5
Minutes	Gold Concentration in Solution μg/mL		
0	9.2	10.0	9
15	7.7	8.2	8
30	7.1	7.6	8
60	5.5	5.9	6
90	4.5	4.5	5
180	3.1	3.2	4
300	2.2	2.0	2
1440	1.0	0.7	1
D/δ, μm/s	20	23	87

A plot of $\ln (C_b/C_{bo})$ versus time yields a straight line indicating that the loading is film-diffusion controlled. Diffusion coefficients divided by boundary layer thickness (D/δ) can be calculated from the slope of the line, Figure 2 and Table 3. The carbon value is probably higher due to its shape being non-spherical, its high density, and its large particle size. The large size and higher density would result in higher terminal settling velocity and thus smaller boundary layer thickness. The small resin size favors film-diffusion controlled kinetics. Thus the resin kinetics will need to be re-tested as the resin size is increased to the greater than 20 US mesh size (841 μm) required for pulp circuits.

PROCESS DESIGN CALCULATIONS

While the experimental polyamine resins look promising, one can better define their potential by doing a preliminary plant design for both resin and carbon. For a design basis the following typical plant parameters will be selected: a) 100 dry tons ore/h, 200 wet tons/h, b) 8 μg/mL gold in pregnant solution, c) 10-11 pH of pulp stream and d) <0.005 μg/mL gold in barrens.

The minimum flow rate absorbent can be calculated from a mass and equilibrium balance. The minimum flow rates are: a) 10.6 kg/h for carbon, b) 14.9 kg/h for DAMP resin and c) 19.9 kg/h for DAP resin. The minimum flow rates are directly related to the equilibrium capacity, Figure 1. The absorbent flow rates, 10-20 kg/h, are much slower than the ore flow rate, 100000 kg/h.

The minimum number of contact stages can be calculated from mass and equilibrium balances once the absorbent flow rate is specified. A flow rate of 3 times the minimum are here specified for both carbon and resin. The minimum number of stages are : a) 2 stages for carbon, b) 3 stages for DAMP resin and c) 4 stages for PDA resin.

The absorbent stage inventory and gold hold-up can be calculated from Fick's first law, D/δ, and mass balances once the flow rates and number of stages are specified. Enough absorbent must be present to absorb the required gold in the required time. If one sets the flow rate at 3 times the minimum and the number of contact stages at the minimum, one finds the stage inventory and gold hold-up is excessive. The gold held up is greater than a months production or 6.5×10^6. Typically the stage residence time is 48 h and the carbon/resin is turned over at gold loadings much lower than equilibrium, 6000 mg/kg gold compared to an equilibrium loading of 60000 mg/kg. Calculations done assuming a kinetically limited process: a) 6000 mg/kg gold

loading, b) residency time of 48 h, and c) 5 stages give the
results in the next paragraph.

Figure 2, KINETICS of LOADING

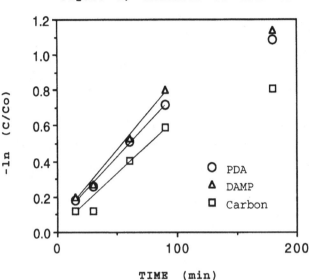

TIME (min)

The carbon or resin flow rate can be calculated from an
overall mass balance to be 133 kg/h. The carbon or resin
inventory can be calculated from the flow rate and residence
time to be 6380 kg or 1280 kg/stage. Stage-to-stage
calculations based on mass balance and kinetics show that the
gold loading reaches 6000 mg/kg loading in 4 stages for both
carbon and resin. This confirms that the plant is kinetically,
not equilibrium limited. As long as resins absorb gold as fast
as carbon and have adequate capacity, it does not matter that
they have lower equilibrium capacity.

CONCLUSIONS

Data have been presented to show that experimental polyamine
resins have good equilibrium capacities for gold. The kinetics
of initial gold loading for both resin and carbon are film-

diffusion controlled. Calculations show that plants are not under equilibrium control but under kinetic control.

The preferred system, carbon or resin, needs to determined from testing on real mine streams. Factors such as fouling resistance, elution cost, attrition rate, absorbent cost, selectivity, and operating capacity will play an important role in absorbent selection.

REFERENCES

1) Fleming, C.A., and Cromberge, G., Journal of the South African Institute of Mining and Metallurgy, pp. 125-137, May 1984.

2) Harris, W.I., and Stahlbush, J.R., U.S. Patent 4,758,413 (1988).

3) Harris, W.I.,Stahlbush, J.R., Pike,W.C., and Stevens, R.R., "The Extraction of Gold......Ion Exchange Resins", presented at 6th Symposium on Ion Exchange, Lake Balaton, Hungary, Sept. 3-7,1990. To be published in the journal of Reactive Polymers.

4) Harris, W.I.,Lindy, L.B., and Dixit, R.S., Reactive Polymers, **4** (1986) 99-112.

SEPARATION OF MOLYBDENUM AND TUNGSTEN BY A POLYSTYRENE RESIN FUNCTIONALIZED WITH IMINODIACETATE LIGAND

TOSHISHIGE M. SUZUKI, MASATOSHI KANESATO, TOSHIRO YOKOYAMA
AND MOHAMMED H. H. MAHMOUD
Government Industrial Research Institute, Tohoku
4-2-1, Nigatake, Miyagino-ku, Sendai, Japan

ABSTRACT

A chelating polymer resin having iminodiacetic acid (IDA) as the functional group has been examined for the selective separation of Mo(VI) and W(VI). The ^{1}H and ^{13}C NMR spectroscopies in the homogeneous system suggest the much more favorable complexation of IDA with Mo(VI) than with W(VI). The mechanism has been described as follows on the basis of NMR and the adsorption equilibrium studies:

$$MO_4{}^{2-} + RNH^+(CH_2CO_2{}^-)_2 + H^+ \rightleftharpoons [RN(CH_2CO_2)_2\text{-}MO_3]^{2-} + H_2O$$

A selective separation of Mo(VI) over W(VI) has been demonstrated by the IDA resin in a column system.

INTRODUCTION

The chemical properties of molybdenum and tungsten are very similar to each other (1,2). They tend to be found as mixtures in rocks and minerals. Due to the remarkable similarity, their mutual separation usually requires tedious procedures. For the separation of metals there is a growing interest in the use of hydrometallurgical techniques including solvent extraction (3) and ion-exchange processes (4,5). Ion-exchange methods are being applied to the analytical separation of molybdenum and tungsten (5) however the selectivity of the conventional anion-exchangers toward oxo-molybdenum(VI) and oxo-tungsten(VI) anions is not sufficient to allow the clear separation of them.

Chelating polymer resins are attractive materials to apply to the metal separation processes because of their high selectivity and operational convenience (6). Iminodiacetic acid (IDA) is the most popular functional group for a chelating resin which can form stable complexes with metal ions of high valency. The NMR study in homogeneous solution has suggested the higher selectivity of IDA toward Mo(VI) over W(VI). This paper deals with the complexation properties of a chelating resin having IDA groups and its application to the column separation of Mo(VI) and W(VI).

EXPERIMENTAL

Chelating Resin

Macroreticular type chloromethylated styrene-10%-divinylbenzene copolymer beads (30-60 mesh) with surface area, 7.3 m^2g^{-1} and mean pore diameter, 720 Å were used for the resin matrix. The polystyrene resin having IDA groups was prepared by treatment of the chloromethylated polystyrene beads with diethyliminodiacetate, followed by hydrolysis of the ester with sodium hydroxide. The nitrogen content of the IDA resin was 4.2% which corresponds to a ligand content of 3.0 mmol/g dry resin.

Capacity and Distribution Coefficient

Equilibrium capacity for metal ion uptake (amount of metal ion retained on one gramme of the resin) was determined by a batchwise technique with the metal ion always being in excess over the resin capacity. The distribution coefficient is defined as, K_d = amount of metal ion extracted in one gramme of resin / amount of metal ion remaining in 1 ml of solution. A 1 g portion of dry resin was added to a 100 ml buffered metal ion solution (1 mM). After shaking for seven days at 25 °C, the equilibrium pH and the amount of metal ion remaining in the solution was determined. The K_d value was then calculated.

Column Breakthrough Experiments

A buffered solution (pH 5) containing an equimolar amount (5 mM) of Mo(VI) and W(VI) was continuously supplied to the column packed with the IDA resin (5 g, wet volume: 19.6 ml). The column effluent was fractionated into 20 ml portions and the amount of metal ions determined. The retained molybdenum and /or tungsten was liberated from the column by elution with 1M sodium hydroxide solution.

Measurements

The metal ion concentration was determined with a SEIKO ICP-atomic emission spectrometer, Model SPS-1200A. The wavelengths of analytical lines were 202.03 nm and 248.75 nm for molybdenum and tungsten, respectively. The ^{1}H-NMR and ^{13}C-NMR spectra were measured in D_2O with a Varian VXR 300 spectrometer.

RESULTS AND DISCUSSION

Metal Complex Formation of Iminodiacetic Acid

Iminodiacetic acid (IDA) gives two ^{13}C-NMR peaks corresponding to the acetate carbons in D_2O. Upon addition of equimolar amount of Mo(VI) to IDA at pH 6, these two peaks shift 6.9 and 6.5 ppm down the field. The peaks for free IDA almost disap-

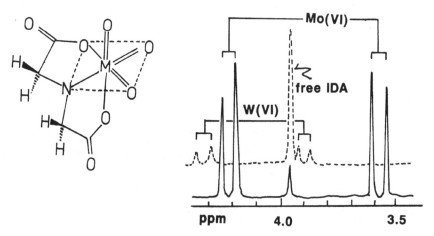

Fig. 1 ^{1}H NMR spectra of IDA in the presence of equimolar amount of Mo(VI) and W(VI)

peared presumably due to complete complexation. The ^{1}H-NMR spectrum of the IDA-Mo(VI) species shows new quartet peaks assignable to AB pattern of the coordinated acetate protons (Fig. 1). The spectrum does not change remarkably for pH 4.5 to 6.0. These observations suggest the formation of 1 : 1 complex in which two acetate arms of IDA coordinate symmetrically to the Mo(VI) core (7). Under similar conditions, the ^{13}C and ^{1}H-NMR spectra of IDA in the presence of W(VI) were measured. In this case, four ^{13}C peaks corresponding to free and coordinated acetate groups were observed. In agreement with the ^{13}C-NMR spectra, free IDA protons were observed along with the small quartet peaks of the coordinated acetate protons (Fig. 1). This indicates that IDA forms a less stable complex with W(VI) than with Mo(VI).

Resin Complexation of Molybdenum(VI) and Tungsten(VI)
The Mo and W capacity vs. pH of the IDA resin is shown in Fig. 2. Apparently the extraction of Mo(VI) is favored over W(VI).

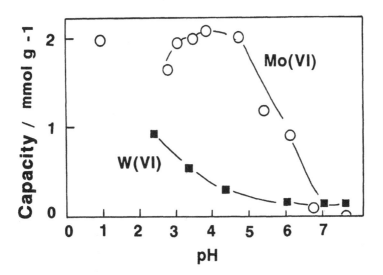

Fig. 2 Complexation capacity of the IDA resin for Mo(VI) (—○—) and W(VI) (—■—).

We observed that the complexation of Mo(VI) is always accompanied by an increase in pH of the solution suggesting that protons are withdrawn from the solution upon complexation. Practically no complexation of molybdenum or tungsten takes place at pH>9. Therefore retained Mo(VI) and W(VI) were readily liberated from the resin on treatment with dilute sodium hydroxide solution. The time-courses of Mo(VI) and W(VI) extraction by IDA resin were examined at various pH. The extraction of Mo(VI) was always faster than that of W(VI), irrespective of the pH. The optimum condition was pH 4.5-6.0, where the Mo(VI) uptake is sufficiently rapid to allow column operation.

Distribution Coefficient

According to the pKa values of N-benzyliminodiacetic acid, the major chemical species of the IDA moiety can be represented as $R-NH^+(CH_2CO_2^-)_2$ at pH 2.5-8.0 (8). The NMR study suggested the

$$R-CH_2-\overset{+}{N}H\begin{matrix} CH_2CO_2H \\ CH_2CO_2^- \end{matrix} \underset{pKa_2=2.18}{\rightleftharpoons} R-CH_2-\overset{+}{N}H\begin{matrix} CH_2CO_2^- \\ CH_2CO_2^- \end{matrix} \underset{pKa_3=8.9}{\rightleftharpoons} RCH_2-N\begin{matrix} CH_2CO_2^- \\ CH_2CO_2^- \end{matrix}$$

1 : 1 complex formation of IDA with Mo(VI). In addition, the complexation of Mo(VI) is accompanied by withdrawal of protons from solution. Therefore the following reaction is assumed to take place for the complexation of Mo(VI).

$$MoO_4^{2-} + RNH^+(CH_2COO^-)_2 + H^+ \rightleftharpoons [RN(CH_2COO)_2-MoO_3]^{2-} + H_2O \quad (1)$$

where R denotes the resin matrix. The equilibrium constant of the above reaction is given by:

$$K = \frac{[(RN(CH_2COO)_2-MoO_3)^{2-}]}{[MoO_4^{2-}][RNH^+(CH_2COO^-)_2][H^+]} \quad (2)$$

The distribution coefficient (K_d) is defined as:

$$K_d = \frac{[(RN(CH_2COO)_2-MoO_3)^{2-}]}{[MoO_4^{2-}]} \qquad (3)$$

From Eqs. (2) and (3), the following relationship can be obtained.

$$\log K_d = \log K + \log [RNH^+(CH_2COO^-)_2] - pH \qquad (4)$$

Equation 4 predicts that the plot of log K_d vs pH gives a straight line with a slope of -1 under excess resin. Figure 3 shows the log Kd vs pH profile where linear plots with the slope approximately -1 were observed for the complexation of Mo(VI). This result is consistent with Eq (4). It is known that both Mo(VI) and W(VI) exist as various species depending on pH. Among them mononuclear MoO_4^{2-} is most likely as the com-

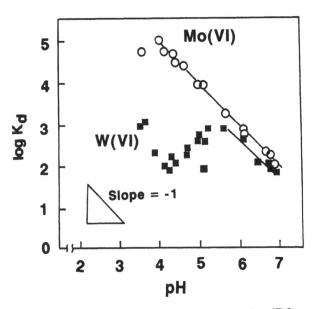

Fig. 3   The distribution coefficients of the IDA resin as a function of pH
Mo(VI) (—O—),   W(VI) (—■—)

388

plexed species. Similarly the complexation of W(VI) as the WO_4^{2-} form may occur, but in a narrow pH range (5.5-7) because the distribution of mono nuclear species is predominant only in a solution of higher pH.

Column Separation of Molybdenum(VI) and Tungsten(VI)

The remarkable difference in the Kd values between Mo(VI) and W(VI) suggests the favorable separation of the two metal ions. We have examined the selective complexation of Mo(VI) by use of the IDA resin in a column system. A solution of given pH containing equimolar Mo(VI) and W(VI) was continuously passed through the column and the metal ions in the effluent were determined. The optimum pH for successful separation was around pH 5; the amount of Mo(VI) retained in the resin is high and the leakage of W(VI) is marked (Fig. 4). At much higher pH, neither metal was effectively extracted. Metal ions were quantitatively released from the resin by elution with 1 M NaOH solution. The metal ion ratio in the eluted solution was 100 : 4, Mo(VI) : W(VI).

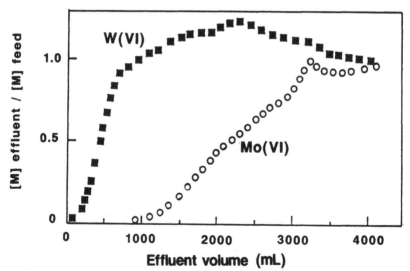

Fig. 4   Column breakthrough profile for the separation of Mo(VI) and W(VI).

Bed resin: 5 g (19.6 ml), pH: 5,   Flow rate: 1 ml min^{-1}

CONCLUSION

The NMR spectroscopic studies showed that the coordination of IDA with Mo(VI) is much more favorable than with W(VI). The chelating resin having an IDA group revealed a higher selectivity for the extraction of Mo(VI) over W(VI) at pH 4-6. A mechanism is proposed in which mononuclear oxo-metal ion (MoO_4^{2-}) is the major complexation species

Mo(VI) and W(VI) can be achieved using a column packed with the IDA resin.

REFERENCES

1. Cotton F. A. and Wilkinson G., Advanced Inorganic Chemistry, John Wiley & Sons, (1988), 804-847.
2. Parish R. V., The Metallic Elements, Longman Group Ltd., (1977).
3. Coca J., Diez F. V. and Moris M. A., Hydrometallurgy, (1990), 25, 125-135.
4. Zhou Z., Ion Exchange for Industry, ed by M. Streat, Ellis Horwood (1988), 373-384.
5. Morhol M., Comprehensive Analytical Chemistry, Elsevier, (1982), 14, 291.
6. Sahni S. K. and Reedijk J., Coord. Chem. Rev., (1986), 59, 1-139.
7. Freeman M. A., Schultz F. A. and Reilly C., Inorg. Chem., (1982), 21, 567-576.
8. Martell A. E. and Smith R. M., Critical Stability Constants" Plenum Press (1974), 1, 135.

USE OF NEW FUNCTIONALISED MATERIALS WITH ORGANOPOLYSILOXANE STRUCTURE FOR PURIFICATION OF ARSENIC(III) IN HYDROCHLORIC ACID MEDIA

GERARD COTE, FENMING CHEN and DENISE BAUER
Laboratoire de Chimie Analytique (Unité CNRS n° 437)
E.S.P.C.I., 10, rue Vauquelin 75005 Paris, France.

ABSTRACT

This study shows that impurities such as bismuth(III) and antimony(III) can be efficiently removed from arsenic(III) solutions in highly acidic media (e.g. 3 to 6 mol/L HCl) by using a new functionalised material possessing an organopolysiloxane structure and diphenylphosphine functions.

INTRODUCTION

Classical procedures for chemical purification of arsenic are based on the separation of arsenic(III) halo-complexes by distillation or solvent extraction. Highly concentrated chloride media (e.g. 8 to 12 mol/L HCl) have proved effective for the extraction of the chloro-complexes of arsenic(III) into carbon tetrachloride (1,2), chloroform (1), benzene (1 - 4), bis(2-chloroethyl) ether (1,5) or Solvesso 150® (solvent constituted of 99% aromatics) (6). However, distillation and solvent extraction of arsenic(III) halo-complexes are not completely selective against various impurities such as antimony(III) or bismuth(III). In the present paper, we have investigated the removal of antimony(III) and bismuth(III) from moderately concentrated (e.g. 3 to 6 mol/L) hydrochloric acid solutions containing arsenic(III) by using new materials possessing an organopolysiloxane structure and monophenyl [=PPh] (material 1) or diphenylphosphine [-PPh$_2$] (material 2) groups as functional groups. Such materials belong to a series of functionalised materials which have been synthesized at Degussa (Germany) by the polycondensation of suitable bifunctional silane monomers (7). They appear as spherical or egg-shaped particles with a grain size ranging between 0.1 and 1.5 mm (mean value : 0.6 mm). The main physical advantages of functionalised organo-polysiloxanes lie in the absence of swelling or shrinking during ion sorption and in their high thermal stability.

EXPERIMENTAL

The two functionalised organopolysiloxanes have been kindly supplied by Degussa Company and used as received. Their functional group concentration was as follows : material 1 (monophenylphosphine groups) : 1.5×10^{-3} mol/g (dry mat.); material 2 (diphenylphosphine groups) : 1.2×10^{-3} mol/g (dry mat.). Arsenic trioxide of analytical (Prolabo) or technical (Metaleurop) grade quality was used for the preparation of arsenic(III) solutions. The other reagents from various suppliers were all of analytical grade.

Metal determinations in aqueous phases were carried out by inductively coupled plasma emission spectrometry with an ICP 1500 Plasma Therm instrument coupled with a Video 11 A/A Instrumentation Laboratory Spectrophotometer. The batch sorption experiments were carried out by vigorously shaking measured amounts (typically 0.5 g of wet material) of each of the two functionalised organopolysiloxanes with definite volumes of aqueous metal solutions of known concentration for 24 h at 20 ± 2 °C. The experiments in dynamic conditions were performed at 20 ± 2 °C with a column filled with a given quantity of functionalised organopolysiloxane.

RESULTS AND DISCUSSION

The equilibrium data for sorption isotherms of As(III), Bi(III) and Sb(III) from 5 mol/L HCl on materials 1 and 2 are given in Figures 1 and 2, respectively.

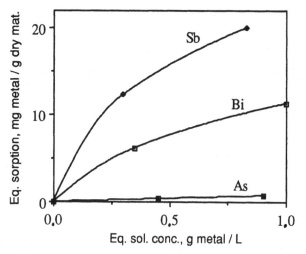

Figure 1. Equilibrium sorption of As(III), Bi(III) and Sb(III)
from 5 mol/L HCl on material 1 (monophenylphosphine)
at 20 ± 2 °C.

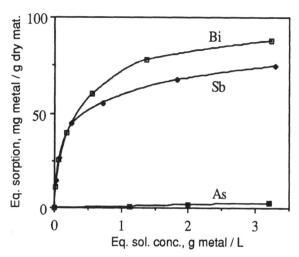

Figure 2. Equilibrium sorption of As(III), Bi(III) and Sb(III)
from 5 mol/L HCl on material 2 (diphenylphosphine) at 20 ± 2°C.

Examination of Figure 1 shows that organopolysiloxane 1
has almost no affinity for As(III), but can extract Bi(III) and
Sb(III) to a small extent. Examination of Figure 2 shows that
organopolysiloxane 2 has also almost no affinity for As(III),
but can efficiently retain Bi(III) and Sb(III). It is noted

that the Sb/P and Bi/S molar ratios in material 2 at saturation are close to 0.5 and 0.4, respectively. Such values suggest that the sorption of Sb(III) and Bi(III) on organopolysiloxane 2 may be as a result of the formation of the $SbCl_3(-PPh_2)_2$ and $BiCl_3(-PPh_2)_n$ (n = 2 or 3) complexes, respectively. It is of interest that several complexes of this type have already been reported with various metals and triphenylphosphine (PPh₃) : $SnCl_4(PPh_3)_2$, $FeCl_3(PPh_3)_2$, $PdCl_2(PPh_3)_2$, etc. (8 - 10). The properties of organopolysiloxane 2 have been investigated further in dynamic conditions.

The breakthrough curves obtained by passing an aqueous solution containing As(III), Bi(III) and Sb(III) in HCl (3 to 6 mol/L) media through a column filled with organopolysiloxane 2 have typically the shape represented in Figure 3.

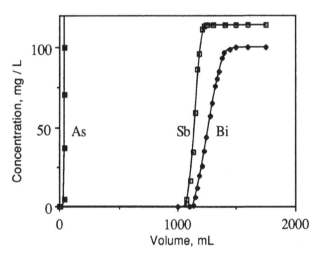

Figure 3. Breakthrough curves of As(III), Bi(III) and Sb(III) for a column filled with polysiloxane 2 (diphenylphosphine), at 20 ± 2°C. Experimental conditions : bed height = 37.5 cm; column diameter = 0.8 cm; mass of dry polysiloxane = 7.0 g; flow rate = 50 mL/h. Feed solution composition = 11 g/L As(III) + 100 mg/L Bi(III) + 114 mg/L Sb(III) + 6 mol/L HCl.

Examination of this figure shows that arsenic(III) (initially 11g/L) appears slightly after the dead volume, whereas Bi(III) and Sb(III) (initially 100 mg/L and 114 mg/L, respectively) are efficiently retained on the column until a volume of solution of about 1100 mL (58 bed volumes) has percolated through the system. Typical results obtained with arsenic(III) solutions prepared with technical grade samples of As_2O_3 are given in Table 1.

TABLE 1

Concentration of Bi(III) and Sb(III) in the effluent solutions of arsenic(III) prepared with technical grade samples of As_2O_3.

Solution	Volume of solution through the column expressed in bed volumes	Mass of As through the column per gramme of dry organopoly-siloxane 2	Bi(III) mg/L	Sb(III) mg/L
A	0 to 247	6.7	< 0.2	< 0.2
	277	7.5	0.2	< 0.2
	293	7.9	0.5	2.6
	311	8.4	1.0	4.2
	331	8.9	2.2	4.7
	364	9.8	5.1	4.9
	400	10.8	7.5	4.7
B	0 to 136	3.7	–	< 0.2
	151	4.1	–	0.2
	172	4.6	–	11.7
	191	5.2	–	29.2
	209	5.6	–	40.7
	252	8.8	–	47.3

General conditions : Bed height = 17.2 cm; column diameter = 0.8 cm; mass of dry organopolysiloxane 2 = 3.2 g; flow rate = 50 mL/h; feed solution composition = 10 g/L As(III) + x mg/L Bi(III) + y mg/L Sb(III) + 3 mol/L HCl; temperature = 20 ± 2°C. Specific conditions : Solution A : x = 8.0 mg/L, y = 4.7 mg/L; solution B : x < 0.2 mg/L (limit of the analytical method used in this work), y = 47 mg/L.

These results show that 4.1 to 7.5 g of technical arsenic(III) can be efficiently purified per gramme of dry organopolysiloxane 2 depending on the origin of arsenic trioxide used.

The influent and effluent solutions have the same composition at the equilibrium. Thus, the determination of the concentrations of As(III), Sb(III) and/or Bi(III) in the solid phase (which can be performed after stripping of these elements) allows the determination of the selectivity coefficient $\beta_{M/As} = [M(III)]_{org} \cdot [As(III)]_{aq} / ([M(III)]_{aq} \cdot [As(III)]_{org})$ where M = Bi or Sb and org and aq refer to the species in the solid organopolysiloxane and aqueous phases, respectively. The values obtained for $\beta_{M/As}$ are given in Table 2.

TABLE 2

Values of the selectivity coefficients $\beta_{Sb/As}$ and $\beta_{Bi/As}$

Aqueous phase composition at equilibrium	As(III) : 10 g/L Bi(III) : 8.0 mg/L Sb(III) : 4.7 mg/L HCl : 3 mol/L	As(III) : 11 g/L Bi(III) : 100 mg/L Sb(III) : 114 mg/L HCl : 6 mol/L
$\beta_{Sb/As}$	237	45
$\beta_{Bi/As}$	346	49
$\beta_{Sb/As} \times (a_{Cl^-})^2$ (*)	1034	1184
$\beta_{Bi/As} \times (a_{Cl^-})^2$ (*)	1510	1289

(*) The values of a_{Cl^-} were taken from reference (6).

Examination of this table shows that the selectivity coefficients $\beta_{Bi/As}$ and $\beta_{Sb/As}$ vary significantly depending upon the experimental conditions. Among the various parameters which may influence the values of $\beta_{Bi/As}$ and $\beta_{Sb/As}$, the activity of chloride ions is likely to be the most important. Indeed, the exchange reaction can be written as

$$MCl_{n,\,aq}^{(n-3)-} + \left[As(OH)_pCl_{(3-p)}(-PPh_2)_q\right]_{org} \Leftrightarrow$$

$$\left[MCl_3(-PPh_2)_q\right]_{org} + \left[As(OH)_rCl_s\right]_{aq} + (n-p-s)\,Cl_{aq}^- \quad (1)$$

where M = Bi or Sb. The constant $K_{M/As}$ of reaction (1) is given by

$$K_{M/As} = \frac{\left|\left[MCl_3(-PPh_2)_q\right]_{org}\right| \cdot \left|\left[As(OH)_rCl_s\right]_{aq}\right| \cdot \left|Cl_{aq}^-\right|^{(n-p-s)}}{\left|MCl_{n,\,aq}^{(n-3)-}\right| \cdot \left|\left[As(OH)_pCl_{(3-p)}(-PPh_2)_q\right]_{org}\right|}$$

but can be rewritten as

$$K_{M/As} = \beta_{M/As} \cdot \left|Cl_{aq}^-\right|^{(n-p-s)}$$

The values of the parameters n, p, r and s appearing in reaction (1) depend on the aqueous concentration of hydrochloric acid. For instance, for 3 to 6 mol/L HCl, n ≈ 6 (Bi and Sb) (11,12), r = 2 (6) and s = 0 to 1 (mixture of As(OH)$_2$ and As(OH)$_2$Cl) (6). The parameter p is equal to or less than r. Thus, the term (n-p-s) is greater than 1, which qualitatively explains the negative mass effect of chloride ions on the selectivity coefficients $\beta_{Sb/As}$ and $\beta_{Bi/As}$ when the hydrochloric acid concentration is increased (Table 2). It is of interest that $\beta_{Sb/As} \times (a_{Cl^-})^2$ and $\beta_{Bi/As} \times (a_{Cl^-})^2$ remain roughly constant between 3 and 6 mol/L HCl, which suggests that $n - p - s \approx 2$.

Several methods of elution of Bi(III) and Sb(III) from the column and regeneration of organopolysiloxane 2 have been investigated. In particular, it has been found that Bi(III) and Sb(III) cannot be eluted from the column merely by using a highly acidic aqueous solution (e.g. 8 - 10 mol/L HCl). On the other hand, Bi(III) and Sb(III) can be stripped from the column with a solution of EDTA as a result of the formation of highly stable complexes with Bi(III) and Sb(III) (13,14).

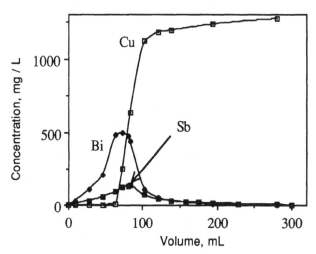

Figure 4. Stripping of Bi(III) and Sb(III) from a column filled of organopolysiloxane 2 (diphenylphosphine) with a 1.27 g/L Cu(I) solution in 6 mol/L HCl at 20 ± 2°C. Experimental conditions : bed height = 17.2 cm; column diameter = 0.8 cm; mass of dry polysiloxane = 3.2 g; flow rate = 74 mL/h.

For instance, in the case of the experiment performed with solution A (Table 1), the percolation of 50 mL (6 bed volumes) of a 0.2 mol/L EDTA solution (flow rate = 50 mL/h) allows the elution of 90% and 80% of Bi(III) and Sb(III), respectively.

Bi(III) and Sb(III) can also be eluted from the column with a solution of copper(I) in hydrochloric acid medium as presented in Figure 4. Copper(I) was selected as stripping agent because of its great affinity for organopolysiloxane 2. In the conditions of Figure 4, the elution yield of Bi(III) and Sb(III) is greater than 99% after the percolation of 200 mL (23 bed volumes) of Cu(I) solution through the column. Copper can be subsequently removed from the column by oxidative stripping in complexing media (e.g. 0.1 mol/L NH_4^+ + 0.9 mol/L NH_3) since organopolysiloxane 2 has almost no affinity for copper(II). It is of interest that the two methods of elution presented above allow efficient regeneration of the column.

ACKNOWLEDGEMENTS

The authors gratefully acknowledge the support of Metaleurop Recherche.

REFERENCES

1. Brink, G.O., Kafalas, P., Sharp, R.A., Weiss E.L. and Irvine, J.W., Jr., J. Am. Chem.Soc., 1957, 79, 1303.
2. Olszer, R. and Siekierski, S., J. Inorg. Nucl. Chem., 1966, 28, 1991.
3. Beard, H.C. and Lyerly, L.A., Anal. Chem., 1961, 33, 1781.
4. Tanaka, K., Japan Analyst, 1960, 9, 574.
5. Myron Arcand, G. , J. Am. Chem. Soc., 1957, 79, 1865.
6. Sella, C., Navarro Mendoza, R. and Bauer, D., Hydrometallurgy, 1991, 27, 179.
7. Cote, G., Chen, F.M. and Bauer, D., Solvent Extr. Ion Exch., 1991, 9, 289.
8. Ohkaku, N. and Nakamoto, K., Inorg. Chem., 1973, 12, 2440.
9. Walker, J.D. and Poli, R., Inorg. Chem., 1989, 28, 1793.
10. Mojski, M., Talanta, 1980, 27, 7.
11. Haight, G.P., Jr., Springer, C.H. and Heilmann, O.J., Inorg. Chem., 1964, 3, 195.
12. Haight, G.P., Jr. and Ellis, B.Y., Inorg. Chem., 1965, 4, 249.
13. Sillen, L.G. and Martell, A.E., Stability Constants of Metal-Ion Complexes, Supplement N°1, Special publication N°25, The Chemical Society, London, 1971.
14. Perrin, D.D., Stability Constants of Metal-Ion Complexes, Part B : Organic Ligands. IUPAC Chemical Data Series N°22, Pergamon Press, Oxford, 1979.

A WEAK BASE COPOLYMER FOR THE SEPARATION OF GOLD(III) FROM BASE METALS

MICHAEL J. HUDSON
Department of Chemistry, University of Reading, Box 224,
Whiteknights, Reading, RG6 2AD.

BILLY KAR-ON LEUNG
Department of Biochemistry, University of Oxford, South Parks
Road, Oxford OX1 3QU.

ABSTRACT
A new copolymer has been prepared by reaction of chloromethylated poly(styrene) (Amberlite XE 305) with 1,3,4-thiadiazole-2-amino -5-thiol in which the pendant group is covalently bound to the polymer through the thiol group. The capacity of the copolymer for gold and the base metals varied over the range of hydrochloric acid concentration, but at 2 M HCl the capacity for gold was 1.9 ± 0.1 mmol g^{-1} and negligible for the base metals. The loaded gold may be quantitatively eluted with acidic thiourea.

INTRODUCTION

In previous studies we have shown that copolymers containing pendant N-dithiocarboxylate (1), N-oxide (2) and thiol (3) groups can be used to separate gold from base metals. Such separations are useful in the preconcentration of gold prior to analysis and also for industrial separations. Chelating resins may be typified by the N-dithiocarboxylates and thiols (4,5). The strong bonds which are formed between the resins and the chloro-anions severely restrict the eluton of the analyte. With strong base amines there is also a difficulty with elution because the ion pairs formed with precious metal chloro-anions are unreactive (6). Consequently, it was decided to investigate the properties of a copolymer carrying a weak base amine such as 1,3,4-thiadiazole-2-amino-5-thiol, which was covalently bound to the polymer matrix through the sulphur atom.

EXPERIMENTAL

Synthesis of the P-SNTD resin

The chloromethylation of poly(styrene-co-divinylbenzene) (Amberlite XE-305) was performed according to a previous method (8) with adjustments to the ratio of the chloromethylating reagent. The analytical figures approximated to 93% substitution of the available rings with 4% crosslinkage leaving 3% of the rings unsubstituted (C 71.8 %; H 6.0 %; Cl 22.1 %).

Chloromethylated XE-305 (61 g) was swollen in DMF (400 mL) for 1 h with gentle stirring and warming. 1,3,4-thiadiazole-2-amino-5-thiol (66.5 g; i.e. 0.5 mol) was dissolved in a mixture of triethylamine (70 mL; 0.5 mol) and DMF (35 mL) and slightly warmed for 30 min. The golden coloured solution was added slowly to the swollen chloromethylated beads. On heating, the solution turned green and this was accompanied by the crystallisation of the triethylamine hydrochloride. The resin mixture was agitated as before and maintained at 85 °C for two hours. The product was allowed to cool and was filtered, and then rinsed thoroughly with DMF; dioxane; dioxane-water (1:1), methanol and finally with de-ionized water. Calculated for P-SNTD with 3.4 mmoles of thiadiazole groups per gramme of dry resin (found): C 56.6 (56.6); H 4.6 (4.6); N 14.4 (14.2); S 22.0 (21.6); Cl 2.4 (2.5).

RESULTS AND DISCUSSION

Structure and properties of the P-SNTD resin

The principal concern was the nature of the binding between the thiadiazole ligand to the poly(styrene) matrix as there was clearly a possibility that the two may be joined through the nitrogen or sulphur atom or both. The reaction was carried out under basic conditions in which the thiol group was ionized and, therefore acts as a strong nucleophile. The absence of the signal at 2580 cm-1 for the free thiol group in the spectrum of P-SNTD indicated that this group has been incorporated to the main polymer backbone, leaving the nitrogen atom adjacent to the aromatic ring free as shown in (1) below. The physical characteristics of P-SNTD include a water regain value of 0.81 g water per gramme of dry resin and a density of 12000 ± 100 kg m^{-3}; the apparent pKb of 8.7 is comparable to that of aminobenzene with pKb of 8.9, suggesting that the P-SNTD resin is also weakly basic. In some physical properties P-SNTD is comparable to Amberlite IRA-400.

(1)

Extraction of gold(III) from chloride media

Resins or solvent extraction reagents, which are capable of separating gold from base metals, are required to form ion pairs with the anion $[AuCl_4]^-$ (9). This was achieved with the related thiadiazole resin (3) but the subsequent elution proved to be very difficult as the gold was covalently bound to the sulphur atom. Consequently, it was anticipated that the above weak base resin could be capable of extracting gold(III), which could also be eluted since the ionic interactions were reduced to a minimum.

The variation of the distribution coefficients for gold(III) in hydrochloric acid and lithium chloride solutions and the corresponding values for some base metals in hydrochloric acid solutions are shown in Figure 1. With respect to the base metals, these are largely rejected by the resin.The exceptions are iron(III) and copper(II) in acid concentrations above 3 and 4 M HCl respectively because the $[FeCl_4]^-$ and $[CuCl_4]^{2-}$ anions were formed and could be extracted by the protonated P-SNTD. However, the extraction of copper(II) was much less than was the case with the corresponding thiol resin (3). With respect to gold(III), the Kd values show a significant increase from 0.05 M HCl (187 ± 2) to 0.5 M solutions (766 ± 2) and then a decline as the concentration of the acid was increased to a value of 253 ± 2 in 6 M HCl. The anion exchange equation for concentrations of acid above 0.1 M is related to that described in equation (2) as the gold is present entirely (10) as the tetrachloroanion.

$$RNH_3^+Cl^- + H^+[AuCl_4]^- ====== RNH_3^+[AuCl_4]^- + HCl$$ (2)

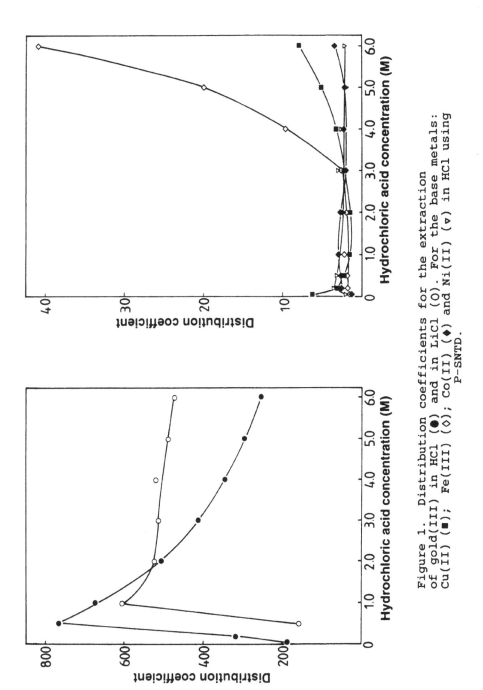

Figure 1. Distribution coefficients for the extraction of gold(III) in HCl (●) and in LiCl (O). For the base metals: Cu(II) (■); Fe(III) (◇); Co(II) (◆) and Ni(II) (▽) in HCl using P-SNTD.

Below this concentration, there are significant amounts of another anion [AuCl$_3$OH]$^-$ which is less well extracted (11). Above 1 M HCl, there is pronounced competition between the free chloride ion and the chloro-anion for the active site on the resin. It is interesting that when lithium chloride was used in place of hydrochloric acid, there was reduced competition between the chloride and the chloroanion. There have been many attempts to explain this phenomenon in other systems (12-15). However, one important factor is that in lithium chloride there is a significant amount of covalent character, as is the case with magnesium chloride so that there are fewer free chloride ions compared with hydrochloric acid.

Elution of gold from the resin

The gold can be quantitatively eluted from the resin. The times to elute 50% (t$_\frac{1}{2}$) of the loaded metal batchwise are 3 and 9 min, or 15 and 38 min in a column operation using acid thiourea (5% in 0.1 M HCl) or alkaline sodium cyanide (0.1 M in 1% NaOH) respectively. The elution of gold from pre-loaded columns is illustrated in Figure 2. The resultant acidic thiourea eluate is unstable with respect to decomposition by sulphide precipitation followed by chlorination or by direct oxidative chlorination. For example, a sample (41.5 mg) of loaded gold could be eluted with acidic thiourea and subsequently decomposed to yield 41.45 mg of metal, a recovery of over 99.9%.

Since the gold can be quantitatively removed from the thiourea complex thereby regenerating the resin, a continuous sorption-desorption process for gold recovery is feasible. The results for such an operation are summarised in Table 1. The capacity of the resin remains unchanged after several cycles of loading and elution.

Cycles	Amount of metal (mg)		Amount eluted	% recovery (column capacity)
	in effluent	in resin		
I	2.9	297.1	296.0	99.6
II	2.9	297.1	298.0	100.0
III	2.7	297.3	297.1	100.0
IV	2.6	297.4	297.8	100.3
V	2.9	297.1	297.2	100.0

Table 1. The cyclic recovery of gold by elution with acidic thiourea.

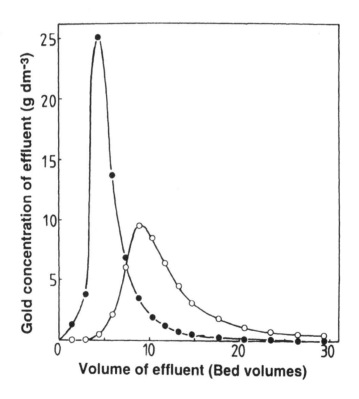

Figure 2. The rate of elution of gold from P-SNTD using acid thiourea(●) or alkaline sodium cyanide (O).

404

X-Ray Microprobe Analysis

The cross-sectional elemental scan (Figure 3) for sulphur, gold and chlorine in the gold-loaded resin sample shows an even distribution for the three elements across the bead. This indicates that the gold was able to migrate to the core of the beads with maximum utilisation of the resin capacity. There was also a similar distribution of the elements after five cycles of extraction and stripping.

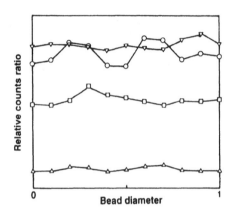

Figure 3. Cross-sectional elemental scans of P-SNTD after gold loading. Elements are sulphur ($\triangledown$), chlorine ($\square$), gold (K-shell) ($\triangle$) and gold (L-shell) (O).

From the above data, it appears that both the chloromethylation and the functionalisation with the thiadiazole groups proceeded uniformly throughout the bead. The extraction of gold also occurred evenly as would be expected for an anion exchange process with the regularly distributed free amine groups.

ACKNOWLEDGEMENTS

We are particularly grateful to Mr. R.D. Hancock of British Petroleum Plc (BP) for help and advice; BP and the SERC for the provision of a CASE Award for BK-OL; Rohm and Haas for supplying Amberlite XE-305 and INCO (Europe) Ltd. for loan of gold metal.

REFERENCES

1. Giwa C.O. and Hudson M.J., *Hydrometallurgy*, 1982, **8**, 65-75.

2. Drew M.G.B.D., Glaves L.R. Hudson M.J., *J. Chem. Soc. Dalton Trans.*, 1985, 771.

3. Hudson M.J. and Shepherd M.J., *Hydrometallurgy* 1983, 9, 223-234.

4. Hudson M.J., Trace Metal Removal from Aqueous Solution, Symposium Proceedings, Royal Soc. Chem. Warwick, 1986.

5. Hudson M.J., Ion Exchange Science and Technology, NATO ASI, Troia, Portugal, 1986.

6. Ellis A.F., Hudson M.J. and Tomlinson A.A.G., *J. Chem. Soc. Dalton Trans.*, 1985, 1655.

8. Belfer S., Glozman R., Deshe A. and Warshawsky A., *J. Appl. Polym. Sci.*, 1980, **25**, 241.

9. Warshawsky A., Ion Exchange and Sorption Processes in Hydrometallurgy, Wiley, New York, 1987.

10. Kraus K.A. and Nelson F., *J. Am. Chem. Soc.*, 1954, 76, 984.

11. Kraus K.A., Michelson D.C. and Nelson F. *J. Am. Chem. Soc.*, 1959, **81**, 3204.

12. Rieman W., Ion Exchange in Analytical Chemistry, Wiley, New York, 1970.

13. Kertes A.S. and Marcus Y., Ion Exchange and Solvent Extraction of Metal Complexes, Wiley-Interscience, New York, 1969.

14. Mirza M.Y., *Talanta*, 1980, **2**, 101.

15. Marcus Y., *J. Phys. Chem.*, 1959, **63**, 1000.

HYDROLYSIS OF CHELEX 100 CHELATING RESIN
AND ITS EFFECT ON NICKEL ION EXCHANGE

JUKKA LEHTO, AIRI PAAJANEN, RISTO HARJULA and HEIKKI LEINONEN
University of Helsinki, Department of Radiochemistry,
Unioninkatu 35, SF-00170 Helsinki, Finland

ABSTRACT

Chelex 100 chelating resin in sodium form was found to be significantly hydrolysable. Contact with de-ionised water in solution volume to exchanger weight ratios of 10-4000 yielded a conversion to hydronium form of 1 to 10%. Hydrolysis brought sodium ions into solution with concentrations of 0.05-1.5 mM and increased the pH from 6.1 to 8.7-10.0. Hydrolysis has a marked effect on the nickel ion exchange on Chelex 100 in dilute nickel solutions (<0.1 mM): hydrolysis controls the sodium ion concentration in the solution and the increasing pH diminishes the absorption of nickel.

INTRODUCTION

Chelex 100, a chelating resin consisting of polystyrene-DVB matrix and iminodiacetic acid functional groups, is a highly selective exchanger for divalent transition metal ions, especially copper and nickel (1,2). Owing to the weakly acidic nature of the carboxyl groups this exchanger greatly prefers hydronium ions. Hydrolysis of the exchanger causes its partial conversion to hydronium form. When an exchange of divalent transition metal ion on Chelex 100 in sodium form takes place, the concurrent hydrolysis results in the formation of a three-component ion exchange system. Hydrolysis may accordingly have a great influence on the ion exchange, especially in dilute solutions.

EXPERIMENTAL

Hydrolysis of Chelex 100 in sodium form (BIO-RAD, 200-400 mesh, water content 77.5%) was studied by shaking samples of the exchanger with 150 mL of ultrapure de-ionised water. The ratio of solution volume to dry exchanger weight (BF) varied between 11 and 4444. The initial pH of the water was 6.1. After shaking for 24 h the phases were separated by centrifugation and the following parameters in the solution were determined: 1) pH with a standard pH meter, 2) sodium ion concentration with an atomic absorption spectrophotometer, 3) alkalinity by titration with nitric acid, 4) total carbon dioxide with a CO_2-sensing electrode and an ion chromatograph and 5) iminodiacetic acid with an aminoacid analyzer.

Ion exchange of nickel on Chelex 100 was studied by shaking for three days 0.35 g (dry weight 0.0675 g) samples of the exchanger with 35 mL of $5 \cdot 10^{-7}$-10^{-2} M $Ni(NO_3)_2$ solution containing ^{63}Ni tracer. After separation of the phases by centrifugation the following species in the solution were determined: 1) pH, 2) sodium ion concentration and 3) nickel ion concentration. The nickel ion concentration was determined by measuring the ^{63}Ni activity of the solution with a liquid scintillation counter.

RESULTS AND DISCUSSION

Hydrolysis of Chelex 100 can be described by the equation

$$RN(CH_2COO)Na_2 + 2H_2O \rightleftharpoons RN(CH_2COO)H_2 + 2Na^+ + 2OH^- \quad (1),$$

where R is the polystyrene-DVB matrix. Chelex 100 was found to be significantly hydrolysed (Fig. 1). Owing to the hydrolysis the sodium ion concentration in the solution was fairly high (0.05-1.5 mM) and the pH increased considerably (from 6.1 to 8.7-10.0).

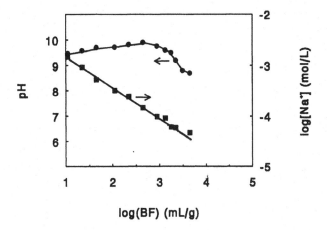

Figure 1. Sodium ion concentration [Na⁺] (■) and equilibrium pH (●) of the solution after shaking samples of Chelex 100 with deionised water as a function of the ratio of solution volume to exchanger weight, BF (mL/g).

According to Eq. 1 the concentrations of sodium and hydroxyl ions should be identical. Owing to the presence of carbonate species in the solution this was not valid, but the sodium concentrations corresponded fairly well to the alkalinity ($[OH^-]$ + $[HCO_3^-]$ + $2[CO_3^{2-}]$) of the solution (Table 1). Carbonate is assumed to originate from both the air in the free space of the shaking vials and from the exchanger itself, the concentration of the latter being estimated to be 0.01 mmol/g of dry exchanger.

The amino acid analysis revealed that no iminodiacetic acid was released from the exchanger, which means that sodium ions in the solution did not originate from the dissolution of the exchanger. The ratio R in Table 1 is close to unity, which value would mean that the sodium ions in the solution originated entirely from the hydrolysis.

Table 1. Equilibrium pH, OH⁻, sodium ion, carbonate and bicarbonate concentrations and the ratio of the sodium concentration to the alkalinity (R) as a function of the ratio of solution volume to exchanger weight, BF (mL/g). Concentrations are $\mu M \cdot 10$.

BF	pH	[OH⁻]	[Na⁺]	[CO₃²⁻]	[HCO₃⁻]	R
11	9.54	3.5	147	13.6	83.9	1.28
22	9.70	5.0	92	10.6	45.3	1.28
44	9.85	7.1	57	7.2	21.8	1.20
111	9.87	7.4	32	3.6	10.4	1.28
222	9.96	9.1	24	2.2	5.1	1.30
444	9.72	5.3	15	1.2	4.9	1.17
888	9.35	2.2	9.8	0.4	3.6	1.48
1776	9.30	2.0	6.1	0.2	1.9	1.42

Hydrolysis of Chelex 100 results in a rather high conversion of the exchanger from sodium form into hydronium form (Fig. 2). This means that whenever the sodium form exchanger is rinsed with water or immersed in water for swelling, a considerable proportion of the sodium ions are exchanged for hydronium ions. Thus, when the exchanger pretreated with water is used for the separation of other ions, it must be borne in mind that the exchanger is more or less heterogeneous in nature.

A corrected selectivity coefficient (K_G) for the hydronium exchange reaction can be calculated from hydrolysis data with the equation

$$K_G = \frac{[\overline{H^+}] \quad [Na^+]}{[H^+] \quad [\overline{Na^+}]} \times \frac{\gamma_{Na}}{\gamma_H} \qquad (2)$$

where the barred symbols are the concentrations of the ions in the exchanger and γ_{Na} and γ_H are the activity coefficients of the ions in the solution. Because the solutions were very dilute the activity coefficients were estimated to be 1.0.

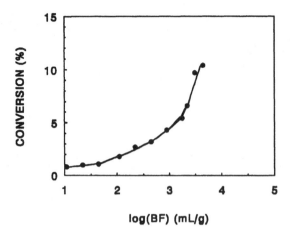

Figure 2. Conversion of sodium form Chelex 100 to hydronium form in deionised water as a function of the ratio of solution volume to exchanger weight, BF (mL/g).

Figure 3 shows K_G as a function of the equilibrium sodium ion concentration. In the concentration range 0.2-1.5 mM the $\log K_G$ has a very high constant value of 4.7 indicating high selectivity for hydronium ions over sodium ions. Below the sodium ion concentration of 0.2 mM the selectivity coefficient falls off markedly. This decrease is in contradiction to the present theory of ion exchange (3,4). It may be explained partially by the increasing conversion of the exchanger to hydronium form, but it is improbable that an increase in hydronium conversion from 3 to 6% (exchanger 94-97% in sodium form) would cause K_G to decrease by approximately one decade. We have observed this anomaly with other ion exchangers and the reason for it is being sought (5).

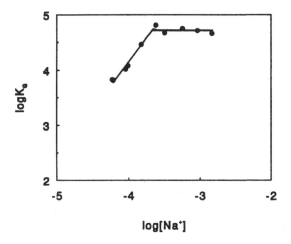

Figure 3. Corrected selectivity coefficient for sodium/hydronium ion exchange in Chelex 100 as a function of equilibrium sodium ion concentration.

The ideal ion exchange of nickel on Chelex 100 can be described by the equation

$$RN(CH_2COONa)_2 \quad + \quad Ni^{2+} \rightleftharpoons RN(CH_2COO)Ni \quad + \quad 2Na^+ \qquad (3).$$

This ideal behaviour was observed only in a rather narrow range of nickel concentration, 0.5-2.5 mM (Fig. 4). At concentrations lower than 0.5 mM the hydrolysis results in the formation of a three-component system of $Ni^{2+}/Na^+/H_3O^+$. The sodium ion concentration in the solution at initial nickel concentrations lower than 10^{-4} M is controlled almost entirely by the hydrolysis and has a constant value of about $3 \cdot 10^{-4}$ M. The separation factor between nickel and sodium is so high, $4.4 \cdot 10^7$ (6), however, that the increased sodium ion concentration caused by the hydrolysis has only a minor effect on the nickel absorption. A more important factor is that at low nickel concentrations the hydrolysis increases the equilibrium pH. Chelex 100 has an absorption maximum for nickel at pH 4-6 and the increasing pH diminishes the nickel absorption.

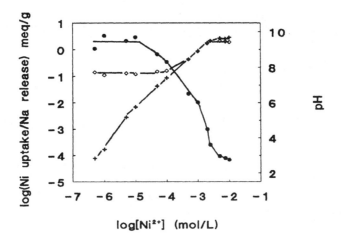

Figure 4. Nickel absorption on Chelex 100 (+), released sodium (◊) and equilibrium pH (●) as a function of initial nickel concentration. BF 444 mL/g(d).

Figure 5 shows the equivalent ionic composition of Chelex 100 as a function of initial nickel concentration. As can be seen, nickel has only a minor effect on the ionic composition below initial concentration of 10^{-4} M, but the hydrolysis causes a conversion of about 6% to hydronium form.

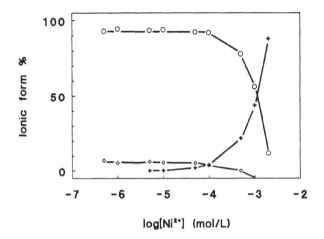

Figure 5. Equivalent ionic composition of Chelex 100 as a function of initial nickel ion concentration. o = Na$^+$, + = Ni^{2+}, ◊ = H$_3$O$^+$.

REFERENCES

1.   Loewenschuss H., Schmuckler G., Talanta, 1964, **11**, 1399.
2.   Leyden D.E., Underwood A.L., J. Phys. Chem., 1964, **68**, 2093.
3.   Barrer R.M., Klinowski J.K., J. Chem. Soc. Faraday Trans., 1974, **70**, 2080.
4.   Kraus K.A., in: Trace Analysis, Yoe J.H. and Koch H.J., Eds., pp. 42-102, John Wiley & Sons, New York, 1957.
5.   Harjula R., Lehto J., Pothuis J.H., Dyer A., Townsend R.P., in: New Developments in Ion Exchange, Kodansha Ltd., Tokyo, 1991, p. 157.
6.   BIO-RAD Bulletin 2020, Bio-Rad Laboratories, Richmond, USA, 1983.

THE REMOVAL OF DIVALENT ANIONS AND CATIONS FROM FEED BRINE FOR CHLORALKALI-ELECTROLYSIS CELLS BY UTILIZATION OF ION EXCHANGE TECHNOLOGY

R.M.KLIPPER, H.HOFFMANN , T. AUGUSTIN
Lewatit Research and Applications Department
Bayer AG, 5090 Leverkusen, Germany

ABSTRACT

The production of commercial quantities of chlorine gas and alkali-metal-hydroxides is performed by electrolysis of brine solutions in a chloralkali cell.
The most advanced technology for this process requires membrane electrolysis cells. For good performance of these cells there is a demand for high purity brines with extremely low levels of divalent cations (magnesium, calcium, strontium) and low levels of sulphate anions.
The following discussion describes how ion exchangers with chelating functional groups meet the required specifications.

INTRODUCTION

The advanced membrane electrolysis of brine solutions requires high purity brines for good performance. The stringent specifications of 20 ppb (µg/kg brine) of calcium and magnesium, as well as 50 to 100 ppb of strontium, can only be met by a two step process. In the first step most of the alkaline earth ions are removed with the conventional carbonate precipitation, followed in a second step by treatment with selective ion exchangers, producing high purity brine (1).

To reduce the level of sulphate anions below 5 g sulphate per litre brine, it is necessary to treat the sulphate rich brine in a bypass process with a sulphate selective chelating ion exchanger (2).

SYNTHESIS OF THE ION EXCHANGER RESIN

The highly specific ion exchangers for the removal of alkaline earth ions are either styrene/DVB polymers with amino alkylene phosphonic acid groups (3, 4, 5) -see Scheme 1- or styrene/DVB polymers with iminodiacetic acid groups (6) -see Scheme 2-.

Scheme 1.Synthesis of ®Lewatit OC 1060

Scheme 2.Synthesis of ®Lewatit TP 208

Polymers based on acrylic acid amides are used for the sulphate removal. They are functionalized with polyamine groups synthesized as shown in Scheme 3 (7).

Scheme 3.Synthesis of ®Lewatit E 304 / 88

RESULTS

All commercially available chelating resins used for the removal
of alkali earths are characterized by the following selectivity
sequence: Mg> Ca >> Sr > Ba > alkalines.The calcium and magnesium ions are removed preferentially, regardless of whether the
functional group is of the iminodiacetic acid type (TP 208) or
the aminomethylphosphonic acid type (OC 1060). This means that
if strontium rather than calcium is selected as the breakthrough
criterion, considerably smaller volumes of brine solutions are
processed.

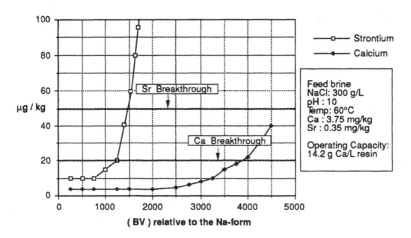

Figure 1.Strontium and calcium operating capacity of OC 1060

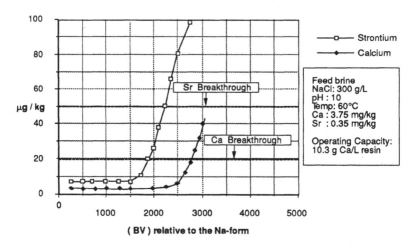

Figure 2. Strontium and calcium operating capacity of TP 208

Figures 1 and 2 show the operating capacity of ®Lewatit
OC 1060 and ®Lewatit TP 208 for calcium and strontium in a
precipitated NaCl brine.
Under the conditions of calcium breakthrough, ®Lewatit OC 1060
achieves a 25 - 30 % higher operating capacity (based on BV)
than ®Lewatit TP 208 and other resins of the same functional
group.
On the other hand ®Lewatit TP 208 is the superior product to the
aminomethylphosphonic acid resin or similar products if strontium is the breakthrough criterion (see conditions above).
The reason is the extremely high total capacity of ®Lewatit
TP 208 in conjunction with a more marked selectivity for
strontium.

The operating capacity of the chelating resins depends on the
ratio of calcium, magnesium and strontium in the feed brine.

After the membrane electrolysis cell the brine contains about
200 g / L NaCl and more than 5 g / L sulphate. For good performance it is necessary to reduce the sulphate amount below
5 g / L. To achieve this, the brine is treated in a bypass process with a sulphate selective ion exchanger e. g. ®Lewatit
E 304/88.The bypass brine is therefore adjusted to a pH of about
3 and a NaCl-concentration of about 100 to 150 g / L.
Figure 3 shows the sulphate capacities of ®Lewatit E 304/88 in
varying brine concentrations.

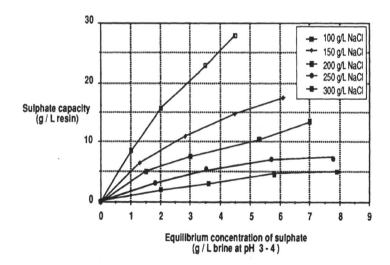

Figure 3. Sulphate capacity of ®Lewatit E 304 / 88 with
varying NaCl- and sulphate concentration

In the exhaustion step brine with 125 g NaCl / L at pH 3 is used. Figure 4. shows a typical exhaustion run of ®Lewatit E 304 / 88. The exhausted resin is regenerated with a brine containing 300 g NaCl / L.

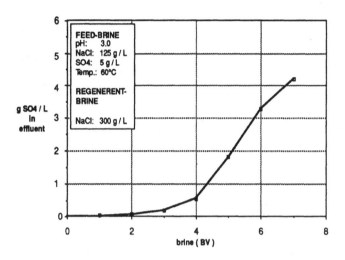

Figure 4.Exhaustion run of ®Lewatit E 304 / 88

CONCLUDING REMARKS

Several tests on   pilot plant scale have shown that the application of highly specific chelating resins can meet the demands of modern membrane technology.

REFERENCES

(1) Klipper, R.M., Hoffmann, H., Mitschker, A., Wagner, R., The influence of morphology and degree of substitution on the selectivity of chelating resins,Ion exchange for Industry, Ellis Horwood Limited, p.243 (1988).

(2) German Offenlegungsschrift, DE 3345898.

(3) Fields, E.K., J.Amer .Chem. Soc.,74, p.1528 (1952).

(4) Kabachnik, M.J., Medved, T.V., Dokl. Akad. Nauk SSSR,
       83, p.689 (1952).

(5) Moedritzer, K., Irani, R.R., J. Organ. Chem., 31,
       p.1603 (1966).

(6) Ullmanns Enzyklopädie der technischen Chemie, Band 13,
       p.295-308.

(7) Krauß, D., Sabrowski, E., Schwachula, G., Funktionali-
       sierte Polymere auf Acrylatbasis I, Plaste und Kau-
       tschuk, 29, p.449-451 (1982).

INDEX OF CONTRIBUTORS

Printed in the United States
By Bookmasters